IR Part-66

Aircraft Maintenance Licence

Distance Learning Modules

Module 3 – Electrical Fundamentals

Copyright © Cardiff and Vale College 2015
Issue 1-15

Cardiff and Vale College

ICAT

International Centre
for Aerospace Training

Cardiff Airport Business Park
Port Road
Rhoose
Vale of Glamorgan
CF62 3DP
+44 (0) 1446 719821
info@part66.co.uk

www.part66.com

Copyright © Cardiff and Vale College 2015

All rights reserved. No part of this publication may be reproduced or transmitted in any form or by any means, electronically or mechanically, including photocopying, recording or any information storage or retrieval system, without permission in writing from the publisher or a license permitting restricted copying.

In the United Kingdom such licenses are issued by the Copyright Licensing Agency:

90 Tottenham Court Road, London W1P 4LP.

Liability

Whilst the advice and information in this book are believed to be true and accurate at the date of going to press, neither the author nor the publisher can accept any legal responsibility or liability for any errors or omissions that may be made.

IR Part-66   MODULE 3

Electrical Fundamentals

Knowledge Levels

The basic knowledge requirements for categories A, B1 and B2 certifying staff are indicated in the contents list of the module notes by knowledge level indicators 1, 2 or 3 against each chapter.

Level 1
A familiarisation with the principal elements of the subject

Objectives

The student should be familiar with the basic elements of the subject.

The student should be able to give a simple description of the whole subject using common words and examples.

The student should be able to use typical terms.

Level 2
A general knowledge of the theoretical and practical aspects of the subject. An ability to apply that knowledge.

Objectives

The student should be able to understand the theoretical fundamentals of the subject.

The student should be able to give a general description of the subject using, as appropriate, typical examples.

The student should be able to use mathematical formulae in conjunction with physical laws describing the subject.

The student should be able to read and understand sketches, drawings and schematics describing the subject.

The student should be able to apply knowledge in a practical manner using detailed procedures.

Level 3

A detailed knowledge of the theoretical and practical aspects of the subject. A capacity to combine and apply the separate elements of knowledge in a logical and comprehensive manner.

Objectives

The student should know the theory of the subject and interrelationships with other subjects.

The student should be able to give a detailed description of the subject using theoretical fundamentals and specific examples.

The student should understand and be able to use mathematical formulae related to the subject.

The student should be able to read, understand and prepare sketches, simple drawings and schematics describing the subject.

Contents

Chapter Zero – Foreword

Foreword 1

Chapter One – Electron Theory

Knowledge Levels: A – 1, B1 – 1, B2 – 1

Introduction	1
Matter, Energy & Electricity	1
Matter	2
Elements & Compounds	2
Atoms	2
Molecules	4
Atom Energy Levels	5
Atomic Shells and Sub-shells	7
Ionisation	8
Conductors, Semiconductors and Insulators	9
Conductors	9
Insulators	10
Semiconductors	10
Revision	12

Chapter Two – Static Electricity and Conduction

Knowledge Levels: A – 1, B1 – 2, B2 – 2

Introduction	15
Static Electricity	16
Triboelectric Series	17
Charged Bodies	18
Gold Leaf Electroscope	20
Coulomb's Law of Charges	21
Electric Fields	21
Conduction of Electricity	22
Conduction of Electricity in Solids	23
Conduction of Electricity in a Liquid	24
Water Solutions that Conduct Electricity	26
Conduction of Electricity in a Gas	26
Conduction of Electricity in a Vacuum	26
Revision	27

Chapter Three – Electrical Terminology

Knowledge Levels: A – 1, B1 – 2, B2 – 2

Introduction	29
Electric Current (I)	29
Electromotive Force (EMF)	31
Potential Difference (PD)	31
Voltage (V)	32
Resistance (R)	32
Conductance	33
Electrical Charge	34
Revision	37

Chapter Four – Generation of Electricity

Knowledge Levels: A – 1, B1 – 1, B2 – 1

Introduction	39
Voltage Produced by Light	40
Voltage Produced by Heat	41
Voltage Produced by Friction	42
Voltage Produced by Pressure	43

Piezo Accelerometers	44
Types	44
Voltage Produced by Chemical Action	45
Voltage Produced by Magnetism and Motion	47
Revision	48

Chapter Five – DC Sources of Electricity

Knowledge Levels: A – 1, B1 – 2, B2 – 2

Introduction	51
Batteries	51
The Cell	53
Primary Cell	54
Secondary Cell	55
Primary Cell Chemistry	55
Secondary Cell Chemistry	56
Cell Types	59
Primary Dry Cell	59
Manganese Dioxide-Alkaline-Zinc Cell	60
Lithium Cells	60
Secondary Cells	64
Lead-Acid Cell	64
Nickel-Cadmium Cell	68
Aircraft Nickel-Cadmium Batteries	69
Silver-Zinc Cell	71
Silver-Cadmium Cell	71
Lithium Ion Cell	71
General Maintenance	73
Battery Safety Precautions	73
Battery Charging	74
Capacity and Rating of Batteries	74
Capacity Checks	74
Cells Connected in Series and Parallel	75
Series-Connected Cells	76
Parallel-Connected Cells	77
Series-Parallel-Connected Cells	78
A Battery's Internal Resistance	79
Thermocouples	80
Operation of Photocells	82
Revision	85

Chapter Six – DC Circuits

Knowledge Levels: B1 – 2, B2 – 2

Introduction	89
The Source	90
The Load	90
The Transmission System	90
Control	91
Ohm's Law	91
Application of Ohm's Law	93
Kirchhoff's Laws	94
Kirchhoff's First Law	95
Kirchhoff's Second Law	96
Branch Current Method	99
Loop Current Method	99
Significance of a Supply's Internal Resistance	100
Revision	103

Chapter Seven – Resistance & Resistors

Knowledge Levels: B1 – 2, B2 – 2

Introduction	105
What Affects Resistance?	106
Specific Resistance	107
Resistor Rating Colour Code	110
Wattage Ratings	116
Resistor Construction	117
Solid Carbon Resistors	117
Wirewound Resistors	118
Film Type Resistors	119
Resistors in Series	120
Rules for Series DC Circuits	121
Resistors in Parallel	122
Rules for Parallel DC Circuits	125
Series-Parallel Resistor Networks	125
Open and Short Circuits	130
Open Circuit	130
Short Circuit	131
Operation & Construction of Potentiometers and Rheostats	132
Thermistors	134
Voltage Dependent Resistors	135
Operation and Construction of a Wheatstone Bridge	137
Temperature Coefficient of Resistance	140
Revision	143

Chapter Eight – Power

Knowledge Levels: B1 – 2, B2 – 2

Introduction	149
Energy	149
Kinetic Energy	151
Potential Energy	151
Electrical Energy	152
Worked Example	155
Power Rating	155
Revision	156

Chapter Nine – Capacitance & Capacitors

Knowledge Levels: B1 – 2, B2 – 2

Introduction	159
What is Capacitance?	160
Description of a Capacitor	161
Dielectric Materials	163
Capacitance Measurement	164
Energy Stored in a Capacitor	164
Factors Affecting Capacitance Value	165
Plate Area	165
Plate Gap	165
Dielectric Constant	166
Capacitor Voltage Ratings	167
Capacitor Symbols	168
Capacitor Construction	168
Paper Construction	168
Mica Construction	169
Glass Capacitors	169
Tantalum Capacitors	170
Ceramic Capacitors	170
Film Capacitors	171
Electrolytic Capacitors	172
Variable Capacitors	173
Identification of Capacitors	175
Capacitor Colour Coding	175
Charging of a Capacitor	177
Discharging of a Capacitor	179
Resistive-Capacitive Series Circuit Time Constant	179
Circuit Discharging	181

Capacitors in a Parallel Circuits	182
Capacitors in Series Circuits	183
Testing Capacitors	183
Some Example Capacitor Applications	184
Condenser Microphones	184
Revision	185

Chapter Ten – Magnetism

Knowledge Levels: B1 – 2, B2 – 2

The Theory of Magnetism	191
What is Magnetism?	193
Properties of a Magnet	195
Magnetic Field Lines	195
Magnetic Flux Density	196
Magnetic Field Strength & Direction	197
Magnetic Flux Line Characteristics	197
Magnetic Fields due to Electric Current	198
Magneto Motive Force (MMF)	203
Magnetic Field Strength	203
Permeability or Magnetic Constant	203
Permeability of Free Space	204
Absolute permeability of a material	205
Relative permeability	205
Reluctance	206
Further Comparisons of Electric & Magnetic Circuits	208
B/H Characteristics	208
Hysteresis Loss	210
Soft Magnetic Materials	211
Para-Magnetic Materials	211
Dia-Magnetic Materials	212
Eddy-Current Loss	212
Retentivity	213
Magnetic Shielding	214
Demagnetising or Degaussing	215
Degaussing Operation for Small Objects	216
Degaussing CRTs	217
Locating Areas of Unwanted Magnetism	217
Precautions for Care & Storage of Magnets	218
Transporting Magnets	218
Revision	219

Chapter Eleven – Inductors & Inductance

Knowledge Levels: B1 – 2, B2 – 2

Introduction	223
Inductor History	224
Faraday's Law of Electromagnetic Induction	224
Lenz's law	228
Induction Principles	229
Self Inductance	231
Inductance in Terms of Flux-Linkages	233
Effects on Circuit Inductance	234
Effects on circuit induced voltage	235
Mutual Inductance	236
Coefficient of Coupling	238
Inductors in Series	239
Series Inductors with Magnetic Coupling	240
Inductors in a Parallel Circuit	240
Rise & Fall of 'I' in an Inductive Circuit	241
Resistive-Inductive Series Circuit Time Constant	243
Energy Stored in an Inductive Circuit	244
Inductor Types	245
Air Core	245
Ferromagnetic Core	245
Inductor Applications	246
Generators	246
Radio Receivers	247
Metal Detectors	248
Moving Coil Meter	248
Moving Coil Microphone & Speaker	249
Revision	251

Chapter Twelve – DC Generator & Motor Theory

Knowledge Levels: B1 – 2, B2 – 2

Introduction	255
Basic Generator Principles	255
The Simple Generator	257
The Simple DC Generator	259
Electromagnetic Poles	262
Commutation	262
Armature Reaction	263
Compensating Windings and Interpoles	265
Generator Motor Reaction	265

Armature Losses	267
Copper Losses	267
Eddy Current Losses	267
Hysteresis Losses	268
The Practical DC Generator	268
Gramme-Ring Armature	269
Drum-Type Armature	270
Field Excitation	271
Generator Classification	272
Series Wound Generator	272
Shunt Wound Generator	273
Compound Wound Generators	273
Generator Construction	275
Voltage Regulation	276
Voltage Control	277
Manual Voltage Control	277
Automatic Voltage Control	278
Basic DC Motor Principles	278
Back Emf	280
Motor Loads	281
Practical DC Motors	281
Series DC Motor	281
Shunt Motor	282
Compound Motor	283
Types of Armatures	284
Gramme-Ring Armature	284
Drum-Wound Armature	285
Direction of Rotation	286
Motor Speed	286
Armature Reaction	287
Manual & Automatic Starter/Generators	289
Typical Start Sequence	289
Start PWR Switch	290
Eng. Starter Switches (2)	290
Power Generation	291
Revision	293

Chapter Thirteen – AC Theory

Knowledge Levels: A – 1, B1 – 2, B2 – 2

Introduction	301
Sinusoid or Sine Wave	302
Amplitude	303
Instantaneous Amplitude	303
Cycle	303
Periodic Time	303
Frequency	303
Wavelength	304
Average or Mean Value	304
Root Mean Square (RMS) Value	304
Peak Amplitude	305
Peak to Peak Amplitude	305
Angular Frequency	305
Instantaneous Values of a Sine Wave	305
Sine Wave Phase Relationships	307
Signal Phase Relationships	308
Adding Phasors	311
Power in AC Circuits	312
Other Alternating Waveforms	313
Square Wave	313
Form Factor	314
Single Phase	315
Three-Phase AC	315
Revision	316

Chapter Fourteen – Resistive, Capacitive & Inductive Circuits

Knowledge Levels: B1 – 2, B2 – 2

Introduction	319
Inductance & Alternating Current	319
Inductive Reactance	321
Capacitors & Alternating Current	322
Capacitive Reactance	325
Reactance, Impedance & Power Relationships in AC Circuits	326
Series RLC Circuits	326
Reactance	327
Impedance	327
Ohms Law for AC	331

Power in AC Circuits	331
Calculating True Power in AC Circuits	334
Calculating Reactive Power in AC Circuits	335
Calculating Apparent Power in AC Circuits	336
Power Factor	336
Power Factor Correction	337
Series RLC Resonant Circuits	338
Parallel R, L, & C Circuits	340
Inductance in AC Circuits	343
Phase Relationships of an Inductor	343
Inductive Reactance	343
Capacitance in AC Circuits	343
Phase Relationships of a Capacitor	343
Capacitive Reactance	344
Total Reactance	344
Phase Angle	344
Ohm's Law Formulas for AC	344
True Power	344
Reactive Power	344
Apparent Power	344
Power Factor	344
Power Factor Correction	345
Resonance	345
Revision	346

Chapter Fifteen – Transformers

Knowledge Levels: B1 – 2, B2 – 2

Introduction	349
Basic Operation of a Transformer	350
A Transformer's Components	350
Core Characteristics	351
Hollow-Core Transformers	352
Shell-Core Transformers	352
Enclosures	354
Transformers Schematic Symbols	354
How Transformers Work	354
No-Load Condition	354
How a Voltage is induced in the Secondary	356
Primary and Secondary Phase Relationship	356
Coefficient of Coupling (K)	357
Turns & Voltage Ratios	357
Load Effects	360
Turns & Current Ratios	361
Power Relationship between Primary & Secondary Windings	362

xi

Transformer Losses	363
Copper Loss	363
Eddy-Current Loss	363
Hysteresis Loss	364
Transformer Efficiency	364
Transformer Ratings	365
Types & Applications of Transformers	365
Power Transformers	365
Autotransformers	367
Three-phase Transformers	368
Audio-Frequency Transformers	371
Radio-Frequency Transformers	371
Current Transformers	372
Impedance-Matching Transformers	373
Basic Transformer	373
Transformer Construction	374
Excitation Current	374
Phase	374
Turns Ratio	374
Power & Current Ratios	374
Transformer Losses	375
Transformer Efficiency	375
Power Transformer	375
Autotransformer	375
Audio-Frequency Transformer	375
Radio-Frequency Transformer	375
Current Transformers	375
Impedance-Matching Transformer	376
Revision	376

Chapter Sixteen – Filters

Knowledge Levels: B1 – 1, B2 – 1

Introduction	379
Low-Pass Filters	381
Inductive Low Pass Filter	381
Capacitive Low Pass Filter	381
High-Pass Filters	383
Inductive High Pass Filter	384
Band-Pass Filters	385
Band-Stop Filters	387
Filter Applications	388
Revision	390

Chapter Seventeen – AC Generators

Knowledge Levels: B1 – 2, B2 – 2

Introduction	393
3-Phase Revisited	393
Star or 'Y' Wound 3-Phase AC Generator	394
Delta Wound Generator	395
AC Versus DC Advantages	395
Basic AC Generators	396
Rotating-Armature Generators	396
Rotating-Field Generators	397
Practical Generators	398
Functions of Generator Components	398
Prime Movers	400
Generator Rotors	400
Generator Characteristics and Limitations	401
Single-Phase Generators	402
Two-Phase Generators	403
The Three-Phase Generator	405
Frequency of Operation	406
Generator Loads	408
No Load Condition	408
Resistive Loads	409
Reactive Loads	409
Voltage Regulation	411
Principles of AC Voltage Control	411
Permanent Magnetic Generator (PMG)	412
Faults	413
Revision	414

Chapter Eighteen – AC Motors

Knowledge Levels: B1 – 2, B2 – 2

Introduction	419
Series AC Motor	420
Rotating Magnetic Fields	421
Single Phase Rotating Magnetic Field	421
Two-Phase Rotating Magnetic Field	422
Three-Phase Rotating Fields	424
Rotor Behaviour in a Rotating Field	427
Synchronous Motors	427
Induction Motors	429
Single-Phase Induction Motors	431

xiii

Split-Phase Induction Motors	432
Capacitor-Start	432
Resistance-Start	433
Shaded-Pole Induction Motors	434
Hysteresis Motor	436
Speed of Single-Phase Induction Motors	437
Induction Motor Slip	437
How to reverse the Direction of a 3 Phase AC Motor	438
Revision	439

Electron Theory

Introduction

In today's modern aircraft, there is hardly anything that functions without using some form of electricfial or electronic signal. Typical areas of use include:

- Lighting
- Power
- Instrumentation
- Remote monitoring
- Intercommunications
- Radio
- Radar
- Navigation

Of course, this is only for starters!

Matter, Energy & Electricity

Modern science's roots stem from the great thinkers that were part of ancient Greece. When Greece was at the height of its influence, only a few other people took any interest in the structure of materials. Most people and civilisations accepted things at face value, they did not know or care how things were structured and accepted solids as continuous, uninterrupted substances. However, some inquisitive Greeks believed that if a substance, such as Gold, was subdivided, it could be done ad infinitum and that only that material, i.e. the Gold, would ever be found. Although some other Greeks felt that there must have been a limit to the number of subdivisions, they all held the premise that there was a basic particle upon which all substances were created.

It was not until almost 2000 years later in the 18th, 19th and 20th centuries that experiments confirmed that this was indeed the case and that there are several basic particles, or ***building blocks*** within all substances.

The rest of this chapter explains how substances are classified as elements and compounds, made up of molecules and atoms.

CHAPTER ONE
ELECTRON THEORY

Matter

Matter is defined as anything that occupies space and has weight; i.e. the dimensions and weight of matter can be identified and measured. Examples of matter include the air we breathe, water, cars, clothing, birds, mammals and even our own bodies; indeed, all that is around us. Therefore, looking at all the things matter can be, it follows then that it may be found in any one of three states:

- Solid
- Liquid
- Gas

Elements & Compounds

An *Element* is a substance that cannot be reduced to a simpler form by chemical means. Some typical examples of elements that are found in everyday life are Gold, Silver, Copper, Aluminium, Oxygen, Iron, etc. There are now over 100 *known* elements and all the various substances we come across in our day-to-day lives are made up of one or more of these.

When two or more elements are chemically combined, the resulting substance is called a *Compound*. A compound is a chemical combination of elements that can be separated by chemical *but not by* physical means. Some typical examples of compounds that you regularly come into contact with are *salt*, made up of the elements Sodium and Chloride and *water* made up of the elements Hydrogen and Oxygen.

Elements and compounds can exist in a third combination called a *Mixture*, defined as a combination of elements and compounds, *not* chemically combined that can be separated by physical means. Some typical examples of mixtures that you regularly come into contact with are air, made up of Nitrogen, Oxygen, Carbon Dioxide, water vapour etc, and seawater, made of water, salts etc.

Atoms

In simple terms, an atom is an element's smallest particle that retains all of its characteristics. However, the atoms of one element differ from the atoms of all others and these give each element its unique characteristic. As already discussed, there are over 100 known elements and consequently, there must be over 100 unique atoms, i.e. a different atom for each element. So if there are only 100 or so distinct atoms, how do we find so many different liquids, solids and gases throughout the world?

The answer is that just as we can combine the 26 proper letters of the English alphabet to make many thousands of words, so elements can combine chemically to form thousands of different materials.

Each element's atoms consist of *Electrons*, *Protons*, and in most cases, *Neutrons*, which are collectively called *sub-atomic particles*. However, the sub-atomic particles of one element are identical to those of any other element, so how are there over 100 of them? The reason is that each element has a different number and arrangement of these protons and electrons.

Physicists have measured the mass and size of the electron and proton, and discovered that the electron is negatively charged while the proton has a positive charge equal and opposite to the electron. This is true even though the mass of the proton is over 1800 times that of the electron.

As mentioned above, most atoms also have another particle called a neutron. The neutron has a mass approximately equal to that of a proton, but it has no electrical charge; i.e. it is neutral. The protons and neutrons form a heavy nucleus of each atom with a positive charge, around which the very light negative electrons revolve or orbit.

One way of looking at this arrangement to understand the concept would be to compare it to our own solar system. Figure 1.1 shows one (1) Hydrogen and one (1) Helium atom, which have the two (2) simplest structures in existence. The Hydrogen atom has only one proton in the nucleus with one electron orbiting around it, while the Helium atom has a nucleus with two protons, two neutrons and two electrons in orbit.

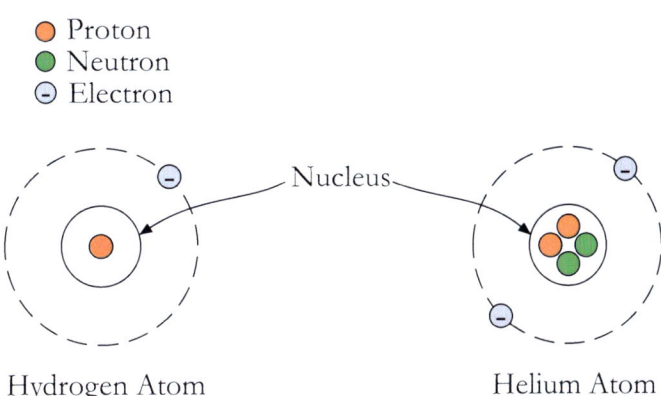

Figure 1.1 – Hydrogen and Helium Atomic Structure

Elements are classified numerically according to the complexity of their atoms with the number of protons in its nucleus determining its **Atomic Number**. Therefore, as illustrated in figure 1.1, Hydrogen has the atomic number of one (1) and Helium two (2). All atoms within an element are identical and in a neutral state, have an equal number of protons and electrons. The **Atomic Mass Number** of an atom is the total number of protons and neutrons it contains. The mass number of an atom is never smaller than the atomic number. It can be the same, as shown with the Hydrogen atom, but is usually bigger. Whilst an electron does have mass, it is considered insignificant and is not included when calculating Atomic Mass. Figure 1.2, shows elements of Carbon and Argon, detailing their Atomic Mass and Atomic Number.

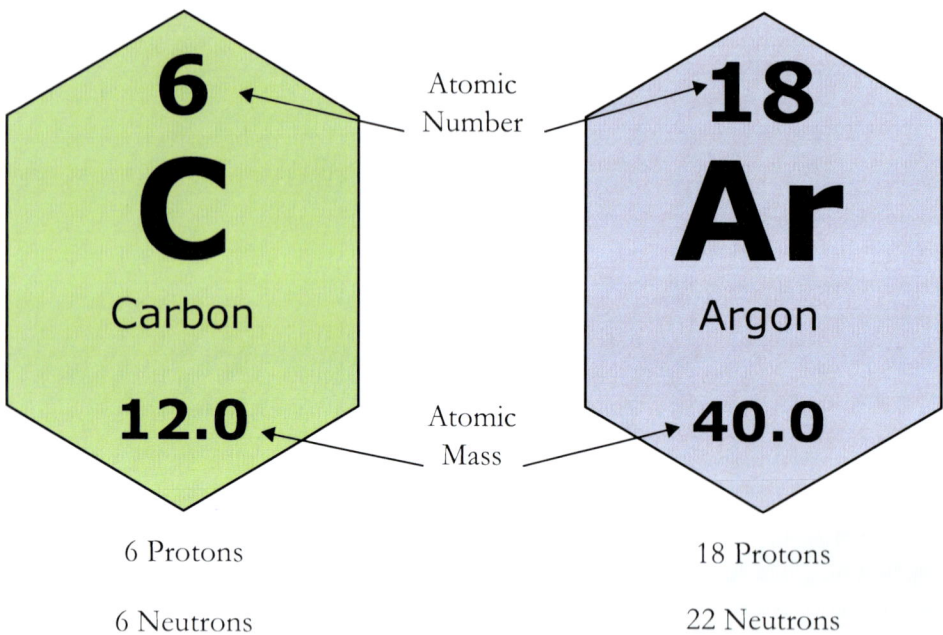

Figure 1.2– Atomic Mass and Atomic Number

From this we can see that the Atomic Number determines the number of Protons, and therefore the number of electrons in an atom. We can also calculate the number of neutrons in an atom by subtracting the Atomic Number from the Atomic Mass.

Molecules

A *Molecule* is a chemical combination of two or more atoms and is the smallest identifiable unit into which a pure substance can be divided whilst still retaining its composition and chemical properties.

In a compound, the molecule is the smallest particle that retains all the characteristics of that original compound.

If we examine one of the commonest forms of matter we know, e.g. water, we know that it is matter as it occupies space and has weight. Depending on the water's temperature, it may exist as a liquid, solid or as a gas, i.e. water, ice or steam respectively. However, regardless of the water's temperature, it always has the same composition.

If we now follow the original Greek idea and start with a quantity of water, and then divide this in half repetitively, we will ultimately end up with a quantity that cannot be divided any further without it ceasing to be water. When we reach this state, the quantity left is called a water molecule. As water is a compound, it can be chemically separated into its element parts and if this were done, we would be left with two (2) parts of the element Hydrogen and one (1) part of the element oxygen. Therefore, we can say that water consists of molecules that are two (2) parts Hydrogen to one (1) part oxygen, written as H_2O.

Molecules may also be made up of only one type of element. For instance, an Oxygen molecule consists of two atoms of Oxygen, and a Hydrogen molecule consists of two atoms of Hydrogen.

Some molecules may be made up of more than two types of elements. A common substance such as sugar is a compound composed of atoms of Carbon, Hydrogen, and Oxygen. These combine to form sugar molecules but as they can be broken down by chemical means into smaller and simpler elements, there cannot be any sugar atoms.

Atom Energy Levels

As an electron in an atom has mass and revolves in motion about the nucleus, following the normal laws of Physics, it has two types of energy. As it is moving around the nucleus an electron has *Kinetic Energy* and due to its position in space also has *Potential Energy*. The electron's total energy, i.e. potential and kinetic, determines the radius of the electron orbit and in order for it to remain in this orbit, by Newton's Laws of Motion, it must not lose or gain energy.

However, life at the sub-atomic level is never that simple and electrons are always losing and gaining energy. To simplify this explanation, we are going to use light as a means of altering the electron's energy level.

It is well known that light is a form of energy that can be converted into another form, e.g. solar panels. However, the physical form in which light energy exists is not so well known. One accepted theory looks upon the existence of light as tiny packets of energy called *Photons*, which can contain various quantities of energy. The energy amount depends upon the colour of the light involved, e.g. ultraviolet, infrared etc.

If light is shone at an atom and a photon of sufficient energy collides with an orbital electron, it will absorb some or even all of the photon's energy, figure 1.3.

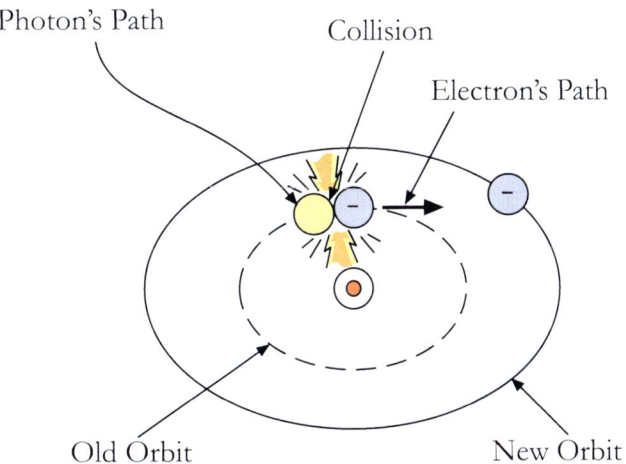

Figure 1.3 – Collision of Photon with an Electron

As the electron absorbs the energy from the photon, it jumps to a new orbit further from the nucleus. However, the new orbit to which the electron jumps is not a random one but has a radius four times larger than the original. Had the electron absorbed even more energy, the next possible orbit to which it could jump is nine (9) times that of the original!

This demonstrates that the electron cannot move randomly, but can only attain orbits that represent several energy levels, i.e. it cannot just jump to any orbit. In reality, the electron will remain in its lowest orbit until it can gain sufficient amounts of energy to allow it to jump to one of a series of permissible orbits. An electron cannot exist in the space between energy levels and the electron will not accept a photon of energy unless it is sufficient to elevate it to one of the higher energy levels.

This explanation has used photons of light as the energy source but heat energy and collisions with other particles can also cause the electron to gain sufficient energy to jump orbits.

Once an electron has been elevated to an energy level higher than the lowest possible, the atom is said to be in an excited state. However, these things are never static and the electron does not remain in this excited state for more than a fraction of a second before it releases the excess energy and returns to a lower energy orbit.

To illustrate this principle, assume that a normal electron has just received a photon of energy sufficient to raise it from energy level one (1) to three (3). In an instant, the electron may jump back to the first level emitting a new photon identical to the one it received. However, this will not always be the case and a second alternative would be for the electron to return to the lower level in two jumps, i.e. from the third to second to first. In this case the electron emits two photons, one for each jump, but each of these would have less energy than the original photon, which excited the electron.

This principle is used in the standard household fluorescent light where ultraviolet light photons, which are not visible to the human eye, bombard a phosphor coating on the inside of a glass tube. The phosphor electrons, in returning to their normal orbits, emit photons of light that are visible, i.e. the light illuminates its surroundings. Any number of colours can be achieved, including white, by using different chemicals for the phosphor coating, as illustrated when used in a TV.

Obviously, this basic principle applies to all elements. However, in atoms with two or more electrons, it gets more complicated as the electrons interact with each other and the exact path of any one electron becomes very difficult to predict. However, we do know that each electron lies in a specific energy band and their orbits are considered as an average of each electron's position.

Atomic Shells and Sub-shells

The difference between the 100+ known elements is dependent on the number and position of the electrons included within the atom. In general terms, when an element is in a stable sate, the electrons reside in a collection of orbits called *shells*. These shells are elliptical and are assumed to be positioned at fixed intervals. In this way, the shells are arranged in steps that correspond to fixed energy levels. The shells, and the number of electrons required to fill them, may be predicted by using *Pauli's* exclusion principle. This principle states that each shell will contain a maximum of $2n^2$ electrons, where 'n' corresponds to the shell number starting with the one closest to the nucleus. For example, by this principle, the second shell would contain $2(2)^2$ or 8 electrons when full.

In addition to being numbered, the shells are also given letter designations, as illustrated in figure 1.4.

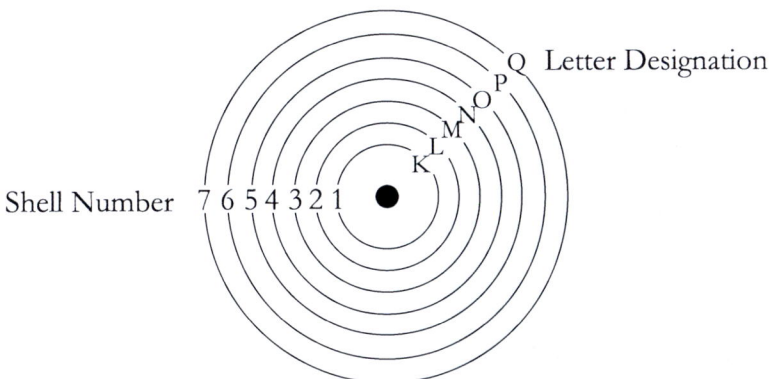

Figure 1.4 – Shell Designations

Starting with the shell closest to the nucleus and progressing outward, the shells are labelled 1 to 7 and K to Q, respectively. The shells are considered full or complete when they contain their full quota of electrons, i.e. two (2) in shell K, eight in shell L, 18 in shell M shell, etc in accordance with the exclusion principle mentioned earlier.

Each of the shells K to Q is a major shell and can be divided into one of four (4) sub-shells, s, p, d, and f. Like the major shells, the sub-shells are also limited to the number of electrons that they can contain, i.e. the 's' sub-shell is complete when it contains two (2) electrons, the 'p' sub-shell with six (6), the 'd' with ten (10), and the 'f' sub-shell with fourteen (14).

However, as mentioned above, the K shell can contain no more than two (2) electrons and so it must have only one sub-shell, i.e. the s sub-shell. The M shell is composed of three sub-shells, s, p, and d and if the electrons in these sub-shells are added together, the total is 18, i.e. the exact number required to fill the M shell.

Let us look at a practical example of an element widely used in electrical and electronic circuits, Copper. The copper atom has 29 electrons that completely fill the first three shells and sub-shells, leaving one electron in the 's' sub-shell of the N shell, illustrated in figure 1.5.

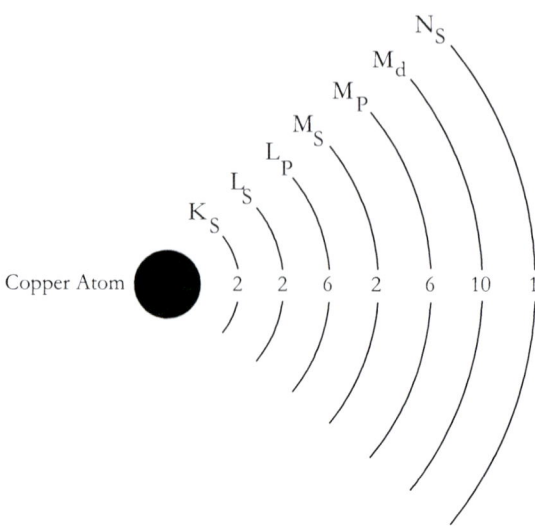

Figure 1.5 – Copper Atom with Electron Orbits

The number of electrons in an element's outermost shell determines the *Valence* of an atom. Therefore, for obvious reasons, an atom's outer shell is called the *Valence Shell* and the electrons in this shell are called *Valence Electrons*.

An atom's valence determines its ability to gain or lose an electron, which in turn determines the chemical and electrical properties of the atom. For example, an atom that is deficient of only one or two electrons from its outer shell will easily gain electrons to complete it, but requires a large amount of energy to free any of its electrons. An atom having a relatively small number of electrons in its outer shell, as illustrated with Copper in figure 1.5, in comparison with the number of electrons it requires to fill it will easily lose these valence electrons. This is because the valence electrons are more weakly attracted to the positive atomic nucleus than the inner electrons as they are a greater distance away. The valence electrons can therefore be shared or transferred in the process of bonding with adjacent atoms. They are also involved in the conduction of electric current in metals and semiconductors.

It should also be noted that the valence shell of an atom will never contain more than 8 electrons.

Ionisation

When an atom loses or gains electrons in this process of electron exchange, it becomes *Ionised*. For ionisation to occur there must be a transfer of energy that results in a change of the atom's internal energy. An atom having more than its usual quantity of electrons acquires a negative charge, and so is called a *Negative Ion*. Conversely, the atom that gives up some of its normal electrons is left with less negative charge and so is called a *Positive Ion*. Therefore, we can then define ionisation as the process by which an atom loses or gains electrons.

Conductors, Semiconductors and Insulators

Although we perhaps have laboured the point a little, the association of matter and electricity is important when discussing electricity, electronics and their respective components.

All elements made of matter may be placed into one of three categories:

- Conductors
- Semiconductors
- Insulators

How elements are categorised will depend on their ability to conduct an electric current. **Conductors** are elements that conduct electricity very readily while **Insulators** have an extremely high resistance to the flow of electricity and all matter between these two extremes are called **Semiconductors**.

As discussed previously, electron theory states that all matter consists of atoms and atoms are composed of smaller particles called protons, electrons, and neutrons. As far as the flow of electricity is concerned, it is the valence electrons that are the most important. These electrons are the easiest to break loose from their parent atom.

As a rule of thumb:

- conductors have three (3) or less valence electrons
- insulators have five (5) or more valence electrons
- semiconductors have four (4) valence electrons

Conductors

The electrical conductivity of matter and therefore by default a conductor, is reliant on the atomic structure of the material from which the conductor is made. In any solid material, atoms making up the molecular structure are bound firmly together. If we take a good conductor such as copper, at room temperature, it contains a considerable amount of heat energy. Since heat energy is one method of moving electrons between and even from their orbits, copper will have many free electrons that can move from atom to atom. When not influenced by an external force, these electrons move in a random and chaotic manner within the conductor, but this movement is equal in all directions so that no part of the conductor loses or gains electrons. However, when an external force is applied, the electrons move generally in the same direction. The effect of this movement is felt almost instantly from one end of the conductor to the other and this electron movement is called an **Electric Current**.

The movement of electrons to form an electric current is sometimes compared to the flow of a liquid. However, this is an over simplification and current flow occurs more due to electron collision, best illustrated by the smartie tube, shown in figure 1.6.

Figure 1.6 - Smarties Effect

The image in figure 1.6 is meant to illustrate electron flow as it really is, i.e. the electron at one end of a conductor that starts the motion or electric current is not necessarily the one that ends up at the other end. Put another way, putting a green smartie in doesn't guarantee a green one will come out.

The structure of material has a direct influence on how current flows. Material made from metals allows current to flow easily, while non-metals allow little, if any, current to flow. However, some metals are better conductors than others. Silver, copper, gold, and aluminium are materials with many free electrons and make good conductors. Gold and silver make the best conductors but are very expensive and so copper and aluminium are the most common materials used. Copper is the better conductor of the two but aluminium is used where weight is a major consideration, such as in aircraft construction.

As a broad statement, we can state that those materials that allow current to flow easily are conductors, while those that do not are insulators.

The ability of a conductor to handle current also depends upon its physical dimensions. Conductors are usually found in the form of wire, but may be formed as bars, tubes, or sheets for specific applications.

Insulators

Non-conductors or *Insulators* have few free electrons and examples of these materials include rubber, plastic, enamel, glass, dry wood, and mica. However, just as there is no perfect conductor, there is no perfect insulator and given enough energy, i.e. high voltage, even insulators will break down eventually.

Semiconductors

At normal room temperature, some materials are neither good conductors nor good insulators and as their electrical characteristics fall between those of conductors and insulators, these materials are classified as *Semiconductors*. As their name implies, semiconductors can be manipulated, under certain circumstances, to be either a conductor or insulator. Germanium and Silicon are two common semiconductors used in solid-state devices.

The conductive state of a semiconductor can be made to vary with temperature and also by adding impurities in the semiconductor crystal. In intrinsic semiconductors, usually made from pure crystals of germanium or silicon as mentioned above, conductivity rises with temperature.

However, the conductivity of extrinsic semiconductors depends on introducing impurities, typically arsenic or phosphorus, into intrinsic semiconductors, a process known as '*doping*'. Typical semiconductor devices such as diodes and transistors each have a different arrangement of impurities in the crystal making up their structure. This is discussed in more detail in Module 4 – Electronic Fundamentals.

Revision

Electron Theory

Questions

1. Matter can exist in one of:

    a) 2 states

    b) 6 states

    c) 3 states

2. Common salt is an example of:

    a) A mixture

    b) An Element

    c) A compound

3. Protons are:

    a) Positively charge

    b) Negatively charged

    c) Neutral

4. An electron moving around an atom has:

    a) Position energy

    b) Kinetic energy

    c) Heat energy

5. **Electron shells that surround an atom are designated outwards from the nucleus with letters and numbers as:**

    a) 1 to 7 & K to Q

    b) 7 to 1 & K to Q

    c) 1 to 7 & Q to K

6. **Atom shells can be further split into four levels that are designated as:**

    a) 1 to 4

    b) s, p, d, f

    c) p, q, r, s

7. **In an atom's L shell, there can be a maximum of:**

    a) 8 electrons

    b) 2 electrons

    c) 18 electrons

8. **An atom's outer shell is called its:**

    a) Valence shell

    b) Extremity shell

    c) Positive shell

9. **If an atom acquires an electron it becomes:**

    a) A positive Ion

    b) A neutral Ion

    c) A negative Ion

10. Elements that are insulator have:

a) 3 or more electrons in their outer shell

b) 5 or more electrons in their outer shell

c) 4 electrons in their outer shell

Revision

Electron Theory

Answers

1. C
2. C
3. A
4. B
5. A
6. B
7. A
8. A
9. C
10. B

Static Electricity and Conduction

Introduction

Static Electricity, also called ***Electrostatics***, is simply the study of electricity at rest. Most, if not all, of us are probably aware of static electricity from the games we played as children. Remember, for example, the way a balloon sticks to objects after it has been rubbed against a woollen jumper or the way a person's hair stands on end after a vigorous rubbing. Although you probably did not know it as such, these are good example of the effect of electrostatics.

The first studies associated with static electricity can be traced back to the Greeks again. A Greek philosopher and mathematician known as Thales of Miletus discovered that when an amber rod was rubbed with fur, the rod attracted some light objects, cloth and shavings of wood.

However, as with other branches of science, it took until the 17th century before anyone studied its affect in detail. At the start of the 17th century an English scientist called William Gilbert, made a study of other substances that were found to have the same attraction qualities as amber. Among these were glass, when rubbed with silk, and ebonite, when rubbed with fur.

As a result of his experiments, William Gilbert classified all substances possessing similar properties to amber as electrics, a word of Greek origin meaning amber.

Because of Gilbert's work with this new subject of electrics, a substance such as ebonite, amber or glass when rubbed vigorously was acclaimed to be ***Electrified***, or ***Charged*** with electricity.

Gilbert's work was taken further by other scientists of that time and beyond and in the year 1733 a French scientist, Charles Dufay, made an important further discovery about electrification. He discovered that when a glass was rubbed with fur, both objects became electrified or charged.

He realised this when he systematically placed the glass rod and the fur near other electrified materials and found that certain substances, which were attracted to the glass rod, were repelled by the fur, and vice versa. From further experiments like this, he concluded that there must be two exactly opposite kinds of electricity.

However it is the American statesman, inventor, and philosopher, Benjamin Franklin, who is credited with first using the terms ***Positive*** and ***Negative*** to describe these two opposite kinds of electricity. He labelled the charge

CHAPTER TWO
STATIC ELECTRICITY AND CONDUCTION

produced on a glass rod when it was rubbed with silk positive, and the charge on the silk negative. He also labelled those bodies that were not electrified or charged, as *Neutral*.

Static Electricity

As already discussed in chapter 1, in a natural or neutral state, each atom of matter will have the correct number of electrons in orbit around it. Therefore, the whole body of matter composed of the neutral atoms will also be electrically neutral and in this state, is said to have a '*zero charge*.' When a substance is neutrally charged, electrons will not be gained or lost if it comes into contact with other neutral bodies.

However, if a substance loses any electrons from its atoms, there will subsequently be more protons than electrons and it will become *Electrically Positive*. If this positively charged body subsequent came into contact with another substance having a normal charge or negative charge, i.e. too many electrons, an electric current will flow between them. This will occur because electrons will leave the more negative substance and enter the positive one to achieve neutrality again. Indeed, electron flow will continue until both bodies have equal charges.

In the same way, when two (2) substance having unequal charges are near one another but not touching, an electric force is exerted between them because of their unequal charges. However, since they are not in contact, their charges cannot equalise. The existence of such an electric force, where current cannot flow, is called *Static Electricity* or an *Electrostatic Force*.

Friction is one of the easiest ways to create a static charge. When two pieces of matter are rubbed together, electrons can be '*wiped off*' one material onto the other. However, if the materials are good conductors, it is quite difficult to obtain a detectable charge on either, as equalising currents can flow easily between the conducting materials and these currents equalise the charges almost as fast as they are created. A static charge is more easily created between non-conducting materials. When a hard rubber rod is rubbed with fur, the rod will accumulate electrons given up by the fur, as illustrated in figure 2.1.

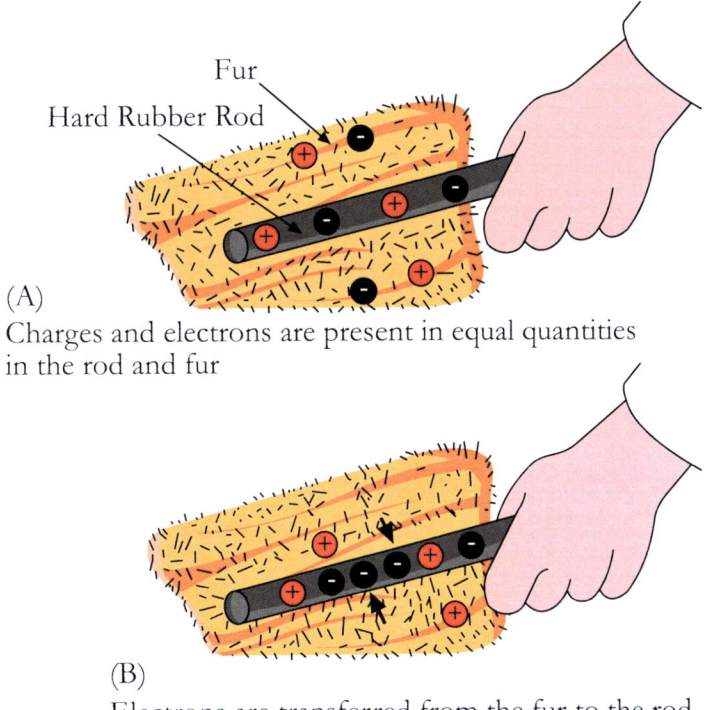

(A) Charges and electrons are present in equal quantities in the rod and fur

(B) Electrons are transferred from the fur to the rod

Figure 2.1 – Build up of Static Electricity

As both materials are poor conductors, very little equalising current can flow, and an electrostatic charge builds up. However, when the charge becomes great enough current will flow regardless of the poor conductivity of the materials and this can cause visible sparks and produce a crackling sound.

Triboelectric Series

Determining which material will acquire the negative charge and which will acquire the positive charge depends on the relative properties of the materials (ie which has a higher affinity for the addition of electrons). A triboelectric series "ranks" different materials according to their ability to gain or tendency to lose electrons when compared to other materials. The conductivity of a material is irrelevant in terms of its ability to lose or gain electrons. Figure 2.2 shows a typical triboelectric series chart.

CHAPTER TWO
STATIC ELECTRICITY AND CONDUCTION

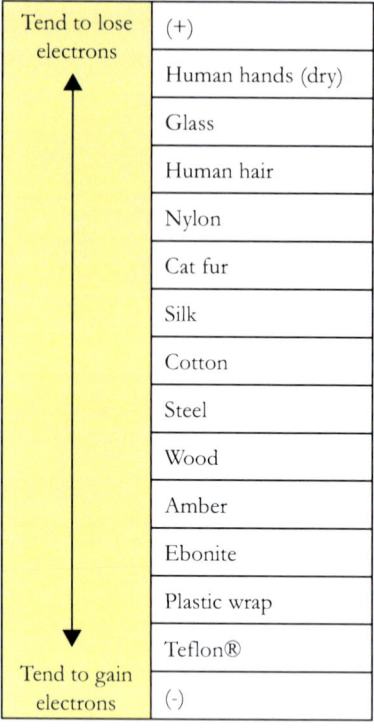

Figure 2.2– Typical Triboelectric Series Chart

Using this chart we can see that cotton is often regarded as the neutral material. If you rub a material that is closer to the negative end of the series with a material that is closer to the positive end of the series, the items will charge negatively and positively, respectively.

Charged Bodies

As mentioned above, when in a natural or neutral state, an atom has an equal number of electrons and protons; i.e. in this balanced state the net negative charge of the electrons in orbit is exactly balanced by the net positive charge of the protons in the nucleus, making the atom electrically neutral.

An atom becomes a *Positive Ion* whenever it loses an electron, and has an overall positive charge and conversely, whenever an atom acquires an extra electron; it becomes a *Negative Ion* and has a negative charge.

Due to normal molecular activity at the sub-atomic level, there are always ions present in any material. If the number of positive ions and negative ions is equal, the material is electrically neutral. However, when the number of positive ions exceeds the number of negative ions, the material is positively charged and is negatively charged whenever the negative ions outnumber the positive ions.

Since ions are actually atoms without their normal number of electrons, it is the excess or the lack of electrons in a substance that determines its charge. In most solids, the transfer of charges is by movement of electrons rather than ions.

One of the fundamental laws of physics and electricity that we probably all know is:

Like charges repel each other/unlike charges attract each other.

Following on from this then, when positive and negative charges come close together, being unlike, they tend to move towards each other. In the atom, the negative electrons are drawn toward the positive protons in the nucleus but this attractive force is balanced by the electron's centrifugal force caused by its rotation about the nucleus. As a result, the electrons remain in orbit and are not drawn into the nucleus. Electrons repel each other because of their like negative charges, and protons repel each other because of their like positive charges.

The law of charged bodies may be demonstrated by a simple experiment that uses two (2) paper balls suspended near one another by threads, figure 2.3.

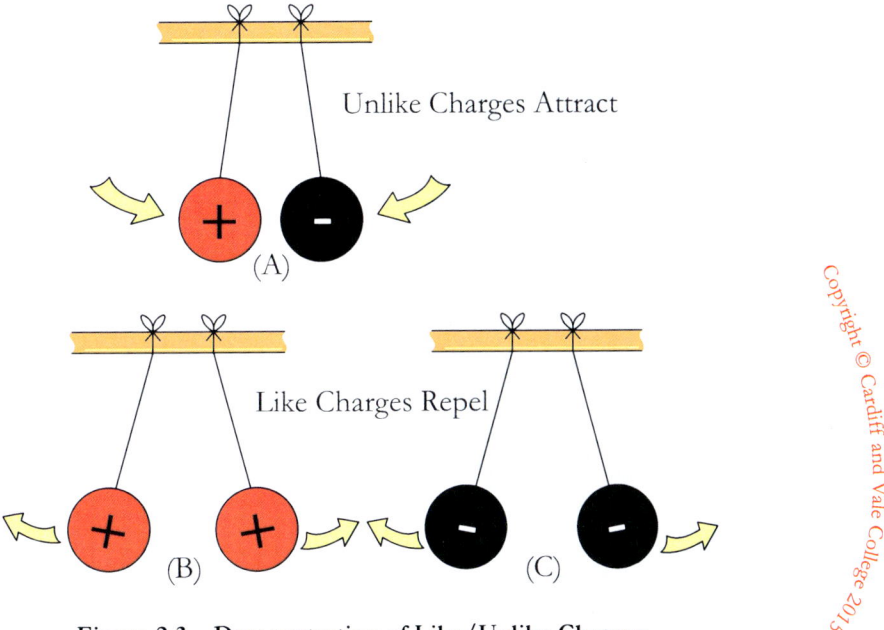

Figure 2.3 – Demonstration of Like/Unlike Charges

If a rubber rod is rubbed with fur to give it a negative charge and is then held against the right-hand ball the rod will give a negative charge to the ball and it will then have a negative charge with respect to the left-hand ball. When released, the two balls will be drawn together, as shown in figure 2.3 (A) and they will touch and remain in contact until the left-hand ball gains a portion of the negative charge of the right-hand ball, at which time they will swing apart as shown in figure 2.3 (C). If a positive or a negative charge is placed on both balls, figure 2.3 (B), the balls will repel each other.

CHAPTER TWO
STATIC ELECTRICITY AND CONDUCTION

Gold Leaf Electroscope

One of the earliest instruments used for testing positive and negative charges was a Gold Leaf electroscope, Figure 2.4.

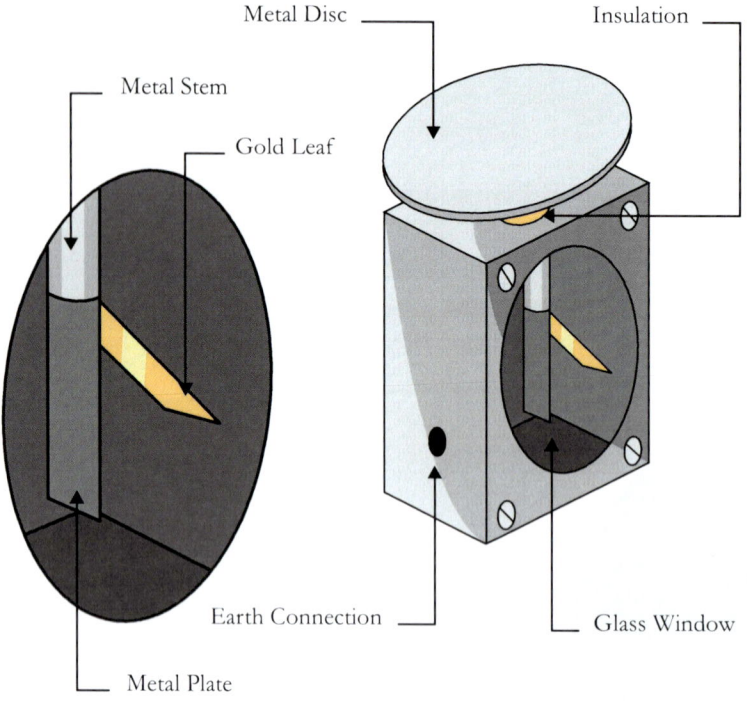

Figure 2.4 – Gold Leaf Electroscope

A metal disc is connected to a narrow metal plate and a thin piece of gold leaf is fixed to the plate. The whole of this part of the electroscope is insulated from the body of the instrument. A glass front allows you to monitor the behaviour of the leaf against a graduated scale and therefore measure the stored charge. When a charge is put on the disc at the top it spreads down to the plate and leaf. This means that both the leaf and plate will have the same charge. Similar charges repel each other and so the leaf rises away from the plate - the bigger the charge the more the leaf rises. Figure 2.5.

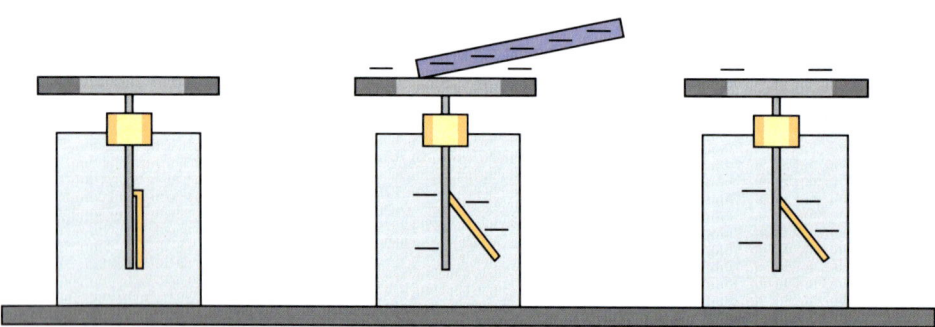

Figure 2.5 – Charging the electroscope

The leaf can be made to fall again by touching the disc - you have effectively earthed the electroscope. The case itself is earthed to prevent it from becoming live.

Coulomb's Law of Charges

The relationship between attracting and/or repelling charged objects was first discovered and written about by a French scientist called Charles A. Coulomb. From 1784 onwards he conducted a series of very delicate experiments, using a torsion balance he had invented himself, capable of detecting forces equivalent to 10^{-5}g. He discovered that the force between two charged poles is inversely proportional to the square of the distance between them and directly proportional to the product of their magnitude. He formalised the relationship, now known as *Coulomb's Law*, which states that:

> *'Charged bodies attract or repel each other with a force that is directly proportional to the product of their individual charges, and is inversely proportional to the square of the distance between them.'*

Put in simple English, this means that the amount of attracting or repelling force which acts between two electrically charged bodies in free space depends on two things:

1. Their charges

2. The distance between the objects

This was a major step forward in the study of electricity and paralleled Newton's law of gravitational attraction. The SI unit of electric charge, the *Coulomb*, SI symbol '*Q*', is named in his honour and is defined as the charge crossing any section of a conductor in which a steady current of one (1) ampere flows for (1) second, commonly abbreviated to 1 A/second.

The charge of one electron might have been used as a unit of electrical charge, as charges are created by displacement of electrons. However, the charge of one electron is so small that it is impractical to use. One coulomb is equal to the charge of 6.24×10^{18} electrons, a big figure by anyone's standards.

Electric Fields

The space between and around charged objects in which their influence is felt is called an *Electric Field of Force*. It can exist in air, glass, paper, or a vacuum. *Electrostatic Fields* and *Dielectric Fields* are other names also used to refer to this area of force.

These fields of force spread out in the space surrounding their point of origin and in general, reduce in proportion to the square of the distance from their source.

CHAPTER TWO
STATIC ELECTRICITY AND CONDUCTION

The field about a charged body is generally represented by lines that are called *Electrostatic Lines of Force*. These lines are imaginary and are used merely to represent the direction and strength of the field. To avoid confusion, the lines of force exerted by a positive charge are always shown leaving the charged body, and for a negatively charged body they are shown entering. Figure 2.6 illustrates the use of lines to represent the field about charged objects.

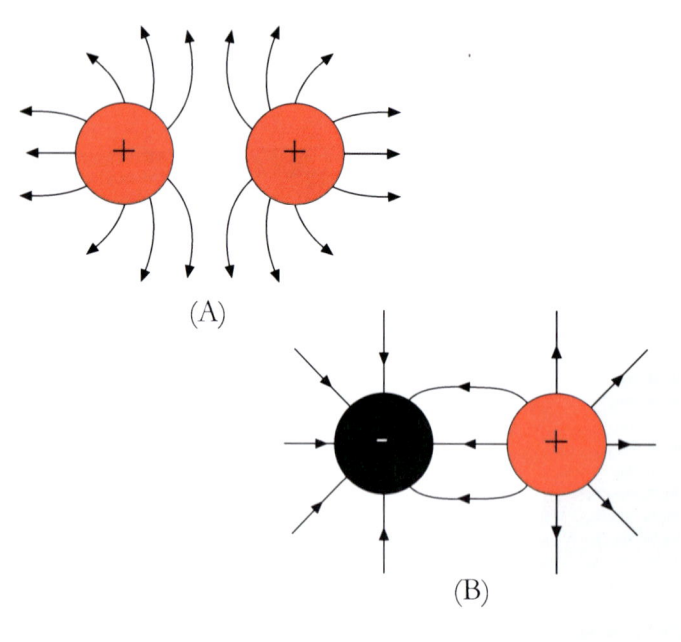

Figure 2.6 – Electrostatic Lines of Force

Figure 2.6 (A) represents the repulsion of like-charged objects and their associated fields while (B) represents the attraction of unlike-charged bodies and their associated fields. This is covered in more detail in Chapter 10 - Magnetism.

Conduction of Electricity in Solids, Liquids, Gases & a Vacuum

As we have already discussed, the first observed artificial electrical phenomenon was static electricity when certain substances become negatively charged when rubbed with a piece of fur or woollen cloth. Amber was one of the first substances used, which became negatively charged, but it was soon discovered that a glass rod rubbed with silk had similar powers and could attract charged objects even more strongly. However, the glass had a *positive* charge, owing to a deficiency of electrons, showing how the electrostatic effect could be demonstrated with different materials.

As we now know, protons lie in the atom's nucleus and are effectively fixed in position in solid material so that when charge moves in a solid, it is carried by its negatively charged electrons. Electrons are easily liberated in some materials, i.e. *conductors*, and all metals, particularly copper, gold and silver, are good examples of these.

Materials in which the electrons are tightly bound to the atoms are called *insulators*, *non-conductors*, or *dielectrics* and good examples of these include glass, rubber, and dry wood.

However, there is a third kind of material where a solid has a relatively small number of electrons that can be freed from their atoms in such a way as to leave a '*hole*' where each electron had been. The hole, representing the absence of a negative electron, behaves as though it were positively charged. An electric field will cause the negative electrons to flow one way through the material and positive holes to move the other way, so producing a *current of electricity*. Such a solid, called a *Semiconductor*, generally has a higher resistance to the flow of current than a conductor such as copper, but a lower resistance than an insulator such as glass. If the negative electrons carry most of the current, the semiconductor is called N-*type* and if most of the current is carried by the positive holes, the semiconductor is said to be P-*type*.

If a material were a perfect conductor, a charge would pass through it without resistance, while a perfect insulator would allow no charge to be forced through it. No substance of either type is known to exist at room temperature. The best conductors at room temperature offer a low, but not zero, resistance to the flow of current. The best insulators offer a high, but not infinite, resistance at room temperature. Most metals, however, lose all their resistance at temperatures near absolute zero, i.e. 0°K; this phenomenon is called *Superconductivity*.

Conduction of Electricity in Solids

The only solids that conduct electricity freely at room temperature are metals, found on the left side of the periodic table, and graphite, which is one form of the carbon element. Nearly all of the other solids in the world, i.e. non-metal elements, solid ionic and covalent compounds, are non-conductors of electricity, i.e. are insulators.

As discussed in chapter one, metal atoms have few electrons in their outer shell, e.g. copper has only one (1), and these are not securely tied in orbit around its nucleus. This outer shell is the highest energy band of any atom occupied by electrons and is called the *Valence* band. As the valence band is only partially filled with electrons in a metal, there are numerous empty levels and electrons move freely between atoms within the structure of the metal, though in a random way. However, under the influence of an electric field, usually created by a *Potential Difference (PD)* across it due to a battery or other power source, the flow is in one direction only as the negative electrons are attracted to the positive side of the PD creating a current carried entirely by electron motion, illustrated in figure 2.7.

CHAPTER TWO
STATIC ELECTRICITY AND CONDUCTION

Figure 2.7 – Current flow in a Metal

Conduction of Electricity in a Liquid

The only liquid elements that conduct electricity are the liquid metals, e.g. at room temperature liquid mercury is a conductor, and other metals will also continue to conduct electricity when melted. Pure covalent liquids, e.g. water, alcohol, propane, hexane etc, are all non-conductors of electricity and even solid covalent substances remain non-conductors when melted.

Covalent bonding is the sharing of electrons between atoms. This type of bonding occurs between two of the same element or elements close to each other in the periodic table. When molecules have the same affinity for electrons, covalent bonds are most likely to occur. Since both atoms have the same affinity for electrons and neither is willing to donate them, they share electrons in order to become more stable. Figure 2.8.

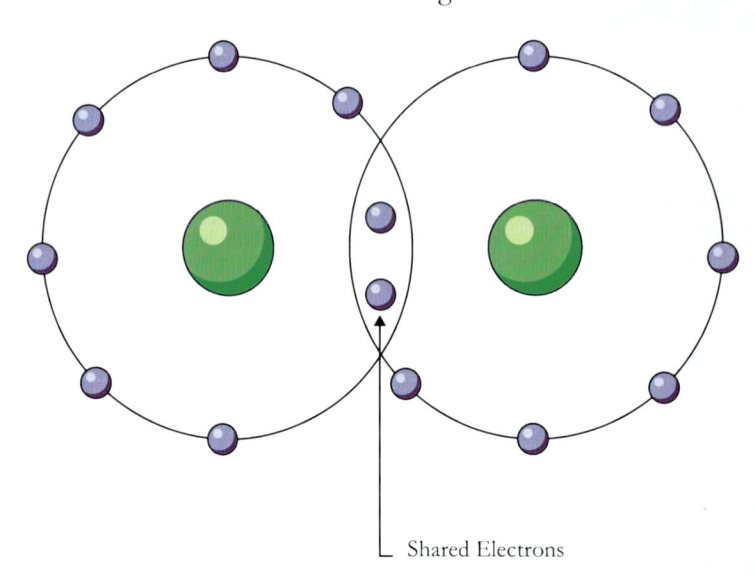

Figure 2.8 - Covalent Bonding

These elements are sharing their electrons because neither one of them has the strength to take the electrons from the other.

Ionic bonding is the complete transfer of valence electron(s) between atoms. It is a type of chemical bond that generates two oppositely charged ions. Figure 2.9.

Electron is given away

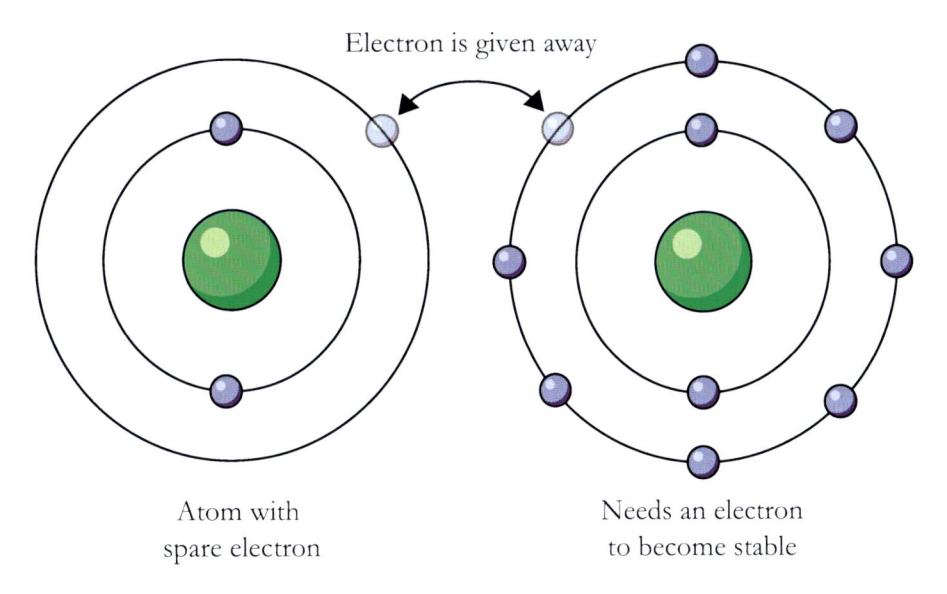

Atom with
spare electron

Needs an electron
to become stable

Figure 2.9 – Ionic Bonding

Ionic substances therefore do conduct electricity when they are melted as they are made of charged particles, i.e. positive and negative ions. In the solid state, ionic substances are held very firmly in place in a lattice structure and the ions cannot move about, figure 2.10.

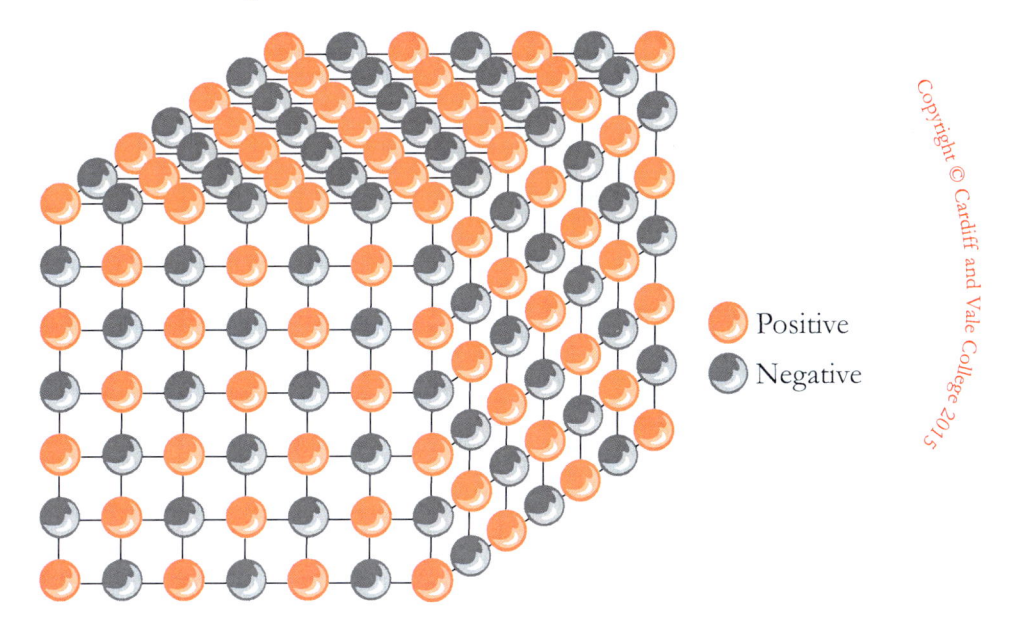

Positive
Negative

Figure 2.10 – Ionic Lattice

When the ionic solid is melted, however, the bonds holding the ions in place in the lattice break and the ions can then move about freely. When an electric current is applied to an ionic melt the positive and negatives ions polarise and electricity is carried by the ions that are now able to move freely.

Water Solutions that Conduct Electricity

As mentioned earlier, water is an insulator so how do we explain the danger of bringing water and electricity into proximity with each other? Pure or distilled water will not conduct electricity, but everyday water we come across, e.g. from a tap, is not pure but contains contaminants, minerals, salts etc, and it is these substances that create ions that allow electricity to flow.

Any ionic solid dissolved in water will allow electricity to flow as the ionic lattice, as mentioned above, breaks up and the ions become free to move around in solution. A liquid that allows current flow is called an *Electrolyte* and all ionic solutions and ionic melts are electrolytes.

Note: Covalent substances do not conduct in solution.

Conduction of Electricity in a Gas

In gases, electrical conductivity is very low and they act as an insulator or dielectric in their neutral state. However, if an electric field is applied to the gas in a confined environment, e.g. a tube, once it reaches a *breakdown value*, which will be different for each gas, it frees the valence electrons from the atoms in an avalanche process and forms a *plasma*. A plasma is an ionised gas and in chemical terms is considered to be a distinct phase of matter like solid, liquids and gases. This fourth state of matter was first identified by the British Chemist *Sir William Crookes (1832 to 1919)* in 1879 and named 'plasma' by the USA chemist *Irving Langmuir (1881 to 1957)* in 1928, as it reminded him of a blood plasma. The free electrons make the plasma electrically conductive and so current will flow as long as there is a potential difference maintained across it.

Although gases in a plasma state do contain ions as the atoms lose electrons, due to their much lower mass, the electrons accelerate quicker than the heavier positive ions in response to an electric field and so carry the bulk of the current.

Conduction of Electricity in a Vacuum

Vacuums under normal conditions are perfect insulators as they usually do not contain any charged particles. However, if a metal electrode is placed in a vacuum and is then heated, electrons are released through a process called *Thermionic Emission*. Thermionic emission is defined as the flow of electrons due to thermal vibration energy overcoming the electrostatic forces holding the electrons in place on the metal's surface. If a potential difference is then placed across the vacuum, current will flow as the electrons are attracted towards the positive potential.

With this arrangement, the metal electrode creating the free electrons is called the *Cathode*, while the positive terminal is called the *Anode*.

Revision

Static Electricity and Conduction

Questions

1. With electrostatics:

   a) Like charges attract

   b) Unlike charges attract

   c) Unlike charges repel

2. Charged bodies attract or repel each other with a force that is:

   a) Directly proportional to the distance between them

   b) With the same force irrespective of the distance between them

   c) Inversely proportional to the distance between them

3. Charged bodies attract or repel each other with a force that is:

   a) The square of the individual charges

   b) The sum of the individual charge

   c) The product of the individual charges

4. The SI unit of electric charge the Coulomb, has the SI symbol:

   a) Q

   b) C

   c) F

5. One Coulomb is equivalent to a current flow of:

   a) 10 A/second

   b) 1 A/second

   c) 0.1 A/second

Revision

Static Electricity and Conduction

Answers

1. **B**
2. **C**
3. **C**
4. **A**
5. **B**

Electrical Terminology

Introduction

As with all walks of life, electricity has its own terms, units and nomenclature. Electricity is a broad name that encompasses the study of electrical charges and currents and their influences. As we know from chapter 2, there is a distinction between positive and negative charges and if a conducting path exists between groups of positive and negative charge, like water, the charges will attempt to become level by electrons flowing from one to the other. This flow of electrons is what constitutes an electric current.

Electric Current (I)

In any electrical conductor there are numerous free electrons. Forcing these free electrons to flow in one direction is what creates an electric current. The unit of measurement of electric current is the *Ampere (A)* and the SI symbol used for current is '*I*'. The textbook definition of current is more complex:

> *'The ampere is that current, which if maintained in two (2) straight parallel conductors of infinite length, or negligible circular cross-section and placed one (1) metre apart in a vacuum, would produce a force between these conductors of 2 × 10^7 Newton/metre of length.'*

This definition is derived from experimentation and mathematic principles that are way beyond what you need to know for this Module, but by using this definition, most other electrical units take on a more suitable scale. In more basic terms, a current of one ampere (1A) equates to the movement of 6.24 × 10^{18} electrons per second.

However, to produce current, as already discussed above, the electrons must be moved by a difference in charge between two (2) points, called a *Potential Difference*. For example, if a copper wire is placed between two charged objects that have a potential difference, such as a battery, all of the negatively charged free electrons will feel a force pushing them from the negative charge towards the positive charge, illustrated in figure 3.1.

CHAPTER THREE
ELECTRICAL TERMINOLOGY

Figure 3.1 - Electron Flow through a Copper Wire with a Potential Difference

The direction of electron flow, shown in figure 3.1 is from the negative (-) side of the battery, through the wire, and back to the positive (+) side of the battery, i.e. the direction of ***Electron Flow*** is from a point of negative potential to a point of positive potential.

As electrons vacate their atoms during electron current flow, the atoms become positively charged, sometimes represented in textbooks as positive 'holes'. Looking at the flow of current in this way, we can say that the flow of electrons in one direction causes a flow of positive charges in the opposite direction. This flow of positive charges is known as ***Conventional Current*** flow and is illustrated in figure 3.2 below as a dashed arrow.

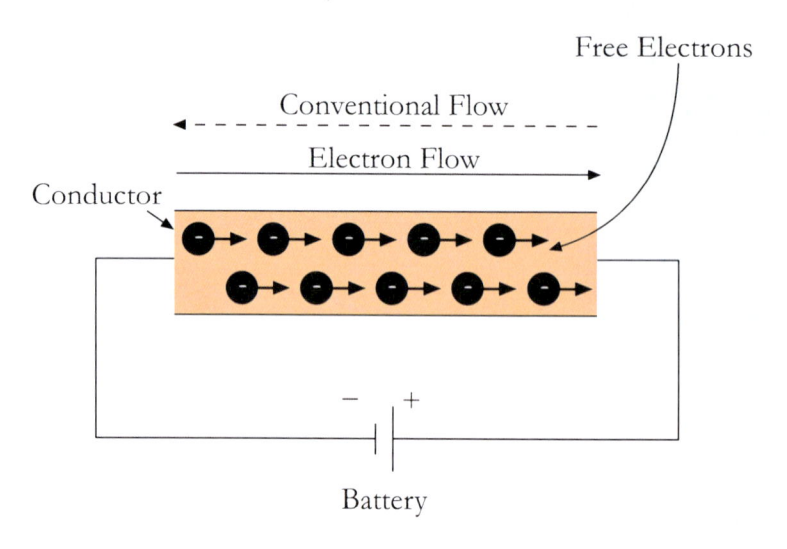

Figure 3.2 – Schematic showing Conventional and Electron Flow

All of the electrical effects of ***Electron Flow***, i.e. from negative to positive, or ***Conventional Flow*** from a higher potential to a lower potential, are exactly the same as those if we considered the flow of positive charges in the opposite direction and it is important to realise that both conventions are in use today to describe current flow. In the USA electron flow is the preferred method of representing current flow while in the UK and Europe, the use of conventional flow is more predominant. However, they are essentially equivalent, i.e. all predicted affects, e.g. heating, mathematical calculations etc, are the same, but this will only result in accurate work if you stick to one or other of the conventions throughout your work.

The confusion came about because in early experimentation with electricity, it was thought that current flowed from positive to negative and this was put forward by the American inventor *Benjamin Franklin (1706 to 1790)*. This idea was widely accepted, until in 1898, when the British physicist *Sir Joseph J Thomson (1856 to 1940)* discovered the electron. However, conventional current flow was by then in such common use that it has prevailed until today, especially within Europe.

Generally, electric current flow can be classified as one of two general types, *Direct Current (DC)* or *Alternating Current (AC)*. A direct current flows continuously in the same direction while an alternating current periodically reverses direction. We will be studying DC and AC current in more detail later in this module, but an example of DC current is that obtained from a battery while an example of AC current is our common household supply.

Electromotive Force (EMF)

As mentioned previously, an electric current is created by forcing free electrons to move around a circuit in one direction. This force is called the *Electromotive Force (EMF)* and can come from many sources, e.g. a battery. However, EMF is technically not really a force and can best be described as the driving influence that causes a current to flow, representing the energy expended in passing a unit of charge. An EMF is perhaps best thought of as always being connected with energy conversion.

From this it is perhaps a little too obvious to state, but for a current to flow in a circuit, it must have an EMF. The SI unit of measurement for EMF is the *Volt (V)* and it has the SI symbol '*E*'.

Potential Difference (PD)

Potential Difference (PD) is a measurement of the difference in voltage between two points in a circuit and can also be a measurement of how large the electrostatic force is between two charged objects. As you already know, according to Coulomb's Law, charged bodies attract or repel each other with a force that is directly proportional to the product of their charges and is inversely proportional to the square of the distance between them.

For a current to flow between any two points in an electrical circuit there must be a potential difference between them. Figure 3.3 shows a resistor with a voltage of 12 volts at point 'A', and a voltage of 6 volts at point 'B'.

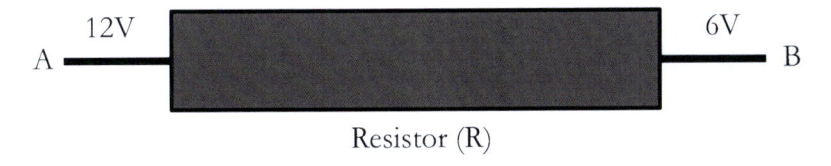

Resistor (R)

Figure 3.3 - Potential difference

The potential difference between these points is:

> 12 Volts - 6 Volts = 6 Volts

The SI unit of measurement of *Potential Difference (PD)* is the *Volt (V)* and has the SI symbol '*V*'.

Note: When discussing a circuit that has a flowing current the terms '*potential difference*' and '*voltage drop*' mean exactly the same thing.

Voltage (V)

As mentioned above, the SI unit of electrical potential difference is the Volt with the symbol V. This was named after *Alessandro Volta*, an Italian Physicist who invented the battery. The technical definition of Voltage is:

If the power dissipated between two (2) points along a wire carrying a current of one (1) ampere is one (1) watt, then the potential difference between the two (2) points equals one (1) volt.

More on this later in this Module when we discuss *Ohm's Law*.

Resistance (R)

From previous chapters, we know that current is the directed flow of electrons. We also know that some materials offer little opposition to current flow, while others significantly oppose current flow. This opposition to current flow in an electric or electronic circuit is defined as *Resistance*, which has the SI symbol '*R*'. Its SI unit of measurement is the *Ohm* represented by the Greek letter omega (Ω). As resistance opposes the flow of electrons, if current flows through a resistance then work is done, usually manifested as heat. The technical definition of resistance is that:

A conductor has a resistance of one ohm (1Ω) when an applied potential of one (1) volt produces a current of one (1) ampere.

Resistance, although an electrical property, is determined by the physical structure of a material and this is governed by many of the same factors that control current flow. The magnitude of resistance is determined in part by the number of '*free electrons*' available within the material as a decrease in the number of free electrons will obviously decrease the current flow. It follows then that resistance is greater in a material with fewer free electrons. Therefore, depending upon their atomic structure, different materials have different quantities of free electrons and so the various conductors used in electrical applications have different values of resistance. From this, we can say, rather obviously perhaps, that the best conductors will have the least resistance.

A material's cross-sectional-area also greatly affects the magnitude of resistance. If the cross-sectional area of a conductor is increased, a greater quantity of electrons is available to move through the conductor and so a larger current will flow for the same amount of applied voltage. An increase in current indicates that as the cross-sectional area is increased, the resistance must have decreased and conversely if it is decreased, the resistance increases and the current decreases. Therefore, *the resistance of a conductor is inversely proportional to its cross-sectional area*.

A conductor's resistance is also influenced by its length. If the conductor's length increases, the amount of energy given up increases as free electrons moving from atom to atom dissipate energy as heat. From this we can see that the longer a conductor, the more energy is lost to heat. This energy loss subtracts from the energy being transferred through the conductor, which results in a decrease in current flow for a given applied voltage. A decrease in current flow indicates an increase in resistance, if the voltage is held constant. Therefore, if the length of a conductor is increased its resistance also increases and so *the resistance of a conductor is directly proportional to its length*.

Temperature changes affect the resistance of materials in different ways. In some materials an increase in temperature causes an increase in resistance, whereas in others, an increase in temperature causes a decrease in resistance. The amount of change of resistance per unit change in temperature is known as the *Temperature Coefficient*. If for an *increase* in temperature the resistance of a material *increases*, it has a *positive* temperature coefficient, while a material whose resistance *decreases* with an *increase* in temperature has a *negative* temperature coefficient.

Conductance

Electricity as a technical subject is frequently explained in terms of opposites. *Conductance* or *Conductivity* is the opposite of resistance and is an indication of the capacity of a material to pass electrons. The factors that affect the amount of resistance are exactly the same for conductance, but obviously affect conductance in the opposite way. Therefore, conductance is directly proportional to area, and inversely proportional to the length of the material. Conductivity is therefore the inverse of resistivity (R) and can be determined from the voltage and current values according to Ohm's law. The material's temperature is also a definite factor, but if we assume a constant temperature, the conductance of a material can be calculated by dividing the current flowing in a circuit by its voltage, i.e. I/V.

The old unit for conductivity was the Mho (G), i.e. ohm spelt backwards, but this has now been replaced by the SI term *Siemens (S)*. From this we can see that the relationship between resistance (R) and conductance (G) or (S) is a reciprocal one, i.e.:

$$R = \frac{1}{S} \quad \text{and} \quad S = \frac{1}{R}$$

For example a circuit with a resistance of 25Ω will have a conductance of:

$$S = 1 \div 25 = 0.04 \text{ Siemens}$$

Electrical Charge

From the earlier chapters on electrostatics, we know that an electrical system generally transmits or transfers energy due to the movement of *Electric Charge*. We also discovered that the strength of the electrostatic field is directly dependent on the force of the charge.

For simplicities sake, we could use the charge of one electron as a unit of electrical charge, as charges are created by dislodging electrons. However, the charge of one electron is so small that it is impractical to use. Therefore, the practical unit adopted for measuring charges is the *Coulomb*, named after the French scientist Charles Coulomb, which has the SI unit '*Q*'. One coulomb is equal to the charge of 6,240,000,000,000,000,000, i.e. six quintillion two hundred and forty quadrillion or 6.24×10^{18} electrons.

The technical definition states that:

> *When a current of one (1) Ampere is maintained for one (1) second, the quantity of electrical charge passing a single point is one (1) Coulomb (C).*

The displacement or movement of electrons creates electrical charges and so there must be an excess of electrons at one point, and a deficiency at another. Consequently, a charge must always have either a negative or positive polarity and a body with an excess of electrons is negative, and one with a deficiency is positive. However, a difference of potential can only exist between two points in a circuit if they have different charges. Figure 3.4.

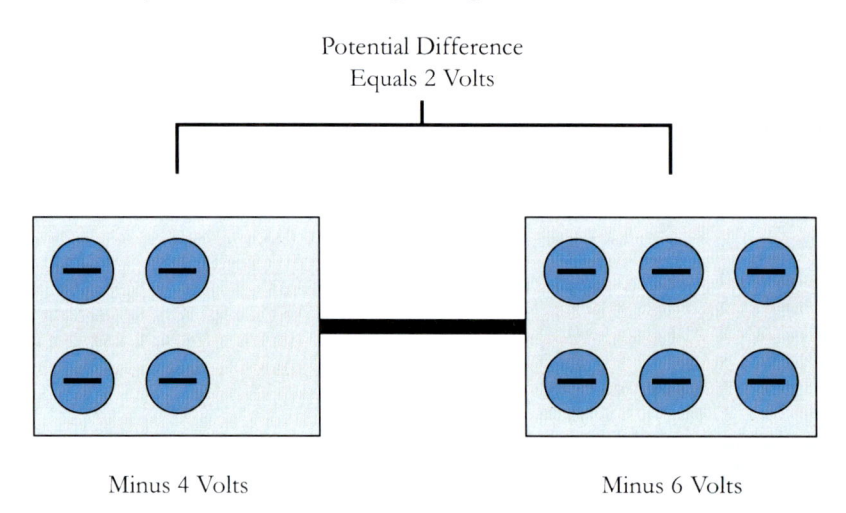

Figure 3.4 – Different Charge States

In other words, there is no difference in potential between two points if both have a deficiency or excess of electrons to the same degree. Figure 3.5.

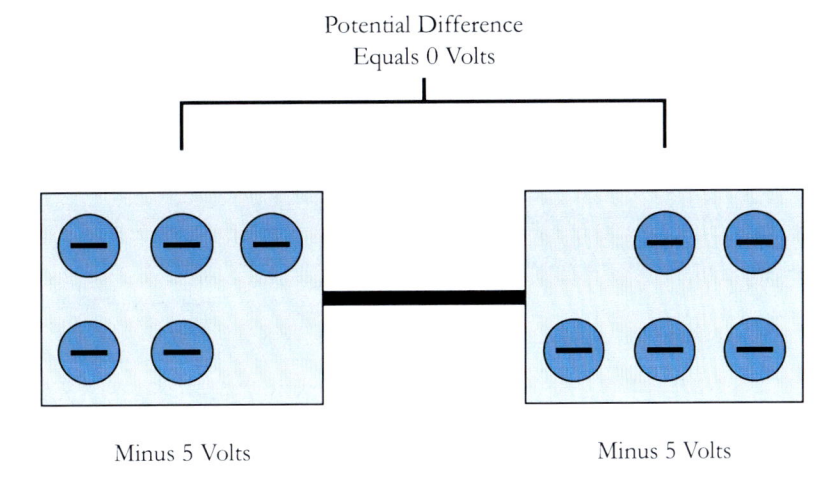

Figure 3.5 – Identical charges

In most electrical circuits only the difference of potential between two points is of importance and the absolute potentials of each point is of little significance. In most, if not all, circuits it is convenient to use one standard reference for all of the various potentials throughout a piece of equipment. For this reason, the potentials at various points in a circuit are generally measured with respect to the metal chassis on which all parts of the circuit are mounted. The chassis is considered to be at zero potential and all other potentials are either positive or negative with respect to it. When used as the reference point, the chassis is said to be at *Ground* or *Earth Potential*. Figure 3.6

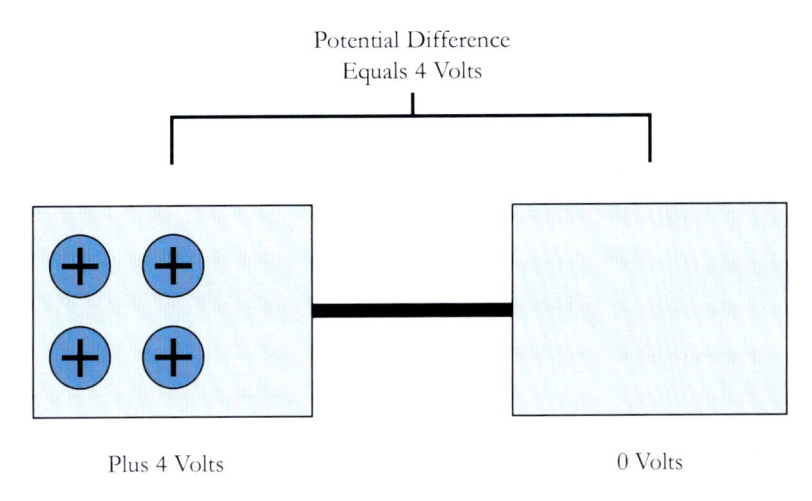

Figure 3.6 – Potential Difference referenced to Earth

When a potential difference exists between two charged bodies that are connected by a conductor, electrons will flow along it in an attempt to equalise charges. As electrons have a negative charge, this flow is from the negatively charged body to the positively charged body, until the two charges are equalised and the potential difference no longer exists. This effect can be simply represented by two water tanks connected by a pipe and valve as shown in figure 3.7.

CHAPTER THREE
ELECTRICAL TERMINOLOGY

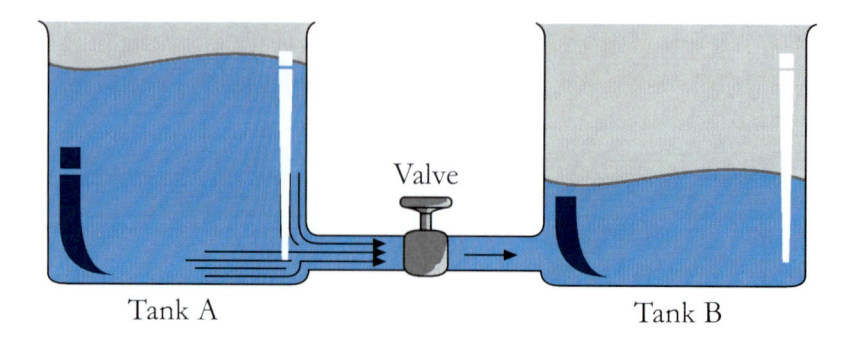

Figure 3.7 – Water Analogy of Potential Difference

At first the valve is closed and all the water is in tank A and so the water pressure across the valve is at maximum. When the valve is opened, the water flows through the pipe from A to B until the water level becomes the same in both tanks. The water then stops flowing in the pipe because there is no longer a difference in water pressure between the two tanks.

Electron movement through an electric circuit is directly proportional to the difference in potential, or EMF, across the circuit, just as the flow of water through the pipe in figure 3.7 is directly proportional to the difference in water level between the two tanks.

A fundamental law of electricity is that the ***electron flow is directly proportional to the applied voltage***, i.e. if the voltage is increased, the flow is increased; if the voltage is decreased, the flow is decreased.

Revision

Electrical Terminology

Questions

1. For current to flow there must be a difference in:

    a) Kinetic energy

    b) Potential energy

    c) Positional energy

2. With conventional flow, current flows from:

    a) Negative to positive

    b) Positive to negative

    c) Either direction

3. In the figure below, the potential difference from A to B is:

    a) -6 V

    b) +18 V

    c) +6 V

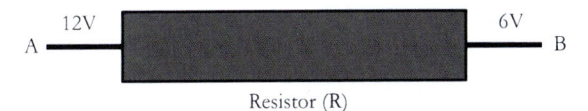

4. The resistance of a conductor is:

    a) Inversely proportional to its cross-sectional area

    b) Proportional to its cross-sectional area

    c) Irrespective of its cross-sectional area

5. If a conductor's length is doubled its resistance:

    a) Halves

    b) Doubles

    c) Remains the same

Revision

Electrical Terminology

Answers

1. **B**
2. **B**
3. **C**
4. **A**
5. **B**

Generation of Electricity

Introduction

We have already looked at how a charge can be produced by rubbing a rubber rod with fur. Because of friction between the rod and fur, the rod acquires electrons from the fur, making it negative, while the fur becomes positive due to the loss of its electrons. This difference in charge between the rod and fur forms a potential difference between them and the electrons making up this up are capable of doing work if a discharge or current flow occurs. Knowing how these charges occur and how they influence their surroundings allows us to use their potential under certain conditions.

However, to be a *practical* source of voltage, any potential difference must be maintained and only be allowed to dissipate under controlled conditions. At the sub-atomic level, this means that as one electron leaves the concentration of negative charge, another must immediately take its place or the charge will eventually shrink to the point where no further work can be done.

Therefore, a *Voltage Source* must be something that is capable of supplying and maintaining voltage while an electrical path exists across its terminals. The source's internal action must be such that as electrons are continuously removed from one terminal, keeping it positive, they must simultaneously be supplied to the second terminal, maintaining a negative charge.

Currently, there are six (6) known methods of producing a voltage or electromotive force (emf), and these are:

1. *Light*, also called *Photo-Electricity* - Voltage produced by light striking photosensitive, i.e. light sensitive, substances

2. *Heat*, also called *Thermo-Electricity* - Voltage produced by heating the joint or junction where two dissimilar metals are joined

3. *Friction* - Voltage produced by rubbing certain materials together

4. *Pressure*, also sometimes called *Piezo-Electricity* - Voltage produced by compressing crystals of certain substances

5. *Chemical Action* - Voltage produced by chemical reaction in a battery cell

6. *Magnetism and Motion* - Voltage produced in a conductor when it has relative motion with a magnetic field, i.e. the conductor cuts the field's magnetic lines of force

Naturally, some of these methods are used more than others, while some are used just for specific applications.

Voltage Produced by Light

In chapter one of this Module, we discussed how light is a form of energy and that when it strikes the surface of a substance, it may dislodge electrons from that material's surface atoms as they revolve in their orbits, figure 4.1.

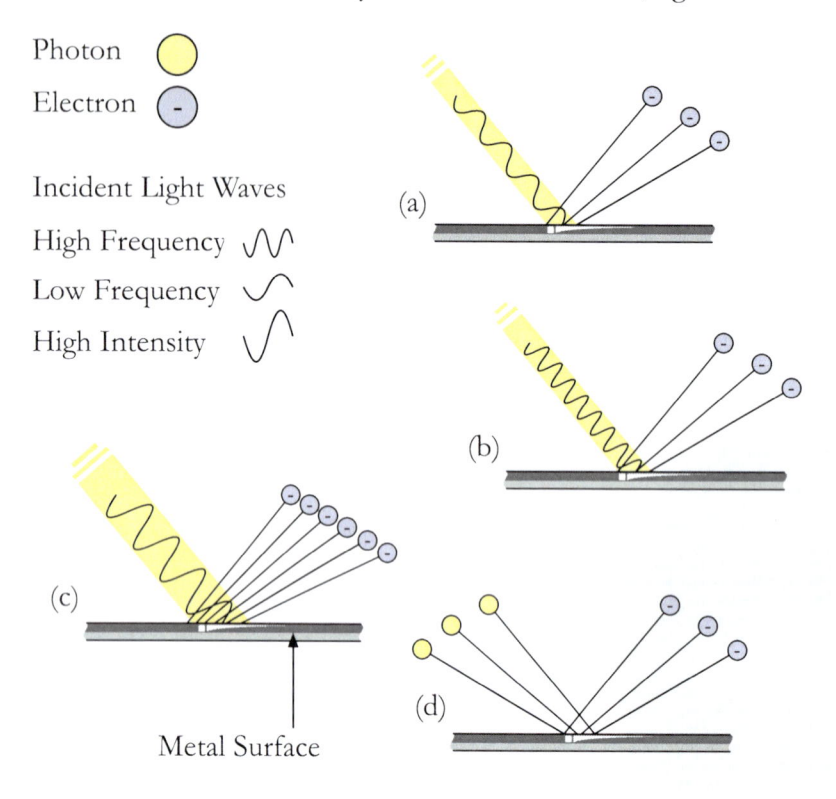

Figure 4.1 – Photoelectric Principle

The photoelectric effect occurs when electrons are released from a metal surface as a result of light shining on that surface as in drawing (a) above in figure 4.1. However, as you are probably aware, light comes in many colours, depending on its frequency, and no electrons are emitted unless the wavelength of the light is less than some critical value. This value will depend on the structure of the material being subjected to the light and the energy of the emitted electrons will depend not on the light's intensity but on its wavelength.

If the *frequency* of the light is increased, the same numbers of electrons are emitted but they now have more energy, (b) in figure 4.1 above. If the *intensity* of the light is increased, more electrons are emitted, (c) in figure 4.1.

These results were not explicable in terms of the classical theory of light waves and it took the great **Albert Einstein (1879-1955)**, to propose that light could be considered to be made up of photons and that each photon collides with an electron, dislodging it from its orbit, (d) in figure 4.1 above, to produce electricity.

As perhaps you would expect, some substances, mostly metallic ones, are far more sensitive to light than others. In other words, more electrons will be dislodged and emitted from the surface of a highly sensitive metal, with a given amount of light, than would be from a less sensitive substance. On losing electrons, the photosensitive, sometimes referred to as light-sensitive, metal becomes positively charged, and an electric force is created. Voltage produced in this way is called *Photoelectric Voltage*.

The most common photosensitive materials used to produce a photoelectric voltage are various compounds of silver oxide or copper oxide. A complete device that operates on the photoelectric principle is called a *Photoelectric Cell*, and there are many different sizes and types in use today. Their operation is discussed in a later chapter of this Module.

Voltage Produced by Heat

When a length of metal such as copper, is heated at one end, electrons tend to move away from the hot end towards the cooler end, which is true of most metals. However, in some metals such as iron, the opposite affect takes place and electrons tend to move *towards* the hot end. This phenomenon in metals, known as the Seebeck effect, was first observed by physicist Thomas Johann Seebeck (1770-1831). These characteristics are illustrated in figure 4.2.

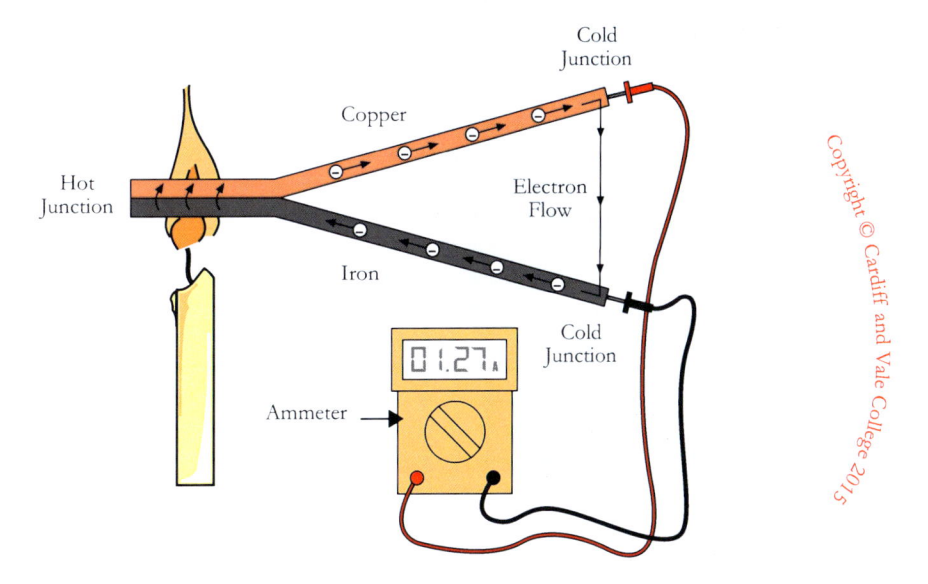

Figure 4.2 - Voltage produced by Heat

In this arrangement, the negative electrons are moving through the copper *away* from the heat and through the iron *towards* the heat. If a circuit is completed, for example by using an ammeter as in figure 4.2, the electrons cross from the iron to the copper through it at the cold junction. This arrangement or device is called a *Thermocouple*.

Thermocouples' power capacities are very small when compared to some voltage sources. The thermoelectric voltage produced depends mainly on the difference in temperature between the hot and cold junctions. Therefore, they are extensively used as temperature measuring devices and as heat-sensing devices in automatic temperature control equipment as they can generally be subjected too much greater temperatures than ordinary thermometers.

CHAPTER FOUR
GENERATION OF ELECTRICITY

Voltage Produced by Friction

As already discussed in earlier chapters, friction was the first method used for creating a voltage and the development of charges by rubbing a rod with a cloth or fur is a good example of how a voltage is generated by friction. However, because of the nature of the materials with which this type of voltage is generated, it cannot be conveniently used or maintained and for this reason, very little practical use has been found for generating voltages by this method.

However, in searching for methods of producing a voltage of larger amplitude and more practical nature, machines were developed in which charges were transferred from one terminal to another by using rotating glass discs or moving belts. The most notable of these is a machine for producing high electrostatic potential differences, invented by **Robert Van de Graaff** in 1931, figure 4.3.

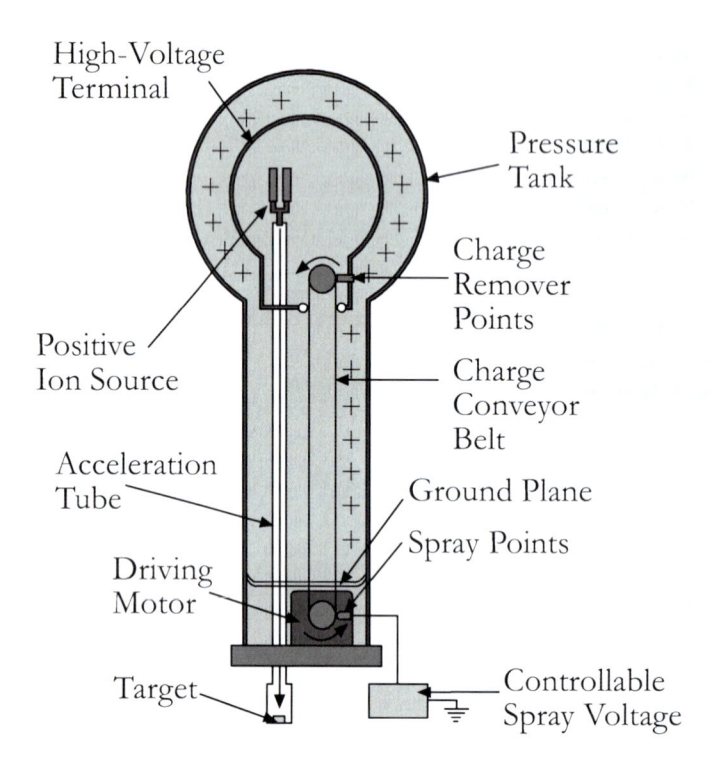

Figure 4.3 - Van de Graaff Generator

In this machine, an electric charge deposited by electrical discharge onto a moving rubber loop is transported to the interior of a hollow metal dome, where it is transferred to the dome and stored. With this device, potential differences of several million volts may be obtained and it was an early example of a particle accelerator. It is used today to produce potentials in the order of millions of volts for nuclear research, but these machines have little value outside the field of research, and so their theory of operation is not discussed in these notes.

Voltage Produced by Pressure

One specialised method of generating an emf uses the characteristics of certain ionic crystals such as **Quartz**, **Rochelle salts**, and **Tourmaline**. These crystals have the extraordinary ability to generate a voltage whenever stresses are applied to their surfaces. When stress is applied it distorts the crystal causing unbalanced electric forces within the material. Therefore, if a crystal of quartz is squeezed, charges of opposite polarity will appear on two opposite surfaces of the crystal. However, if the force is reversed and the crystal is stretched, charges will again appear, but will be of the opposite polarity to those produced by squeezing.

Therefore, if a crystal of this type is stressed with a vibratory motion, it will produce a voltage of reversing polarity between two of its sides. Quartz or similar crystals can therefore be used to convert mechanical energy into electrical energy and this phenomenon is called the **Piezoelectric Effect**, illustrated in figure 4.4.

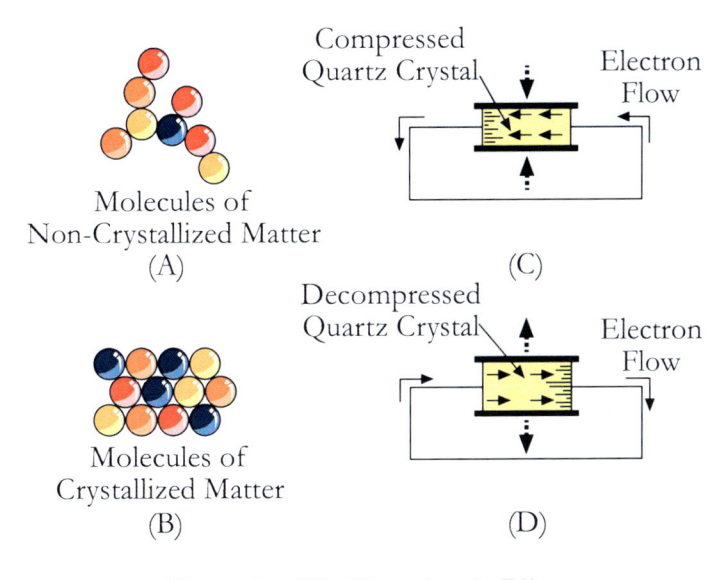

Figure 4.4 - The Piezoelectric Effect

Piezoelectric crystals are commonly used in microphones, record styluses and oscillators used in transmitter/receivers. As you might expect, this method of generating an emf is not suitable for large voltage or power applications, but is widely used in sound and communications systems where small signal voltages can be used effectively.

Piezo Accelerometers

Typical commercial aircraft engines have piezo accelerometers and an associated signal conditioning system to monitor engine health and provide an indication of vibration to the flight crew.

The crystal produces a voltage that is proportional to the accelerative force and is a highly accurate vibration measuring device. The basic arrangement is shown in Figure 4.5.

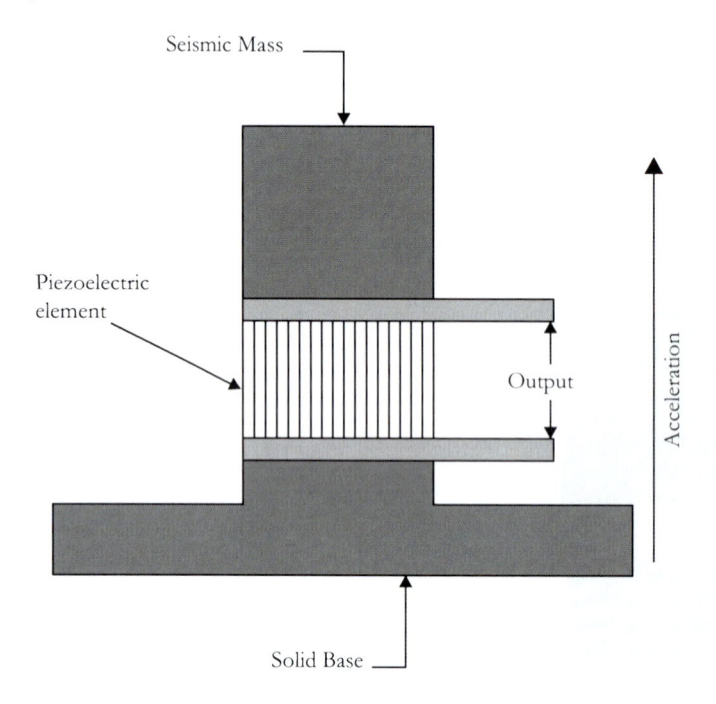

Figure 4.5- Piezoelectric Accelerometer

Types

The most commonly used accelerometers are piezoelectric, and piezoresistive in nature. Piezoelectric devices are more preferred in cases where it is to be used in very high temperatures, and high frequency range up to 100 kilohertz.

Piezoresistive devices are used in sudden and extreme vibrating applications. The piezoresistive effect also involves pressure or stress. The piezoresistive effect only causes a change in electrical resistance and does not alter an electrical charge or voltage. The resistors are normally configured into a Wheatstone bridge circuit, which provides a change in output voltage that is proportional to acceleration.

Piezoelectric crystals also have another interesting property, which is really the converse piezoelectric effect; i.e. they have the ability to convert electrical energy into mechanical energy.

In this case, a voltage applied across the proper surfaces of the crystal will cause it to expand or contract its in response to the applied voltage at a frequency that is specified and proportional to the crystal's shape and dimensions. These are widely used in digital timepieces as they can maintain their natural, i.e. resonant, frequency with significant accuracy.

At first glance, there probably does not appear to be a use for this effect in aviation. However, in 1984, the USA avionics manufacturer Rockwell-Collins used the piezoelectric crystal as a basic gravity detector for their first digital *Attitude and Heading Reference System (AHRS)*, certified first in the SAAB 340 commuter aircraft.

They used two (2) rotating wheel assemblies mounted in a 'T' format and on each rotating assembly there were four (4) piezoelectric crystals mounted in pairs; one pair sensed acceleration perpendicular to the wheel spin axis, while the other pair sensed rate changes about the axis. The outputs from both wheel assemblies are then combined to generate attitude, rate and acceleration information about all three (3) aircraft axis.

The piezoelectric crystals have no output when at rest, but when the assembly was rotated, gravity applied a force on the crystals and they bent so converting the mechanical energy of gravity into an electrical signal. The electrical voltage output was a sine wave under perfect condition, i.e. gravity was acting directly downwards, and its amplitude depended on how much force was applied.

In this way, the AHRS could establish the local gravity vectors during its initialisation phase and would then maintain an attitude and heading output by comparing how far and in what direction the aircraft moved from its initial starting point.

Systems using piezoelectric crystals as the sensing component of an AHRS are in wide use today and can provide outputs with resolutions unheard of in the 1980s, except by using the most expensive gyroscopic devices. AHRS and similar systems are discussed in more detail in Modules 5 & 13.

Voltage Produced by Chemical Action

Voltage may be produced chemically when certain substances are exposed to a chemical action. If two dissimilar substances, which are usually metals or metallic materials, are immersed in a solution that produces a greater chemical action on one substance than on the other, a difference of potential will exist between the two.

If a conductor is then connected between them, electrons will flow through the conductor in order to equalise the charge. This arrangement is called a *Primary Cell*, the two metallic substances are called *Electrodes* and the solution is called the *Electrolyte*.

The diagram in figure 4.6 is a simple example of a primary cell, originally called a *Voltaic Cell*, after its inventor, the Italian physicist *Alessandra Volta*.

CHAPTER FOUR
GENERATION OF ELECTRICITY

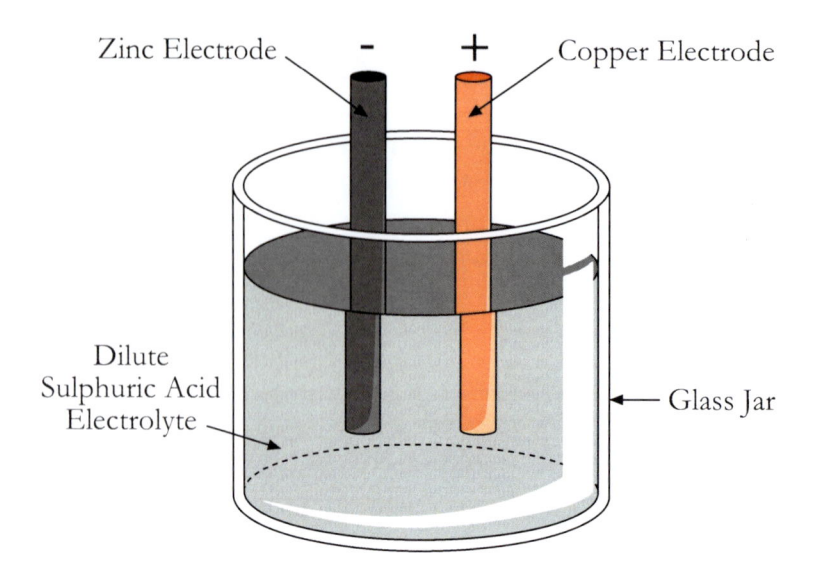

Figure 4.6 – The Voltaic cell, a Simple Primary Cell

When any metal electrode is placed in an electrically conducting liquid (i.e., an electrolyte), the atoms at its surface tend to exchange electrons with the solution, becoming charged in the process. Some of these charged atoms or ions tend to leave the metal and wander around in the solution. This tends to set up a small voltage difference between the metal electrode itself and the electrolyte. The amount of potential difference between the electrodes depends **_predominantly_** on the metals used in their construction. Some metals like lithium and zinc develop a negative voltage, while others like copper, mercury and silver develop a positive voltage. The type of electrolyte and the size of the cell have little or no effect on the potential difference produced.

There are two types of primary cells, the wet cell and the dry cell. In a wet cell, as you might expect, the electrolyte is a liquid. This type of cell must always remain in an upright position and so is not readily transportable except in one position. A car battery is a typical example of this type of cell.

The dry cell is not actually dry but contains an electrolyte mixed with other materials to form a paste. This means it can be stored in any position and is in much more common use than the wet cell. Torches, portable radios and children's toys are typical examples of the use of dry cells.

Batteries are formed when several cells are connected together to increase electrical voltage and current output. These are discussed in the next chapter.

CHAPTER FOUR
GENERATION OF ELECTRICITY

Voltage Produced by Magnetism and Motion

Magnets and magnetic devices are used throughout the World for thousands of different jobs. One of the most useful and widely employed applications of magnets is in the production of electric power from mechanical sources. The mechanical power may be provided by a number of different sources, such as an engine, water action or steam turbines. However, the final conversion of these source energies into electricity is completed by generators, using the principle of *Electromagnetic Induction*.

The types and sizes of generators are discussed in later chapters of this module and so in this chapter we will only discuss the fundamental operating principle of all electromagnetic-induction generators.

To begin with, there are three (3) fundamental conditions that must exist before magnetism can produce a voltage.

1. There must be a *Conductor* in which the voltage will be produced
2. There must be a *Magnetic Field* in the conductor's vicinity
3. There must be *relative motion* between the field and conductor, i.e. the conductor must be moved so as to cut across the magnetic lines of force, or the magnetic lines of force must move to cut across a stationary conductor

If these conditions are met, when relative motion occurs, electrons within the conductor are driven in one direction or another, dependent on the direction of the motion, so generating a voltage or emf. Figure 4.7 illustrates this in graphic form. The emf produced is related to the speed of the relative motion. The greater the relative speed, the greater the voltage.

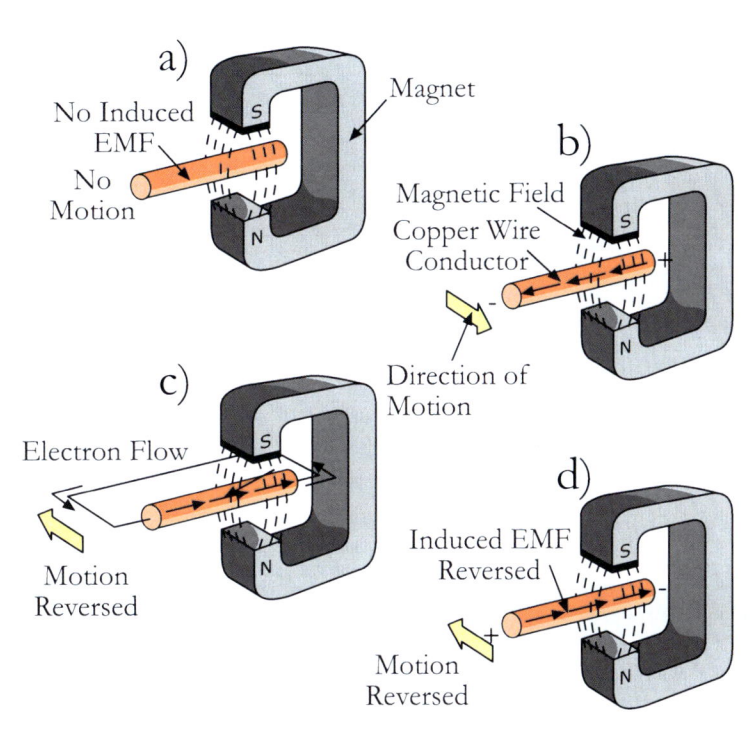

Figure 4.7 - Voltage produced by Magnetism & Motion

47

CHAPTER FOUR
GENERATION OF ELECTRICITY

In figure 4.7, a magnetic field exists between the poles of the C-shaped magnet and the conductor is made of copper wire. The conductor is moved back and for across the magnetic field to produce the relative motion.

In the top right drawing of figure 4.7, the conductor is moving towards the front of the page and the electrons move from right to left. The electrons move in this way due to the magnetically induced emf acting on the electrons in the copper. Consequently, the left-hand end becomes negative, and the right-hand end positive. The conductor is stopped in the top left drawing and so there is no longer an emf produced as the motion has ceased. In the bottom right drawing, the conductor is moving away from the front of the page and an induced emf is created again, but in this case, the reverse motion now causes a reversal of direction in the induced emf.

If the conductor were connected into a circuit of some sort, it would provide a path for electron flow, as illustrated in the bottom left drawing, which would continue as long as the emf continues. However, in looking at figure 4.7, you should realise that an induced emf could also have been created by keeping the conductor steady and moving the magnetic field.

As already mentioned above, the more complex aspects of power generation by use of mechanical motion and magnetism are discussed in later chapters of this Module.

Revision

Generation of Electricity

Questions

1. Electricity can be produced in one of:

    a) 6 ways

    b) 4 ways

    c) 5 ways

2. With electricity produced by light, if the light's intensity increases, the amount of electrons released:

   a) Decreases

   b) Remains the same

   c) Increases

3. With electricity produced by light, if the light's frequency increases, the amount of electrons released:

   a) Decreases

   b) Remains the same

   c) Increases

4. Voltage produced by pressure is called:

   a) The piezoelectric effect

   b) The Rochelle effect

   c) The Quartz effect

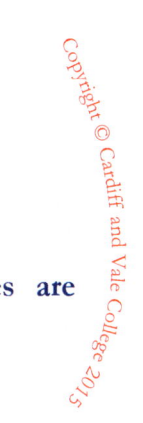

5. To produce voltage by chemical means, two metal plates are required suspended in a solution called:

   a) An electrolyte

   b) A soluble compound

   c) A voltaic solution

6. The amount of potential difference produced by chemical means depends *predominantly* on:

   a) The strength of solution the electrodes are immersed in

   b) The size of the electrodes

   c) The metal used in the electrode's construction

7. In order to produce voltage by electro-magnetism there must:

   a) Parallel motion between conductor & magnetic field

   b) Relative motion of the conductor across the magnetic field

   c) The magnetic field must be stationary

8. The strength of EMF produce by electro-magnetism is:

   a) Inversely proportional to the speed of motion

   b) Irrespective of the speed of motion

   c) Proportional to the speed of motion

Revision

Generation of Electricity

Answers

1. A
2. C
3. B
4. A
5. A
6. C
7. B
8. C

DC Sources of Electricity

Introduction

The purpose of this chapter is to introduce and explain the basic theory and characteristics of batteries and other simple sources of *Direct Current (DC or dc)*. The batteries that are discussed and illustrated are representative of many models and types that are in use throughout the World and those used in today's aircraft industry. It would be impossible to cover every type of battery in use today and we have not attempted to do so. However, after completing this chapter, you will have a good working knowledge of the batteries that are in general use.

Batteries

Batteries are widely used as sources of direct-current electrical energy in cars, boats, aircraft, radios etc. In some instances, they are used as the only source of power, e.g. a glider, while in others they are used as a secondary, standby or emergency power source.

As already discussed in the previous chapter, it was the Italian physicist *Alessandro Volta (1745-1827)*, who first discovered the voltage cell and went on to invent the electric battery. He was born in Como of an aristocratic family devoted to the Church, and became a professor of natural philosophy at Pavia, capital city of the province of Pavia, Italy. He was highly religious, but not prudish as one of his fiends recalled that he '*understood a lot about the electricity of women*'.

Volta became interested in the study of electricity following the discovery of *Galvani*, an Italian anatomist, in the 1780s that an electric spark, or contact with copper and iron, caused a disembodied frog's leg to twitch. Volta became interested in finding the cause of the phenomenon and his experiments showed him that an electric current could be generated by bringing different metals into contact with one another.

In 1799 he succeeded in constructing a battery consisting of metal discs, alternately silver and zinc, with brine-soaked card between them. This *Voltaic Pile*, as it was called, produced a steady electric current and was the first reliable source of electricity. However, he did little further work on the device, but it was to transform the study of electricity and was invaluable to men that followed after him like *William Nicholson (1730-1795)*, who made the first British Battery, *Sir Humphrey Davy (1778-1829)*, who studied and developed electrochemistry and *Michael Faraday (1791-1867)*, who made several major advances in electromagnetism and other sciences.

Volta was given the title of Count by Napoleon, who invaded Italy in the 1790s, and who had become very interested in electricity and saw its importance to science. The SI unit of electric potential, the *Volt (V)*, is named in his honour.

Here is how Alessandro Volta described his new '*pila*', or battery, in a letter to Sir Joseph Banks, written in 1800.

> *"Sixty or more pieces of silver, applied each to a piece of tin or zinc and as many strata of cardboard, soaked in salt solution, interposed between every pair of metal discs, and always in the same order, constitutes my new instrument."*

Once the electric 'battery' had been developed and people discovered they could draw off an electric current that 'flowed' through conductors continuously, the old forms of electricity were largely disregarded. People referred to the *old* electricity as '*static*', as in '*stationary*', electricity, or sometimes as '*frictional*' electricity, while '*proper*' electricity was considered to come from a battery of cells that was soon shorten to just the word battery.

For a long time, the battery was known as a *Voltaic pile*, because it was made of a pile of discs, and the word pile is still used in French, for a battery. It was called this because Volta actually called his battery a 'pila', which is also the Italian word for 'pile'.

However, the Voltaic pile is no longer used as it was superseded by the design of the French Engineer *Georges Leclanché (1839-82)*, who devised a far more convenient Carbon-Zinc electrical cell known as the *Leclanché* Cell. This has since been modified into what we now call a *Dry Cell*.

This cell has a carbon rod as the positive pole, surrounded by a wet paste of carbon black, manganese dioxide and ammonium chloride, with a thickener such as sawdust, inside a zinc container which is the negative pole.

However, before we get too far ahead of ourselves, in the previous chapter, we looked at how voltage can be produced using chemical action. Now we need to put this theory into practice.

Batteries or battery cells are either *primary* or *secondary*. Primary cells can be used only once because the chemical reactions that supply the electrical current are irreversible. Secondary cells, sometimes called *storage* cells or storage batteries in the USA, can be used, charged and reused several times. In these batteries, the chemical reactions that supply electrical current are readily reversed so that the battery is charged, more on this later.

The Cell

A cell is a device that transforms chemical energy into electrical energy. The simplest cell, known as either a *Galvanic* or *Voltaic Cell*, is illustrated in figure 5.1.

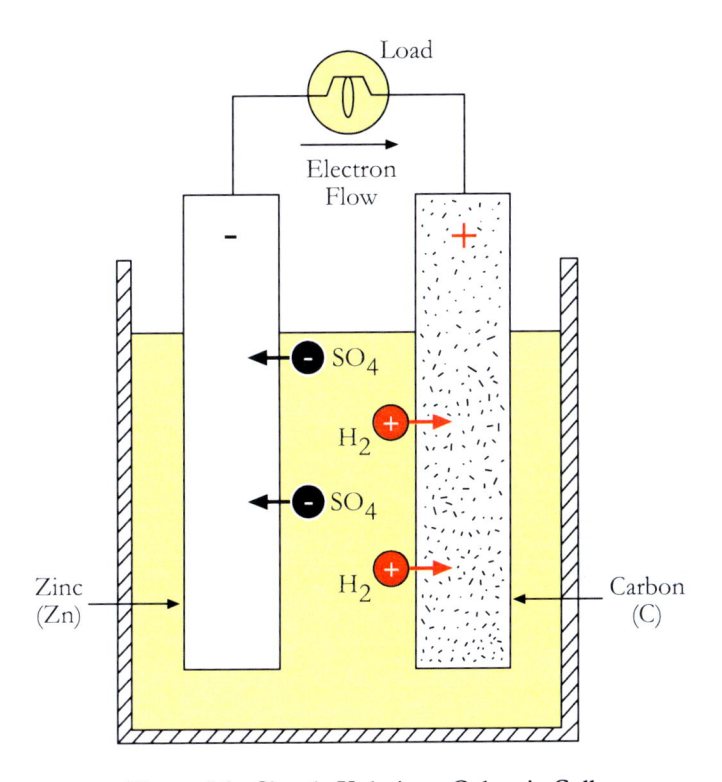

Figure 5.1 - Simple Voltaic or Galvanic Cell

This simple cell consists of a piece of Carbon (C) and a piece of Zinc (Zn), both called *Electrodes*, suspended in a jar that contains a solution of Water (H_2O) and Sulphuric Acid (H_2SO_4) called the *Electrolyte*. The water/acid mix in a fully charged cell is typically 75%/25%.

The cell is the fundamental unit of the battery and all simple cells consist of two (2) electrodes placed in a container that holds the electrolyte. However, in some cells the container acts as one of the electrodes and in this case, is acted upon by the electrolyte.

The electrodes act as conductors to provide the current path by which the current leaves or returns to the electrolyte. In the simple cell above, they are carbon and zinc strips, placed in the electrolyte; while in the dry cell illustrated in figure 5.2, they are the carbon rod, in the centre, and zinc container in which the cell is assembled.

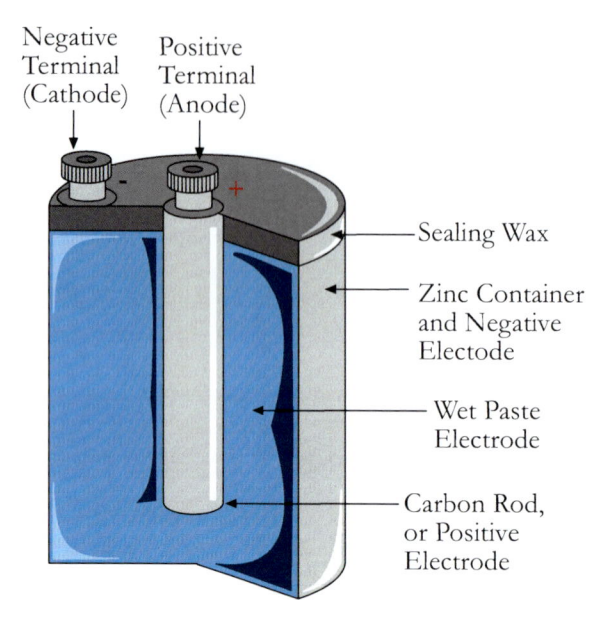

Figure 5.2 - Dry cell, Cross-Sectional View

The electrolyte is a solution that acts upon the electrodes and provides a path for electron flow, and it may be a salt, acid or alkaline solution. As already mentioned earlier, in the simple galvanic cell the electrolyte is in a liquid form, while in the dry cell, it is a paste.

The container, which may be constructed of one of many different materials, provides a means of holding, i.e. containing, the electrolyte and it is also used to mount the electrodes. As perhaps you would expect, although the container shell may sometimes be used as one of the electrodes, it must be constructed of a material that will not be acted upon, i.e. eroded away, by the electrolyte solution.

Primary Cell

A *Primary Cell* is one in which the chemical action eats away one of the electrodes, usually the negative one, and when this occurs, it must either be replaced or the whole cell must be discarded. In the galvanic-type cell, typically the zinc electrode and the liquid electrolyte are replaced when this happens but with a dry cell, it is usually cheaper to buy a new one, especially at today's prices. Indeed, primary cells are so cheap and easy to use the process of reclaiming a cell by replacing one electrode and the electrolyte has all but disappeared.

Secondary Cell

A *Secondary Cell* is one in which the electrodes and the electrolyte are altered by the chemical action that takes place as the cell delivers current but they have the added advantage that they may be restored to their original condition, by forcing an electric current through them in the opposite direction to that of discharge. The car and aircraft batteries are common examples of the secondary cell.

With a primary and secondary cell, if a load, i.e. a device such as a resistor that consumes electrical power, is connected externally to its electrodes, electrons will flow under the influence of a difference in potential across the electrodes, from the negative electrode, commonly called the *Cathode*, through the external conductor to the positive electrode, commonly called the *Anode*. In this way, as we have already discussed, a cell is a device in which chemical energy is converted to electrical energy and consequently, this process is also sometimes called an *Electrochemical* action.

The *voltage* developed by the cell depends on the materials from which the electrodes are made and the composition of the electrolyte. The *current* that a cell can deliver, depends on the resistance of the entire circuit, *including that of the cell itself*. The cell's *internal resistance (r)* depends on the size of the electrodes, the distance between them in the electrolyte, and the resistance of the electrolyte. The larger the electrodes and the closer together they are in the electrolyte, obviously without touching, the lower the internal resistance of the cell and the more current the cell is capable of supplying to the load.

Primary Cell Chemistry

When current flows through a primary cell with carbon and zinc electrodes immersed in an electrolyte solution of sulphuric acid and water, the following chemical reaction occurs.

The current flow through the load is the movement of electrons from the negative zinc electrode to the positive carbon electrode, resulting in fewer electrons in the zinc and an excess of electrons in the carbon. Since the hydrogen ions of the sulphuric acid are positively charged, they are attracted to the negative charge, caused by the excess electrons, on the carbon electrode, illustrated in figure 5.1. When this occurs, the zinc electrode becomes positively charged as it has lost electrons to the carbon electrode. This positive charge attracts the negative ions from the sulphuric acid and these combine with the zinc to form zinc sulphate, eroding the zinc electrode. Zinc sulphate is sometimes visible as a greyish-white substance found on the case of a torch battery that has been left in for too long.

This process of the zinc eroding and the sulphuric acid changing to hydrogen and zinc sulphate, causes the cell to discharge and when the zinc is used up, the voltage of the cell will be reduced to zero (0). Over time, the zinc sulphate then dissolves into the electrolyte and the hydrogen appears as gas bubbles around the carbon electrode. However, in this process, the carbon electrode does not undergo a chemical change but simply provides a return path for the current.

In figure 5.1, the zinc electrode is labelled negative and the carbon electrode positive, which represents the current flow outside the cell from negative to positive.

Secondary Cell Chemistry

As we have already discussed, the differences between primary and secondary cells are that the secondary cell can be recharged and the electrodes are made of different materials. The secondary cell illustrated in figure 5.3, uses **Sponge Lead** as the cathode and **Lead Peroxide** as the anode, hence the name **Lead-Acid Battery** or **Cell**.

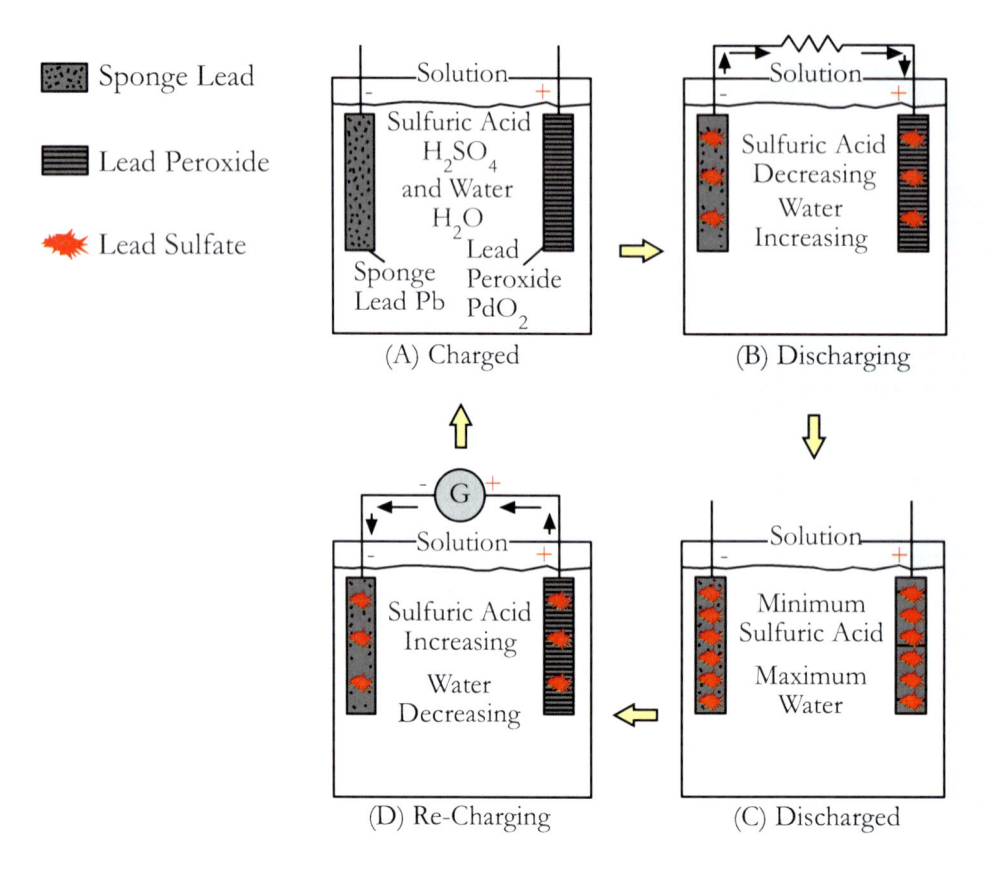

Figure 5.3 – Typical Secondary Cell

(A) Fully charged battery, Cathode is sponge lead (Pb) and Anode is Lead Peroxide (PbO_2).

(B) A load is connected to the battery and it starts to discharge. As it is discharging, the sponge Lead Cathode (Pb) takes Sulphate (SO_4) from the electrolyte, combining with it to form Lead Sulphate ($PbSO_4$). While the anode, which is made of Lead Peroxide (PbO_2), also combines with sulphate from the electrolyte, also to form Lead Sulphate ($PbSO_4$).

(C) When completely discharged both the cathode and anode have become primarily Lead Sulphate with some remnants of the original Lead and Lead Peroxide, ie $PbSO_4+Pb$ and $PbSO_4+ PbO_2$ respectively.

(D) When a generator or other charging device is connected to the battery in a reverse direction to the load, then the chemical processes are reversed; i.e. the cathode and anode give up the Sulphates back into the electrolyte and they revert to their original state of Lead (Pb) and Lead Peroxide (PbO_2) respectively.

The lead-acid battery is one of the most popular batteries in use today and we will use its construction and chemical reactions to describe the general chemistry of the secondary cell. Later in this chapter, when other types of secondary cells are discussed, you will see that the materials that make up the parts of a cell are different, but the chemical action is essentially the same.

Figure 5.3 drawing 'A' shows a fully charged lead-acid secondary with its cathode made of pure sponge lead, and the anode of pure lead peroxide, with an electrolyte that is a mixture of sulphuric acid and water. Figure 5.3 drawing 'B' shows the secondary cell discharging through a load connected between the cathode and anode. The current flows negative to positive as illustrated, creating the same process explained for the primary cell. However, with the secondary cell there are differences in the discharge process to that of the primary cell.

1. In the primary cell, the zinc cathode is eaten away by the sulphuric acid but in the secondary cell, the sponge-like construction of the cathode *retains* the lead sulphate formed by the chemical action of the sulphuric acid and the lead.

2. In the primary cell the sulphuric acid did not chemically act on the carbon anode but in the secondary cell the lead peroxide anode is chemically changed to lead sulphate by the sulphuric acid.

When the cell is fully discharged, figure 5.3 drawing 'C', the cathode and anode retain some sponge lead and lead peroxide respectively, but the amounts of lead sulphate in each is at a maximum and the electrolyte has a minimum amount of sulphuric acid. When it arrives at this state, no further chemical action can take place within the cell and it is fully discharged.

However, as discussed earlier, we know the secondary cell can be recharged, which is the process of reversing the chemical action that occurs as the cell discharges. To recharge the cell, a voltage source, such as a generator, is connected across the cathode and anode as illustrated in figure 5.3 drawing 'D'. In this case, the negative terminal of the voltage source is connected to the cell's cathode and the positive terminal to the cell's anode. With this arrangement, the lead sulphate is chemically changed back to sponge lead in the cathode, lead peroxide in the anode, and sulphuric acid in the electrolyte.

Once all the lead sulphate is chemically changed, the cell is back to being fully charged, figure 5.3 drawing 'A', and the whole cycle can be repeated.

This process makes the secondary cell a useful source of aircraft power, especially in emergencies as its energy can be 'topped-up' by the aircraft generator and will be fully charged until needed. However, while the cell is delivering current, the chemical action does cause hydrogen bubbles to form on the surface of the anode and this is called **_Polarisation_**. Some hydrogen bubbles rise to the surface of the electrolyte and escape into the air, but some do remain on the surface of the anode. If enough hydrogen remains around the anode, the bubbles can form a barrier that increases the cell's internal resistance.

We will be discussing the affect of a cell's internal resistance later in this chapter, but put simply now, when the cell's internal resistance increases, its output current is decreased and the voltage of the cell also decreases.

As perhaps you might expect, a cell that is heavily polarised has no useful output. However, in a practical installation, there are several methods to prevent polarisation or to depolarise the cell if this occurs.

1. One method is to use a vent on the cell to let the hydrogen escape into the air. However, the major disadvantage of this method is that hydrogen is not then available to reform into the electrolyte during recharging. Adding distilled water to the electrolyte, such as you used to be able to do with your own car battery, solves this problem, but is difficult, if not impossible, to accomplish on an aircraft and is certainly not allowed on aircraft flying in the Transport Category.

2. A second method is to use material that is rich in oxygen, such as manganese dioxide, which supplies free oxygen to combine with the hydrogen and form water.

3. A third method is to use a material that will absorb the hydrogen, such as calcium. The calcium releases the hydrogen during the charging process.

All three of these methods remove enough hydrogen so that for practical purposes, the cell is free from polarisation.

Cell Types

There are many different types of primary cells and in the last 10 to 15 years the development of new types has been so rapid to keep up with the demand for sustainable power in portable devices, e.g. portable computers, MP3 players etc, that it is virtually impossible to have a complete knowledge of them all. When choosing what type to use, several factors need to be taken into account such as physical size, voltage, cost, current capacity, ease of replacement, etc. Some of the more popular types are discussed in the following paragraphs.

Primary Dry Cell

The dry cell is the most popular type of primary cell as it is ideal for simple applications where an inexpensive and non-critical source of electricity is all that is needed. However, as we have already discussed, the dry cell is not actually dry, as although the electrolyte is not in a liquid state, it is a moist paste. This is because if the electrolyte were totally dry it would no longer be able to transform chemical energy to electrical energy. Figure 5.4 shows the construction of a common type of Zinc-Carbon dry cell.

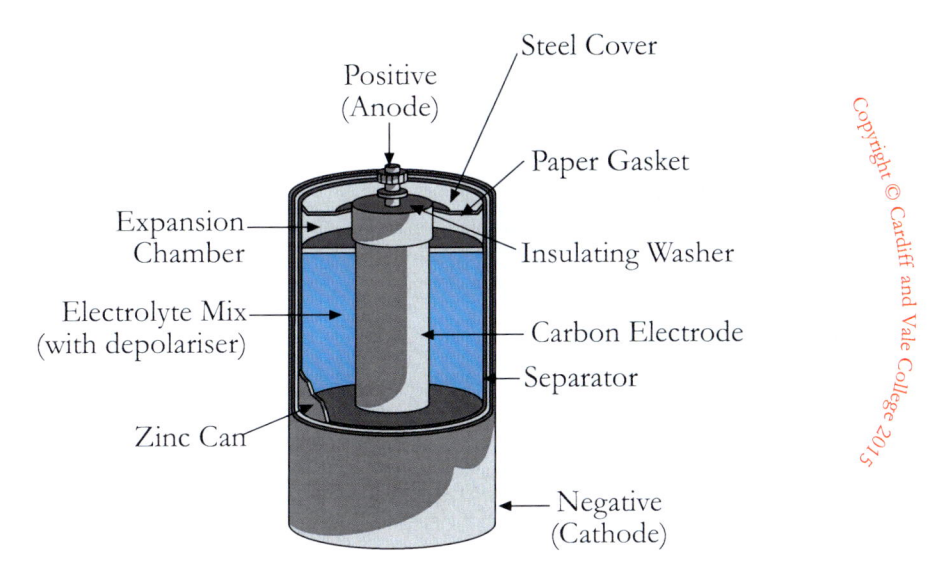

Figure 5.4 – Typical Construction of a Zinc-Carbon Dry Cell

These are also called *Leclanché* cells, and their internal parts are held in a cylindrical zinc container that can be many sizes and shapes, e.g. AA, AAA, C, D, PP3 etc. The zinc container acts as the negative electrode, i.e. cathode, of the cell and is lined with a non-conducting material, such as blotting paper, to separate it from the electrolyte paste. A carbon electrode is then placed in the centre to act as the cell's positive terminal or anode.

The electrolyte paste is a mixture of several substances such as ammonium chloride, powdered coke, ground carbon, manganese dioxide, zinc chloride, graphite, and water. It is packed tightly into the space between the anode and the blotting paper, also allowing it to support the anode rigid in the cell's centre. However, although tightly packed, a small space is left at the top for expansion of the electrolytic paste caused by the depolarisation action. The cell is then sealed with a cardboard or plastic seal.

To complete the construction, as the zinc container is also the cathode, it is protected with an insulating material, e.g. cardboard, to ensure electrical isolation and is then placed in another outer metal skin to give it mechanical stability. Finally, it is sealed at the top to prevent air getting in to dry out the electrolyte.

A dry cell that is not being used, e.g. sitting on a shelf in storage, will gradually deteriorate because of slow internal chemical changes. However, this is usually very slow if cells are properly stored in a cool place.

Manganese Dioxide-Alkaline-Zinc Cell

The *Manganese Dioxide-Alkaline-Zinc Cell* is similar in construction to the zinc-carbon cell except for the electrolyte used. This cell type offers better voltage stability, has a longer operational life and extended shelf life than the zinc-carbon type. In addition, it can also cope with a greater temperature range. It has an output voltage of 1.5 volts and is available in a wide range of sizes. This cell is commonly called an *Alkaline Battery* as seen in many advertisements on TV. Its one disadvantage is that it produces hydrogen when in use.

Lithium Cells

Primary lithium manganese dioxide cylindrical and button cells are widely used for memory backup and portable applications. They are manufactured with lithium metal or lithium compounds as the anode, and are distinctive from other batteries in that, depending on the design and chemical compounds used, lithium cells can produce voltages from 1.5 V (comparable to a zinc–carbon or alkaline battery) to about 3.7 V.

Lithium is the lightest of metals and it actually floats on water, but it also has the greatest electrochemical potential which makes it one of the most reactive of all metals. These properties give Lithium the potential to achieve very high energy and power densities, thus providing batteries with a very long useful life and small cell packages.

As with most batteries they have an outer case made of metal. The use of metal is particularly important because the battery is pressurized. The metal case has a pressure-sensitive vent hole, ensuring that if the battery ever gets so hot that it risks exploding from over-pressure, the vent will release the extra pressure. The vent is only there as a safety measure and the battery will normally be rendered useless after it has operated.

Lithium cells are typically constructed in three ways: bobbin-type, spiral-wound and button type (also known as coin type). Bobbin cells consist of an outer cylinder made of lithium metal and an inner electrode that is reminiscent of a bobbin of thread, Figure 5.5.

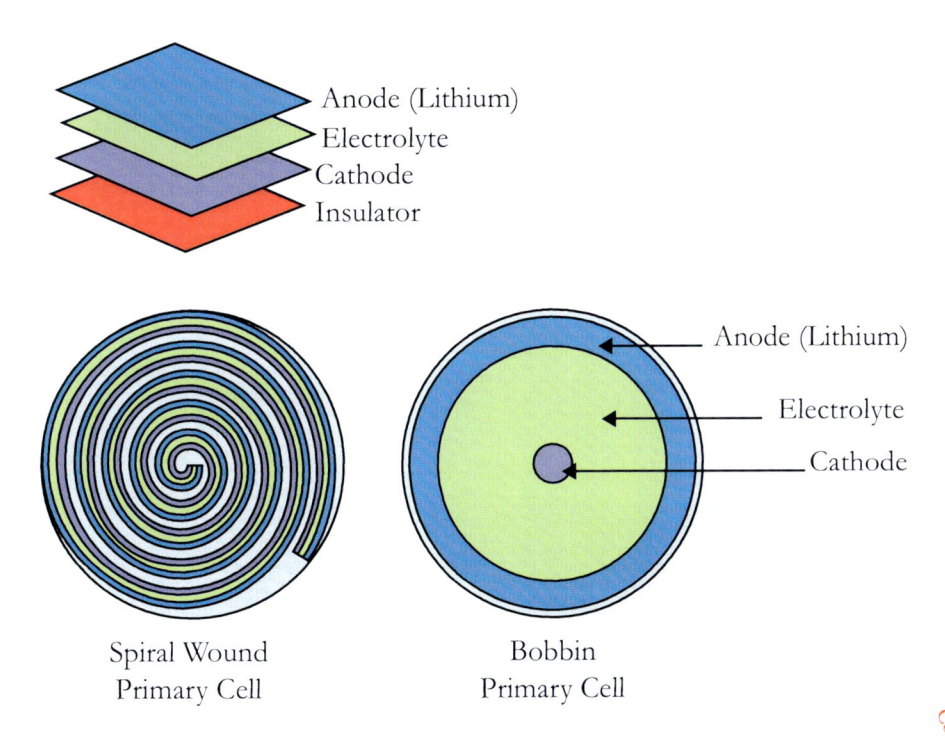

Figure 5.5– Primary Lithium Battery Construction

In a Spiral-wound cell flat sheets of metal are wound around a core, which provides a large surface area that can create high currents. However, the number of layers within the cell reduces the volume of electrolyte the batteries can hold, whereas bobbin cells can hold more electrolyte, enabling them to deliver about 30% more energy than a spiral cell of similar size.

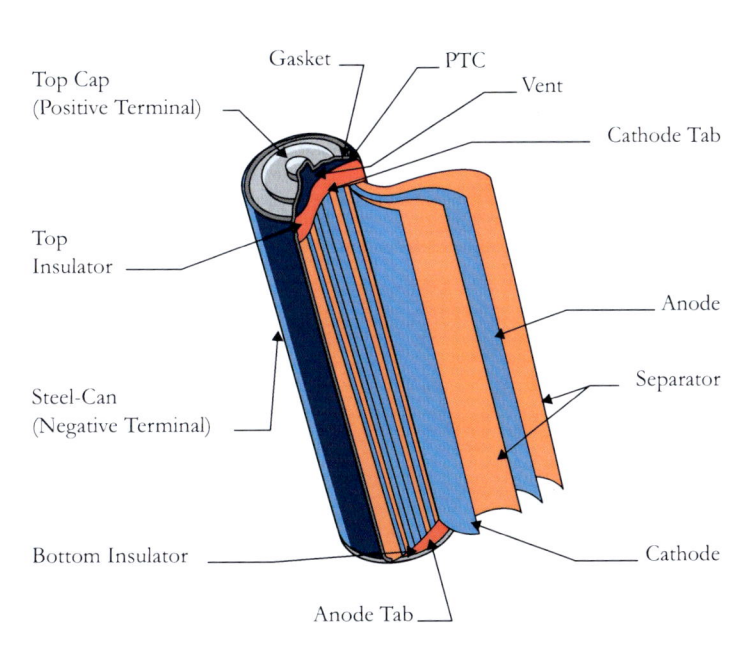

Figure 5.6– Spiral Wound Lithium-ion Battery

CHAPTER FIVE
DC SOURCES OF ELECTRICITY

A Spiral Wound Lithium-ion battery is shown in figure 5.6.

As we can see, the metal case comprises of a long spiral with three thin sheets wrapped together, these are the:

- Positive electrode

- Negative electrode

- Separator

The sheets are submerged in an organic solvent, such as Ether that acts as the electrolyte.

The separator is a very thin sheet of microperforated plastic. As the name implies, it separates the positive and negative electrodes while allowing ions to pass through. Figure 5.7.

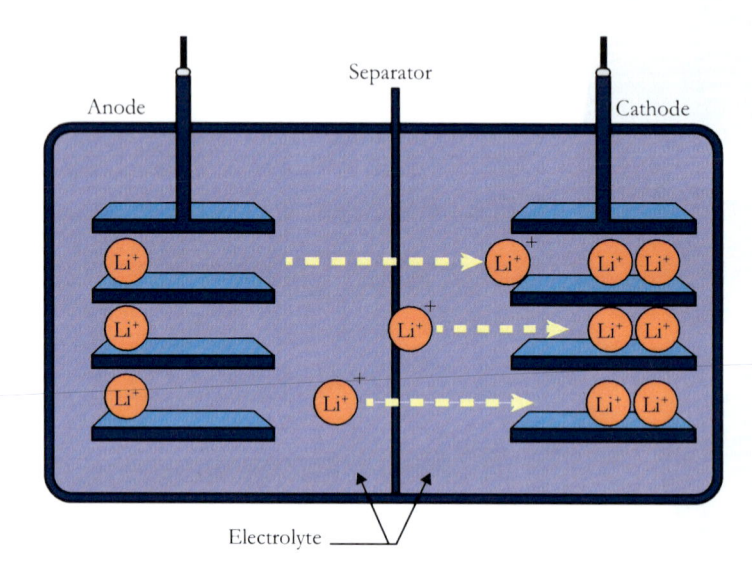

Figure 5.7 – Lithium Ion Cell

During discharge, the energy-containing lithium ion travels from the high-energy anode material through a separator, to the low-energy cathode material. The movement of the lithium releases energy, which is extracted into an external circuit.

In a typical button cell design, the cathode is a pressed piece of active material, and a binder in electrical contact with the cell bottom. A polypropylene separator lies above the cathode while the lithium anode is a foil rolled onto a collector at the top of the cell, Figure 5.8.

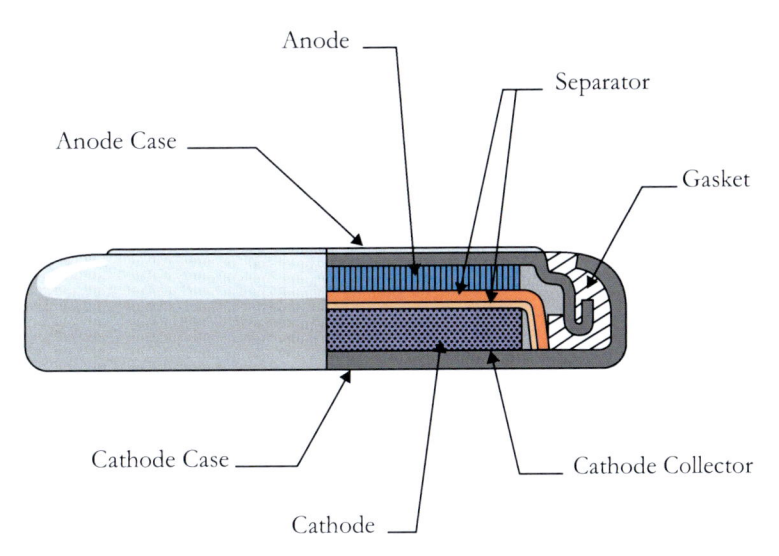

Figure 5.8 – Lithium Button or Coin Battery Construction

All Lithium button cells are rated at 3.0V, which is why they are sometimes called 3V lithium batteries. The four digits in the battery designation actually give the dimensions of the button cell, Figure 5.9.

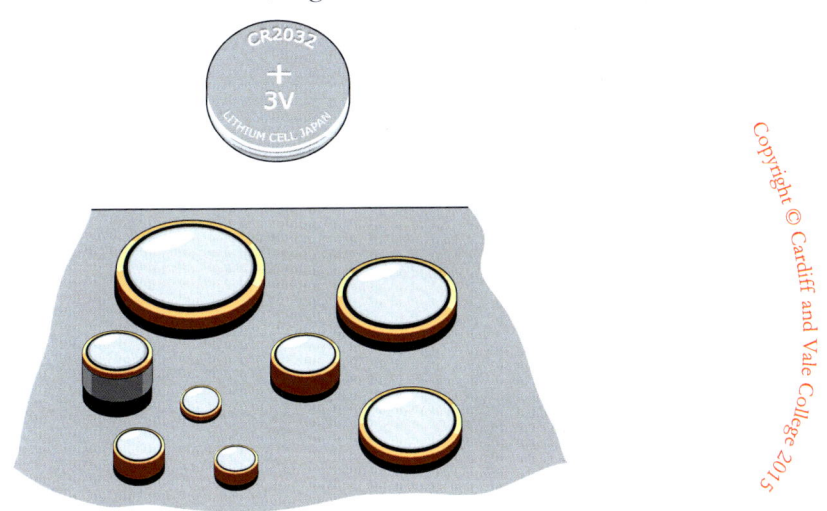

Figure 5.9 – Typical Lithium Button or Coin batteries

The first two digits give the battery diameter in mm. The last two digits give the battery height in mm. Button cells are commonly used in calculators, watches and other small electronic appliances. Most button cells have low self-discharge and hold their charge for a long time if not used.

Secondary Cells

Secondary cells, sometimes called Secondary Wet Cells come in five (5) basic types:

- Lead-acid
- Nickel-cadmium
- Silver-zinc
- Silver-cadmium
- Lithium

Lead-Acid Cell

We have already discussed how the chemical action in a lead-acid cell provides electrical power; i.e. the discharging and charging process - illustrated in figure 5.3 - describes the action of a lead-acid cell.

As a reminder, the lead-acid cell has an anode of lead peroxide; a cathode of sponge lead, and an electrolyte of sulphuric acid and water. There are two (2) basic types of lead-acid battery:

1. Flooded Lead-Acid Batteries
2. Sealed Lead-Acid Batteries

Flooded Lead-Acid Batteries

Flooded cells are those where the electrode plates are immersed in the electrolyte. As gases created during charging are vented to the atmosphere, distilled water must be occasionally added to bring the electrolyte back to its required level. The most familiar example of a flooded lead-acid cell is the old style 12V car battery.

Sealed Lead-Acid Batteries

These battery types confine the electrolyte, but have a vent or valve to allow gases to escape if its internal pressure exceeds a certain safety threshold. During charging, a lead-acid battery generates oxygen gas at the positive electrode and sealed lead-acid batteries are designed so that this oxygen is captured and reabsorbed in the battery. This is called an oxygen recombination cycle and works well as long as the charge rate is not too high. Too high a rate of charge may result in a case rupture, thermal runaway or internal mechanical damage.

Therefore, the valve-regulated battery is the most common type of sealed battery and these have a spring-controlled valve that vents gases at a predetermined pressure, typical 2 to 5 psi, depending on the battery design. Although the term 'valve-regulated' is often used synonymously to describe sealed lead-acid batteries, not all sealed batteries are valve-regulated. Some battery designs use replaceable vent plugs or other mechanisms to relieve excess pressure.

Sealed batteries were developed to reduce maintenance requirements for batteries in active service. Since electrolyte levels are preserved by trapping and reabsorbing gasses, there should not be any need to add distilled water over the life of the battery. These batteries are often misnamed '*maintenance free*' but in reality, all maintenance practices applicable to unsealed type batteries are applicable to sealed type batteries with the exception of electrolyte level replenishment.

Sealed batteries are often used for aircraft backup and emergency power applications, e.g. emergency exit lighting, but as their state of charge cannot be checked by the usual specific gravity measurement, they are subjected to specific time period maintenance checks. However, one disadvantage of the sealed battery is its susceptibility to high temperature and so they are location sensitive, especially in an aircraft environment.

Aircraft Lead-Acid Batteries

The typical aircraft lead-acid battery is constructed of two (2) or more cells housed in a case or container. The container or case housing the separate cells is usually made of hard rubber, plastic, or some other material that is resistant to the electrolyte, mechanical shock and will withstand extreme temperatures. The battery case usually has vent plugs to allow the gases that form within the cells to escape.

As already discussed, the battery plates are the cathodes and anodes and figure 5.10 below shows how these are constructed in more detail.

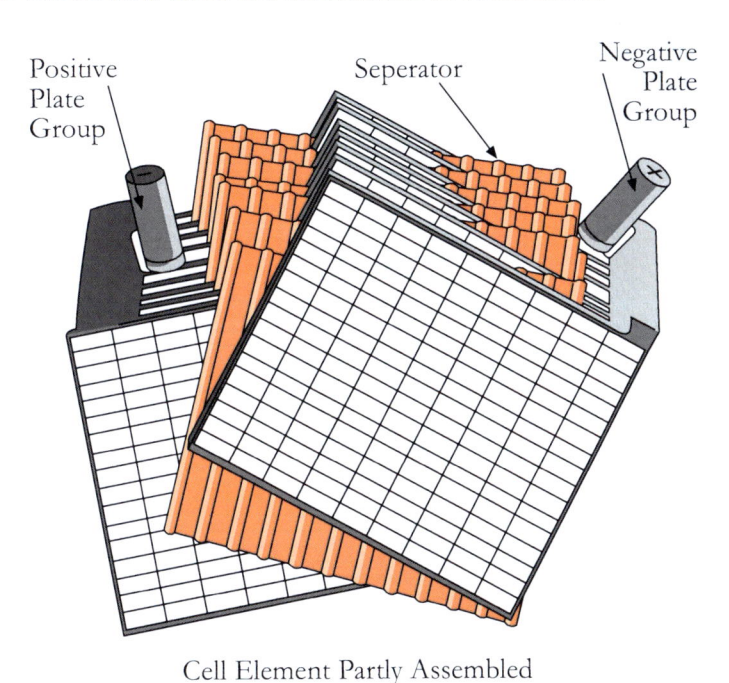

Cell Element Partly Assembled

Figure 5.10 - Lead-Acid Battery Plate Arrangement

CHAPTER FIVE
DC SOURCES OF ELECTRICITY

As illustrated in figure 5.10, the negative plate group forms the cathode of the individual cells and the positive plate group the anode. The plates are interlaced, with a separator maintaining electrical integrity, with a terminal attached to each plate group and the terminals of the individual cells are connected together by link connectors. The cells are connected in series in the battery and the positive terminal of one end cell becomes the positive terminal of the battery while the negative terminal of the opposite end cell becomes the battery's negative terminal.

The terminals of a lead-acid battery are usually identified from one another by their size and markings. The positive terminal is marked with a '+' sign, sometimes coloured *red*, and is usually physically larger than the negative terminal, which is marked with a '−' sign. However, most aircraft batteries are fitted with a quick-release connection that can only be fitted in one way.

Individual cells of a lead-acid battery are not usually replaceable, so in the event of a single cell failure, the battery must be replaced. However, some batteries, typically those used in acrobatic aircraft, are constructed of blocks of cells and these can be replaced on an individual basis.

Lead Acid Battery Maintenance

For a battery to work properly, its electrolyte must contain a certain amount of active ingredient be it acid or alkaline. As the active ingredient is usually dissolved in water, its amount cannot be directly measured and therefore, an indirect method is used, which measures the electrolyte's **Specific Gravity**.

Specific gravity is the ratio of the weight of a substance compared to the weight of the same amount of pure water, with the specific gravity of pure water being 1.0. Any substance that floats has a specific gravity less than 1.0 and conversely, any substance that sinks has a specific gravity greater than 1.0.

The active ingredient in an electrolyte is heavier than water and so has a specific gravity greater than 1.0. A battery's acceptable range of specific gravity is determined at construction by the manufacturer and is determined by using a *Hydrometer*, figure 5.11.

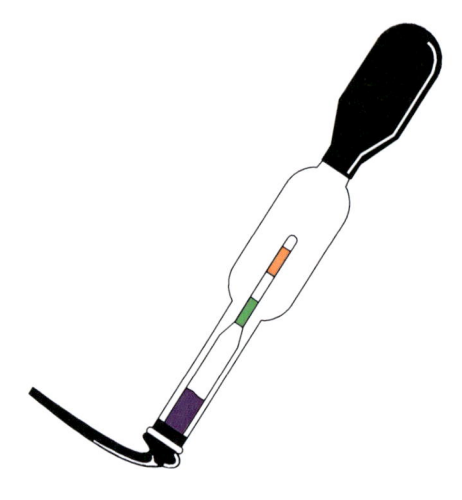

Figure 5.11 – Typical Hydrometer

Put simply, a hydrometer is a glass syringe with a float inside it. The float is a hollow glass tube, sealed at both ends, weighted at the bottom and with a calibrated scale marked on its side.

To test an electrolyte, it is drawn into the hydrometer using the suction bulb in sufficient quantity to make the float rise. The float will rise to a point determined by the electrolyte's specific gravity. If the electrolyte contains a large amount of active ingredient, its specific gravity will be relatively high, i.e. the float will rise higher than it would if the electrolyte contained only a small amount of active ingredient.

The hydrometer must be read in a vertical position where the surface of the electrolyte touches the float and the reading compared with reference to the appropriate Maintenance Manual, which will determine whether or not the battery's specific gravity is within specification.

Note: Hydrometers should be flushed with fresh water after each use to prevent inaccurate readings and must not be used for any other purpose.

With a Lead-Acid Cell, the electrolyte is diluted **Sulphuric Acid (H_2SO_4)** that has a **Specific Gravity** of 1.28 when fully charged, which reduces to 1.15 when fully discharged, allowing it to give an indication of the Cell's charged state. Each cell when fully charged has an open circuit voltage of at least 2.1V.

The voltage of a battery is determined by the number of cells connected in series to form the battery. Although the voltage of a lead acid cell just removed from charging is approximately 2·1 volts it is normally rated at only 2·0 volts, because it soon drops to that value when applied to a load. A Lead Acid battery rated at 12 volts will consist of 6 cells connected in series and a battery rated at 24 volts is composed of 12 cells.

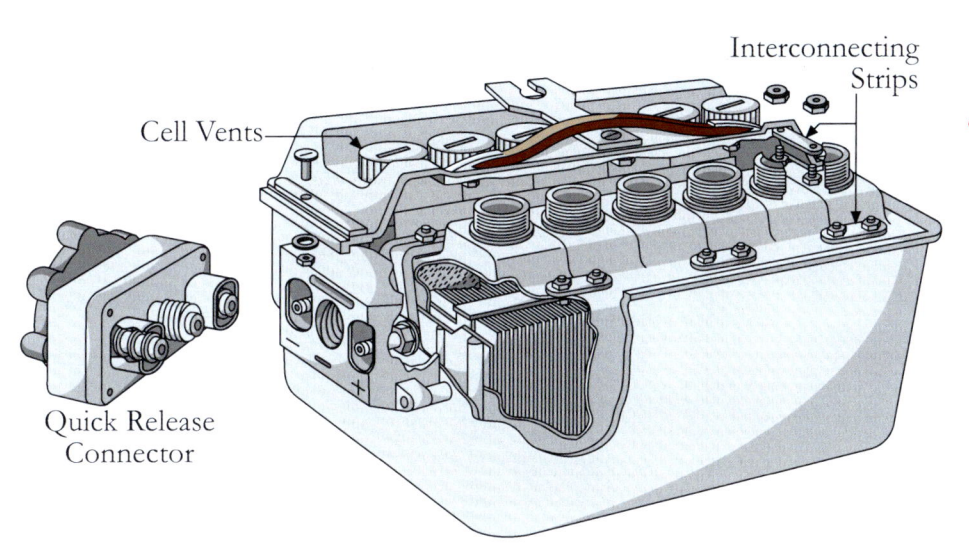

Figure 5.12 - 24 V Lead-Acid Battery

Aircraft with 24 V Lead Acid batteries all have a quick release fastener and are constructed using 12 cells, figure 5.12.

Nickel-Cadmium Cell

The *Nickel-Cadmium* Cell, usually shortened to *NiCad*, is superior to the lead-acid cell in several ways. Their major advantage is that they generally require less maintenance throughout their service life in comparison to lead-acid cells. They are now more popular with manufactures than Lead acid batteries.

The NiCad cell construction differs greatly to the lead-acid cell in that its anode, cathode and electrolyte are made of different materials. The cathode is made of *Cadmium Hydroxide*, the anode *Nickel Hydroxide*, and the electrolyte is *Potassium Hydroxide* and water.

The nickel-cadmium and lead-acid cells have comparable capacities and characteristics at normal discharge rates, but at high discharge rates the nickel-cadmium cell can deliver greater power and can maintain its output for longer. Sealed NiCad cells are equipped with electrodes that allow efficient high current delivery.

All NiCad batteries are capable of:

- Delivering exceptionally high currents
- Being charged and discharged any number of times without any appreciable damage
- Being rapidly recharged hundreds of times
- Tolerance to abuse such as over-discharging or overcharging
- Being charged in a shorter time
- Staying idle longer in any state of charge and keeping a full charge when stored for a longer period of time

However, when compared to many primary batteries and even lead-acid ones, NiCad cells are heavy and have comparatively limited energy density. They last longer and perform better if fully discharged each cycle before being recharge. Otherwise, the cells may exhibit a so-called '*memory*' effect where they behave as if they had lower capacity than was originally delivered.

Larger nickel–cadmium batteries are used for the essential power source in some aircraft and can even start up some aircraft engines. They are also found as backup power systems where very high currents, low temperature conditions, and reliability are important factors. Due to these superior characteristics and capabilities the nickel-cadmium cells are being used extensively in many aircraft applications that require a high discharge rate.

However, these cells do have one major drawback in that they can suffer from *Thermal Runaway*. This condition can exist when the cell's core temperature starts to increase. This causes its *internal resistance*, discussed later in this chapter, to decrease, causing an increase in current flow that causes an increase in temperature, causing a decrease in resistance etc.

Aircraft Nickel-Cadmium Batteries

The nickel-cadmium (NiCad) battery is similar in construction to the lead-acid but it is usually constructed of several smaller cells and these are replaceable on an individual basis.

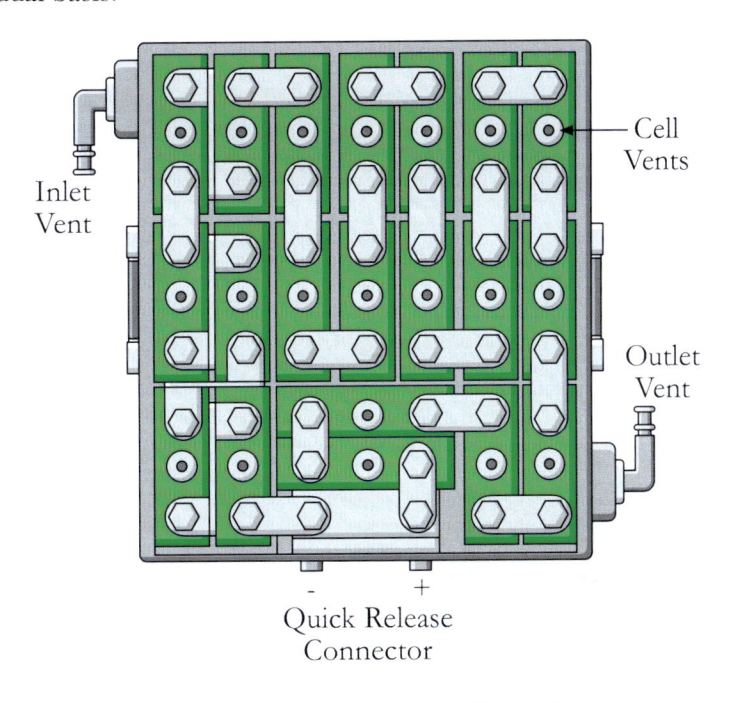

Figure 5.13 - 20 Cell Nickel-Cadmium Battery Interconnect

Therefore, a typical NiCad aircraft battery has twenty (20) cells versus a Lead Acid's twelve (12). Figure 5.13 shows how the individual cells are interconnected.

Figure 5.14 shows the cell of a NiCad battery.

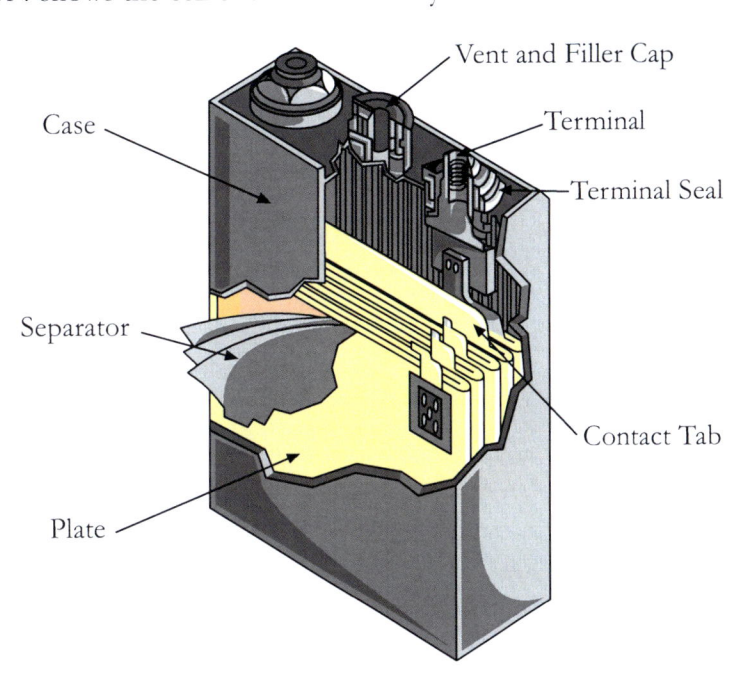

Figure 5.14 - Nickel-Cadmium Cell

CHAPTER FIVE
DC SOURCES OF ELECTRICITY

The construction of secondary cell batteries is so similar, that it is sometimes difficult to distinguish the type by simply looking at it. However, in order to properly check or recharge the battery the type must be known and so each battery should have a nameplate that gives a description of its type and details its electrical characteristics.

With a Nickel-Cadmium Cell, the electrolyte is diluted **Potassium Hydroxide (KOH)**. When fully charged it has an open circuit voltage of 1.3V and the nominal voltage when applied to a load is 1·2V. Therefore, for a nominal 24V, a Nickel-Cadmium battery requires 20 cells. The Specific Gravity is 1.24-1.3, which changes little from a fully charged to fully discharged state.

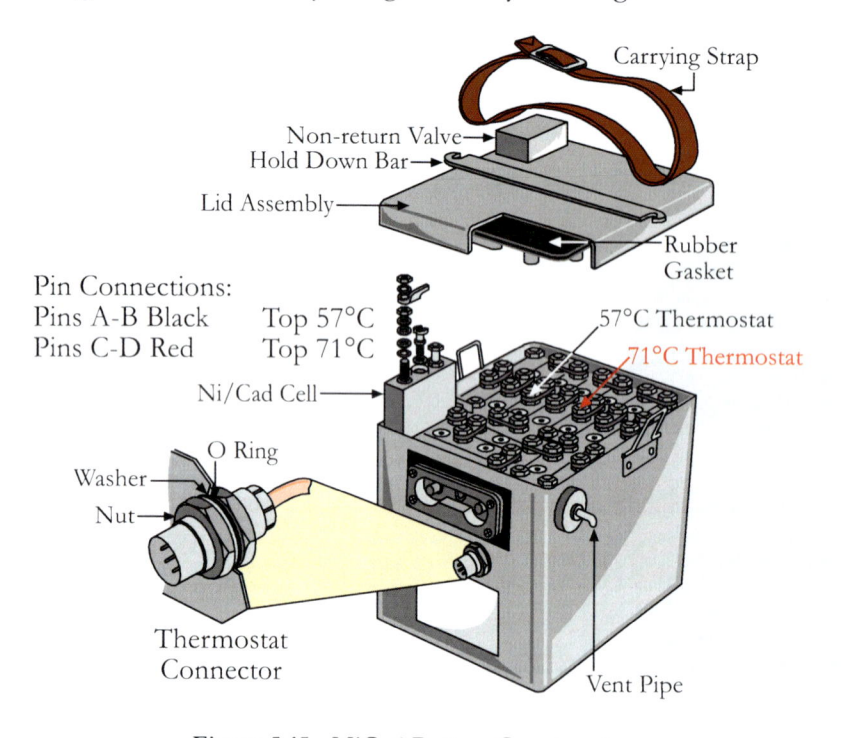

Figure 5.15 - NiCad Battery Construction

To help protect against Thermal Runaway NiCad batteries are usually fitted with two (2) thermostats. As shown in figure 5.15, one thermostat operates at 57°C while the other operates at 71°C. The signals from the thermostats output a warning to the aircraft's **Master Warning System** of **BATTERY HOT** at 57°C and **BATTERY OVERHEAT** at 71°C.

Caution

Lead-acid and NiCad batteries should not be charged or stored together as the fumes from the lead-acid battery will destroy the cells of the NiCad. Consequently, battery maintenance should always be carried out in separate bays for the lead-acid and NiCad.

Silver-Zinc Cell

The *Silver-Zinc* cell is used extensively to power emergency equipment. They are relatively expensive and can be charged and discharged fewer times than other types of cells. However, when compared to the lead-acid or nickel-cadmium cells, these disadvantages are overweighed by the light weight, small size, and good electrical capacity of the silver-zinc cell.

The silver-zinc cell uses the same electrolyte as the nickel-cadmium cell, i.e. potassium hydroxide and water, but the anode is made of *Silver Oxide* and the cathode is made of *Zinc*.

Silver-Cadmium Cell

The *Silver-Cadmium* cell is a fairly recent development and combines some of the better features of the nickel-cadmium and silver-zinc cells. It has more than twice the shelf life of the silver-zinc cell and can be recharged several times, but it is expensive and has a low voltage output.

Its electrolyte is potassium hydroxide and water, as is the nickel-cadmium and silver-zinc cells. However, its anode is made of silver oxide, as in the silver-zinc cell, and the cathode of cadmium hydroxide as in the NiCad cell.

Using different combinations of materials to form the electrolyte, cathode, and anode of cells means that the best features of each can be brought to bear to manufacture cells with different qualities that meet the many varied requirements of today's applications.

Lithium Ion Cell

The high power lithium-ion battery is a new addition to the list of secondary batteries used on aircraft. These batteries are widely used in general aviation, but Boeing was the first large aircraft manufacturer to embrace this technology when installing them on the 787 Dreamliner. They offer several advantages; including the capability to provide optimum power for a longer duration, they are approximately, half the weight, and recharge in a fraction of the time when compared with traditional lead acid and nickel cadmium batteries. According to the manufacturers the battery can charge from 0 to 90% in only 75 minutes.

Charging or discharging a Li-Ion battery involves an exchange of lithium ions between the electrodes. Each cell can provide an output voltage of 3 to 4.2 volts depending primarily on the materials used to construct the cathode. However lithium is also a highly reactive substance; it belongs to the alkali metal group, which also contains sodium and potassium, the volatility of which some of you may remember from school chemistry classes.

Like all batteries, lithium cells consist of two electrodes separated by an electrolyte, typically a solution of lithium salts and organic solvents. The separator is a very thin sheet of microperforated plastic. As the name implies, it separates the positive and negative electrodes while allowing ions to pass through.

Each of the lithium ions acts as a sort of shuttle, carrying an electron from the cathode to the anode. At the cathode, the lithium ions are absorbed, freeing up those electrons to act as current and ionizing the lithium. To recharge the cell, you simply add electricity, which drives the lithium back out of the cathode and into the anode, and it's ready to do it all over again.

As we can see in Figure 5.16, when the battery is charged, lithium ions are driven from the electrolyte into the anode. When the battery is discharged they flow back, creating a balancing flow of electrons in a circuit that powers the device.

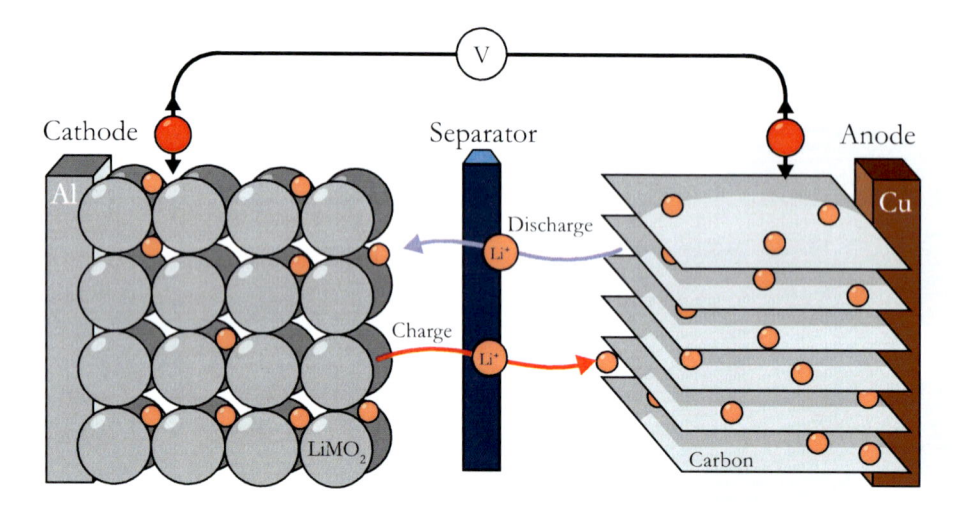

Figure 5.16 – Lithium Ion Battery Charge/Discharge

Lithium batteries can also be damaged by using them in hot environments and by excessive discharging and charging—which is why most lithium batteries contain special circuits to prevent this.

In 2013 rechargeable batteries containing the element lithium were in the headlines for all the wrong reasons, when a lithium-ion battery overheated on a Boeing 787 Dreamliner at Narita airport. Boeing grounded its entire fleet of the next-generation plane after the lithium batteries on two of the aircraft caught fire. The 787s returned to the air after being fitted with a modified system to protect the aircraft against battery fires.

When a fire like this happens, it is usually caused by an internal short in the battery. If the separator sheet that keeps the positive and negative electrodes apart gets punctured and the electrodes touch, the battery heats up very quickly. You may have experienced the kind of heat a battery can produce if you have ever put a normal 9-volt battery in your pocket. If a coin shorts across the two terminals, the battery can get quite hot very quickly.

In a separator failure, that same kind of short happens inside the lithium-ion battery. Since lithium-ion batteries are so energetic, they get very hot. The heat causes the battery to vent the organic solvent used as an electrolyte, and the heat (or a nearby spark) can light it. Once that happens inside one of the cells, the heat of the fire cascades to the other cells and the whole pack goes up in flames.

The two things that will keep lithium-ion batteries relatively safe are continuous improvements in manufacturing techniques and the use of control systems to monitor their temperature and regulate their charging and discharging. An advantage of lithium batteries is that they do not suffer from any "memory effect", which means they can be partially charged and discharged many times without loss of capacity. Running down a lithium battery completely, however, can destroy it. So this too has to be guarded against. As always, researchers are working on alternative chemistries to the lithium-ion battery which could provide even greater energy densities.

General Maintenance

Routine battery maintenance, especially for aircraft, is very simple. Common sense will tell you that the following should be checked as per the Maintenance Schedule:

- Terminals should be periodically checked for cleanliness and good electrical connection
- The battery casing should be checked for cleanliness and evidence of damage

Maintenance procedures for batteries are normally determined by the equipment and aircraft manufacturer, but there are minimum requirements set for all batteries; check the appropriate documentation.

Battery Safety Precautions

All types of batteries should be handled with care to prevent damage to the battery, equipment, aircraft and personnel. Again, battery safety is really a matter of common sense, but the following basic precautions are appropriate:

- Never short the battery terminals
- Always use carrying straps when transporting batteries
- Always wear protective clothing, e.g. rubber apron, rubber gloves, safety goggles etc, when working with batteries
- Do not smoke, allow electric sparks or open flames near charging batteries
- Always prevent electrolyte spillage, but if it does occur, dilute and clear up immediately with large quantities of water and the appropriate neutralising agents

- If electrolyte is spilled or splashed onto skin or into the eyes, *immediately* flush the skin or eyes with large quantities of fresh water for a *minimum* of 15 minutes
- If electrolyte gets into the eyes, make sure the upper and lower eyelids are pulled out sufficiently to allow fresh water to flush under the eyelids and seek immediate medical assistance

Battery Charging

You will no doubt have realised by now that adding the active ingredient to a battery's electrolyte does not recharge it. Adding the electrolyte's active ingredient only increases its specific gravity and has no affect on the active material of the cathode and anode plates, and so therefore does not bring the battery back to a charged condition. The only way to recharge a battery is to apply a charging current in reverse polarity to the discharge current.

Batteries are maintained and tested in specialised CAA approved battery workshops. Although no two workshops are identical the general maintenance and testing procedures are standard and apply to all workshops, as do the safety requirements. These will have specific charging procedures for the various types of batteries in use.

Aircraft batteries are also charged during flight, in a similar way to which a car battery is charged. This will be discussed in more detail in Module 11/13.

Capacity and Rating of Batteries

A cell's capacity is measured in *Ampere-Hours (AH)* and is equal to the product of the current in amperes and the time in hours during which the battery will supply this current. The ampere-hour capacity varies inversely with the discharge current, eg a 400 ampere-hour battery will deliver 400 amperes for 1 hour or 100 amperes for 4 hours, but aircraft batteries are always rated according to a one (1)-hour rate of discharge. Each battery has a low-voltage limit, as specified by the manufacturer, beyond which very little useful energy can be obtained from a battery. This low-voltage limit is normally a test used in battery shops to determine the condition of a battery.

Capacity Checks

A capacity test should be carried out after initial charge, and thereafter at intervals of three months, or at any time the capacity of a battery is in doubt. The battery should be fully charged then connected to a suitable discharge control panel incorporating a variable load resistance, an ammeter and an ampere-hour meter. A voltmeter is necessary to measure voltage at the battery terminals or cell connecting strips.

The battery should be discharged at a rate corresponding to the rating of the battery (e.g. a 25A.H.battery rated at the 1 hour rate would be discharged at 25 amps) until the battery reaches its fully discharged condition. This is denoted by either the main terminal voltage, or the relative density of the electrolyte, falling to the respective fully discharged values for the particular type of battery.

The minimum acceptable capacity for use on aircraft is 80% which, in the case of the example rating quoted, provides a duration of discharge equal to 48 minutes. The result however, should be compared with previous readings to assess rate of deterioration.

$$\% \text{ Efficiency} = \frac{\text{Actual Time of Discharge}}{\text{Expected Time of Discharge}} \times 100$$

Example from above:

If time taken to discharge = 0.8 hours (48 minutes)

then: $\frac{48}{60} \times 100 = 80\%$

Actual capacity of battery would be 20 A.H. because it provided 25 amps for 0·8 hours.

Cells Connected in Series and Parallel

A *Battery* is a voltage source that uses chemical action to produce a voltage; ie it has the same definition as a cell. Indeed, in many cases the term 'battery' is applied to a single cell, such as the series of batteries we use for torches, watches, portable radios, etc. Typically, though not always, a single cell battery has an output of 1.5 volts. Therefore, if a torch, for example, requires six (6) volts to operate, it uses four 1.5 volt batteries in series to make up the six (6) volts. Alternatively, some torch manufacturers may combine four 1.5 volt cells into one unit, creating a six (6) volt battery. In practice, there are three (3) ways of combining cells to form a battery.

1. Series
2. Parallel
3. Series-Parallel

In many cases, a battery-powered device may require more electrical power than one cell can provide in the form of voltage or current, or even both in some cases. Under such conditions, it is necessary to combine, or interconnect, a sufficient number of cells to meet the larger requirements.

Cells connected in series provide a **higher voltage**, while cells connected in parallel provide a **higher current capacity**. To provide adequate power when both voltage and current requirements are greater than the capacity of one cell, then a combination series-parallel network of cells is used.

Series-Connected Cells

Let us assume that a load requires a power supply of 6 volts and a current capacity of 0.125 Amperes. As a single cell normally outputs only 1.5 volts, more than one cell is needed to obtain the higher voltage and these cells are connected in series as shown in figure 5.17.

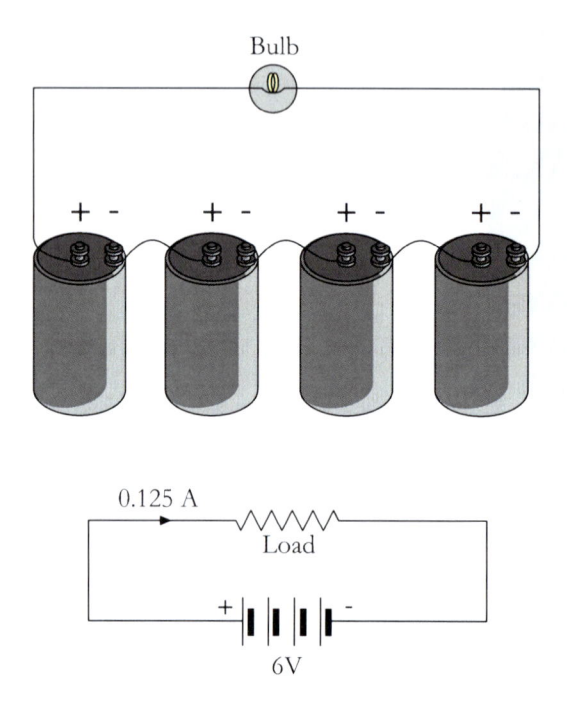

Figure 5.17 – Schematic & Visual Presentation of Series-Connected Cells

The schematic circuit illustrated in figure 5.17 is the equivalent electrical diagram of the physical battery arrangement. The load, which is a simple bulb, is represented by the resistor symbol and the battery is indicated by one long and one short line per cell.

In a series link up like this, the negative electrode, i.e. cathode, of the first cell is connected to the positive electrode, i.e. anode, of the second cell, and so on. The positive electrode of the first cell and negative electrode of the last cell then serve as the terminals of the battery. In this way, the voltage is 1.5 volts for each cell in the series line and as there are four cells, the output terminal voltage is 1.5 × 4 = 6 Volts.

When connected to the load, 0.125 A flows through the load and each cell of the battery, which is well within the capacity of each cell, and so only four (4) series-connected cells are required to supply this particular load.

Caution

When connecting cells in series, always connect alternate terminals together, i.e. negative to positive to negative to positive, etc, and always have only two (2) remaining terminals to connect to the load. Do not connect the two remaining terminals together as this will produce a short circuit across the battery and would not only quickly discharge the cells but could cause some types to explode.

Parallel-Connected Cells

Let us now assume that an electrical load requires only 1.5 volts, but needs 0.5 A of current, making the added assumption that a single cell will supply only 0.125 A. To meet this requirement, the cells are connected in parallel, as illustrated shown in figure 5.18.

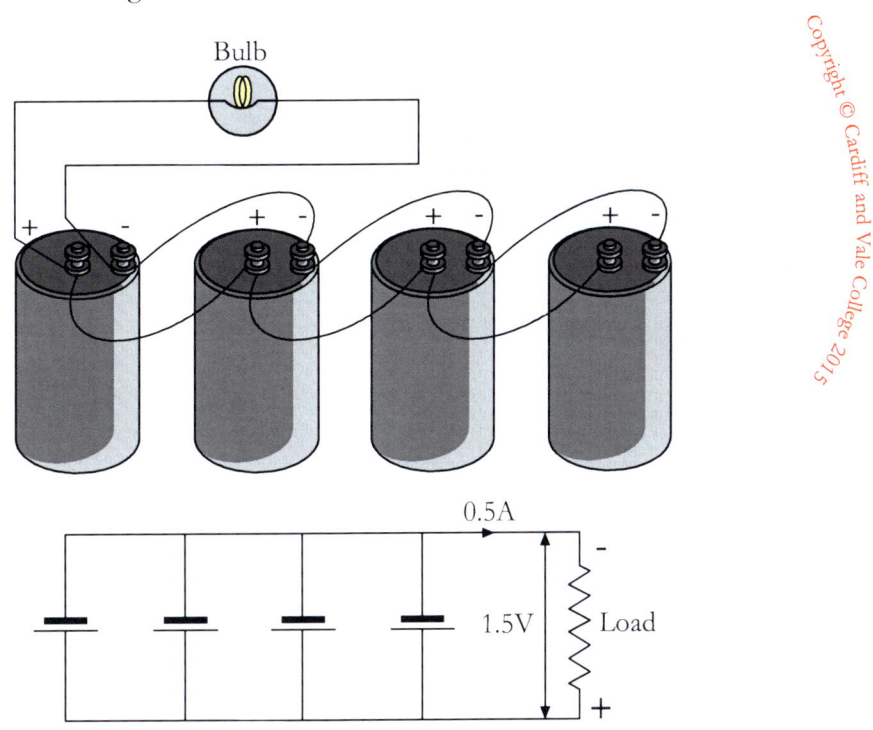

Figure 5.18 – Schematic & Visual Presentation of Parallel-Connection

In a parallel connection, all positive cell electrodes are connected together and all negative electrodes are connected together. No more than one cell is connected between the lines at any one point; so the voltage between the lines is the same as that of one cell, i.e. 1.5 volts. However, each cell may contribute its maximum allowable current of 0.125A to the load and as there are four (4) cells, the total load current is 0.125 × 4 or 0.5 A. In this way, four (4) cells connected in parallel have enough capacity to supply a load requirement of 0.5 A at 1.5 volts.

Series-Parallel-Connected Cells

Figure 5.19 below shows a battery network connected to a load that requires both a voltage and current greater than one (1) cell can provide on its own, ie assuming one (1) cell has an output of 1.5 V at 0.125 A.

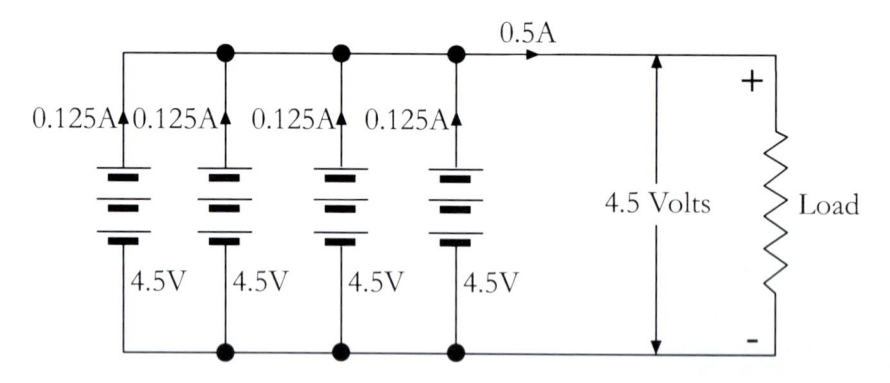

Figure 5.19 - Schematic of Series-Parallel Battery Network

In this case, the load requires 4.5 Volts with a current of 0.5 A. To achieve this, groups of three (3), 1.5 V cells are connected in series to provide the voltage, i.e. 4.5 V, and then four (4) of these series groups are further connected in parallel to provide the required 0.5 A current.

Summary

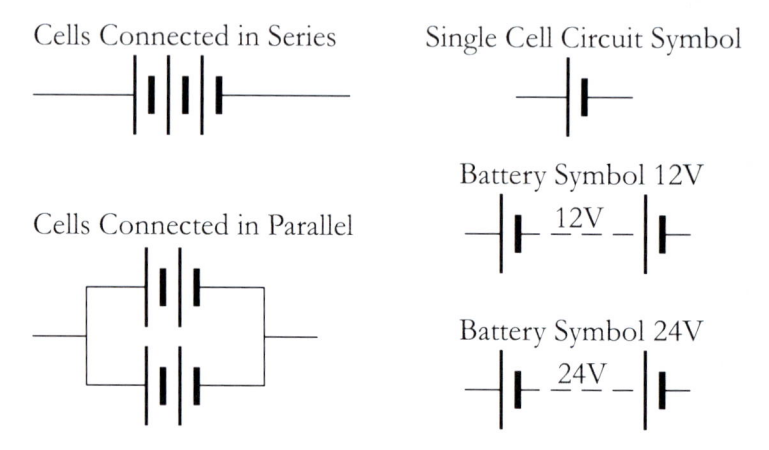

Figure 5.20 - Cell and Battery Circuit Symbology

As shown in figure 5.20, when cells are connected in series their voltages are added, but the overall AH capacity remains the same. When they are connected in parallel, the overall voltage remains the same but their AH capacities are added. In this way, batteries may be created with a combination of cells for any voltage and AH capacity.

A Battery's Internal Resistance

In an ideal world, the total voltage output derived from a cell would be available for the circuit it was introduced to. In addition, since a cell's electrode contains only a limited number of units of chemical energy convertible to electrical energy, it follows that a battery of a given size has a certain capacity to operate devices and will eventually become exhausted having given up all of its energy to the circuit.

Unfortunately we are not living in an ideal world and as we have already discussed, like all matter, the materials used in the construction of a cell has resistance and consequently, these add up to give a cell's *Internal Resistance*. Each cell's internal resistance depends upon several factors and the main ones are:

- The size of the electrodes
- The distance between the electrodes when suspended in the electrolyte
- The resistance of the electrolyte

Now obviously the ideal source would have no internal resistance and would provide the same voltage output for all current conditions. However, if something in a circuit has resistance, and this includes the battery's internal resistance, when a current flows, work is done and is usually shown up as heat loss. Now as the battery is the source of voltage for a circuit, any work done within the battery due to its internal resistance is lost before it even gets to the load. How this affects the circuit's overall performance is discussed in the next chapter, but as far as the battery is concerned, if each cell has internal resistance and there are several cells to a battery, their affect could be cumulative.

Nevertheless, there are several things that can be done in the cell's construction process to minimise its internal resistance. For example, as already mentioned, the larger the electrodes and the closer together they are in the electrolyte without touching, the lower the internal resistance of the cell and the more current the cell is capable of supplying to the load. In addition, the way the cells of a battery are interconnected will affect the battery's overall internal resistance. If the cells are series connected, the battery can supply a large voltage or emf, but this will be at the expense of higher internal resistance as it will be the sum of the individual cell's internal resistance. However, when the cells are connected in parallel, the total internal resistance (r_t) is that of the parallel connected internal resistances. Therefore, if a battery has four (4) cells, each with an internal resistance '*r*', then its total resistance is:

$$\frac{1}{r_t} = \frac{1}{r} + \frac{1}{r} + \frac{1}{r} + \frac{1}{r} = \frac{4}{r}$$

$$\therefore r_t = \frac{r}{4}$$

So in this case, if the cells are connected in parallel the battery's internal resistance is less than ¼ that of an individual cell. Consequently, we can say that for '*n*' cells, if they all have the same internal resistance, when connected in parallel:

$$r_t = \frac{r}{n}$$

Parallel cell connection is therefore a way of getting a lower battery internal resistance for the same emf.

Thermocouples

In chapter 4 we discussed how electricity could be generated by heat at the junction of two (2) dissimilar metals called a *Thermocouple*, as some materials readily give up their electrons and others readily accept electrons. If we look at a practical example of two (2) dissimilar metals like copper and zinc, when they are joined together, a transfer of electrons can take place between them as the copper atoms have excess electrons that the zinc atoms will readily accept. This result in the zinc becoming negatively charged, while the copper takes on a positive charge, creating a voltage potential across the junction of the two metals.

Now as we have already discussed in chapter 4, the heat energy of normal room temperature is enough to make the copper and zinc release and gain electrons, causing a measurable voltage potential. Now if more heat energy is applied to the junction, more electrons are released, and the voltage potential becomes greater, illustrated in figure 5.21.

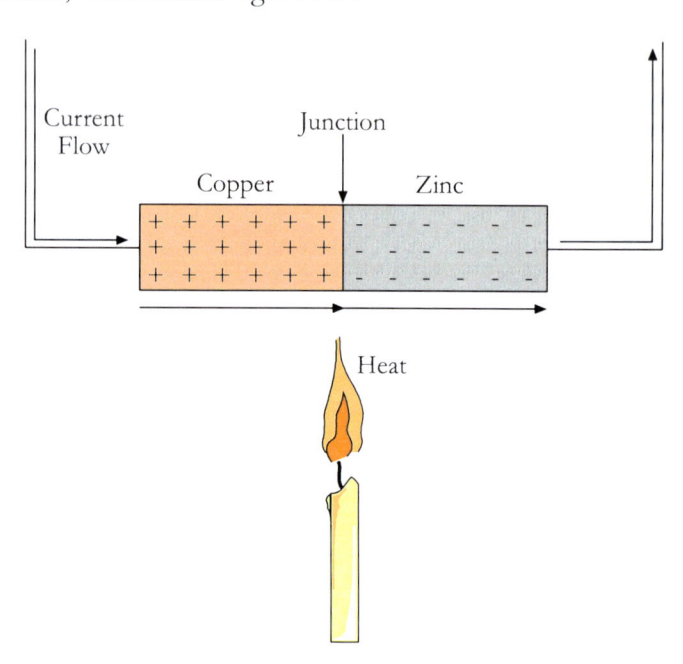

Figure 5.21 - Heat Energy Causes Copper to Release More Electrons to Zinc

When heat is removed and the junction cools, the charges will dissipate and the voltage potential will decrease, and as mentioned before, this process is called *Thermoelectricity*.

The thermoelectric voltage in a thermocouple is dependent upon the heat energy applied to the junction of the two dissimilar metals and to the composition of the metals at the junction. Figure 5.22 shows a graph of the output voltage, in milli-volts, for various junction combinations at varying temperatures.

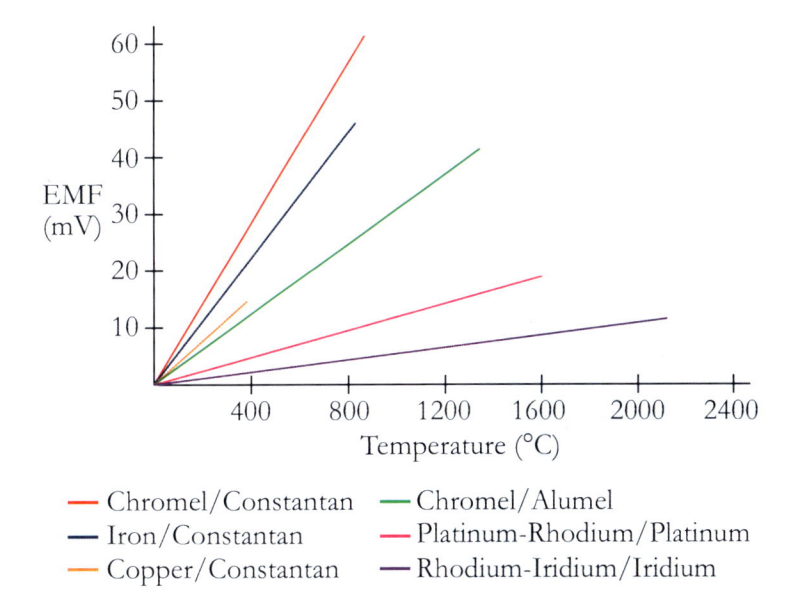

Figure 5.22 – EMF Output for Various Metals at Different Temperatures

Thermocouples are widely used to measure temperature, especially in small or tight areas, as the temperature being measured is only that at the junction of two (2) wires. Their small size also means a small thermal capacity and so they have a terrific response to temperature changes. Depending on the types of material used, as illustrated in figure 5.22, thermocouples can measure temperature in the range of -200°C to +1700°C.

Thermocouple power capacities are very small compared to some other sources, but they are used to measure temperature at higher temperatures than ordinary mercury or alcohol thermometers and are often used as heat-sensing devices in automatic temperature controlled equipment. In addition, they are often used in potentiometer circuits as signal conditioners.

Operation of Photocells

A *Photocell*, also called an *Electric Eye*, *Photoelectric cell*, or *Phototube*, is an electronic device with a photosensitive cathode that emits electrons when illuminated and an anode for collecting these emitted electrons. Various cathode materials are sensitive to specific spectral regions, such as ultraviolet, infrared, or visible light and these include selenium, germanium and silicon, i.e. semiconductors. Under normal circumstances, in darkness there is no potential difference between the anode and cathode and so no current flows as no electrons are emitted. However, illumination in the correct spectral region excites electrons that are attracted to the anode, producing current proportional to the intensity of the illumination.

As we discussed in chapter 4, the most common photosensitive materials used to produce a photoelectric voltage are various compounds of silver oxide or copper oxide and there are many different sizes and types in use. Each type usually serves a special purpose for which it is specifically designed, but nearly all have some of the basic features of the photoelectric cells illustrated in figure 5.23.

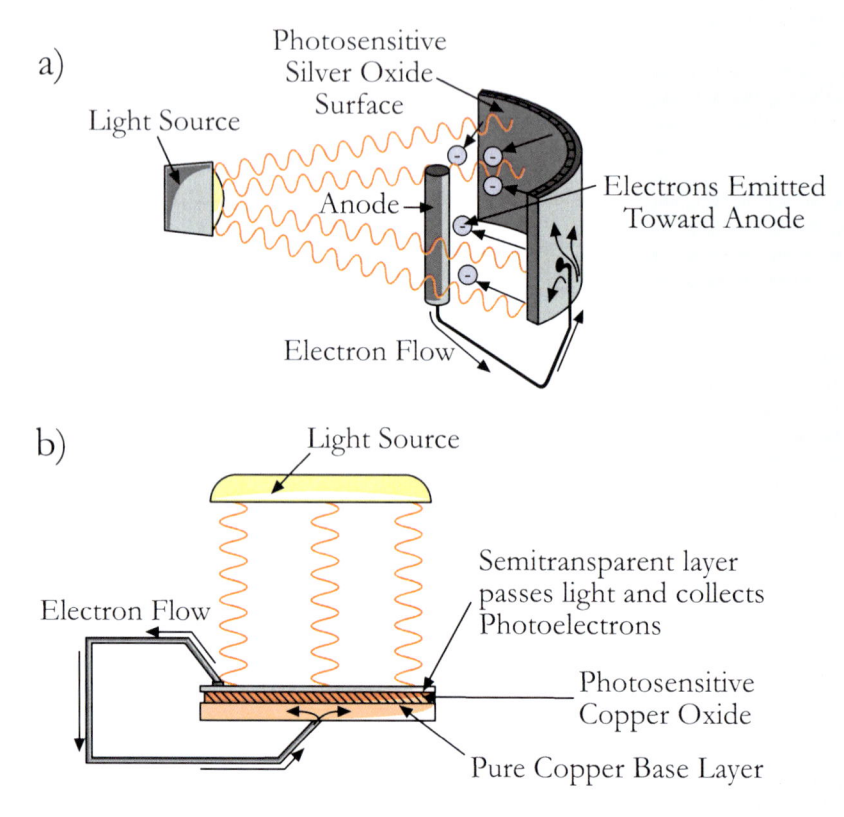

Figure 5.23 – Photoelectric Cell

The cell in figure 5.23 (a) has a curved light-sensitive surface focused on the central anode. When light from the direction shown strikes the sensitive surface, it emits electrons towards the anode and as mentioned above, the more intense the light, the greater the number of emitted electrons. When a wire is connected between the filament and the back, or dark side of the cell, the accumulated electrons will flow to the dark side and will eventually pass through the metal of the reflector and replace the electrons leaving the light-sensitive surface. Therefore, light energy is converted into a flow of electrons, developing a usable current.

The cell featured in figure 5.23 (b) is a more practical and modern design, constructed in layers. The base plate is made of pure copper, coated with light-sensitive copper oxide. An extremely thin semi-transparent layer of metal is placed over the copper oxide, which serves two purposes:

1. It allows the penetration of light to the copper oxide
2. It collects the electrons emitted by the copper oxide

An externally connected wire completes the electron path, the same as in the reflector-type cell.

The output voltage of a photoelectric cell is relatively small and so is usually connected by external wires to some other device that amplifies it to a usable level. Although the power capacity of photoelectric cells is very small, they do react to light-intensity variations in an extremely short time. This characteristic makes them very useful for detecting or accurately controlling a great number of operations.

Modern photoelectric cells are based on a semi-conductor, e.g. selenium, germanium or silicon, which has been suitably prepared, modified or doped. The light radiation releases electrons and depending on the structure of the cell, the result may be a change in conductivity, i.e. photoconductive cells, the production of an emf, i.e. photovoltaic cells, or some other effect. They are used in light meters for photography, light detectors for burglar alarms and aircraft automatic dimming circuits, and aircraft power supplies. They are also used in control systems, where interrupting a beam of light opens a circuit, activating a relay that in turn, supplies power to a mechanism that brings about a desired operation, e.g. remote opening of a garage door.

Another example of their use is in new colour printing techniques where photoelectric cells ensure colour matching by selectively picking up the guide marks printed in each colour as they go by and by reacting to any irregularities in the distance between these guide marks. Any error is automatically corrected by modifying either the speed of one group or the pressure of the rollers that control the tension of the paper between one group and the next. Matching up across the width of the paper is controlled by cells, which react to any lateral digressions as the paper moves along by controlling a lateral-shifting mechanism.

Quality control of colour reproduction is also carried out by photoelectric cells that emit a current whose strength is proportional to the intensity of impression of the guide marks for each colour. By comparing this intensity with a colour scale, a computer determines the continual adjustments needed in the composition of the inks and controls the opening of valves to add pigment if the inks are too light or colourless varnish if they are too dark.

Photocells are useful devices and as their main disadvantage of having a low output is being overcome by ever more sophisticated amplifiers, and so their influence is spreading.

Revision

DC Sources of Electricity

Questions

1. In a Leclanché cell the Carbon electrode is usually:

    a) Positive

    b) Negative

    c) Can be either

2. The water/acid mix in a fully charged cell is typically:

    a) 75%/25% Acid/Water

    b) 50%/50%

    c) 75%/25% Water/Acid

3. In a lead acid cell the electrolyte is usually:

    a) Potassium Hydroxide

    b) Dilute Sulphuric Acid

    c) Distilled water

4. In a secondary lead acid cell, when fully discharged the electrodes become:

    a) Lead sulphate

    b) Pure lead

    c) Lead peroxide

CHAPTER FIVE
DC SOURCES OF ELECTRICITY

5. Secondary Alkaline cells are usually made of:

    a) Nickel-Cadmium

    b) Nickel-Copper

    c) Cadmium-Copper

6. The electrolyte in a NiCad cell is usually:

    a) Sulphuric acid

    b) Potassium Hydroxide

    c) Sulphuric Hydroxide

7. In the figure below, if each cell is 1.5 V and has a current capacity of 0.125A this arrangement will have an output of:

    a) 6V, 0.125 A

    b) 1.5 V, 0.5A

    c) 6V, 0.5 A

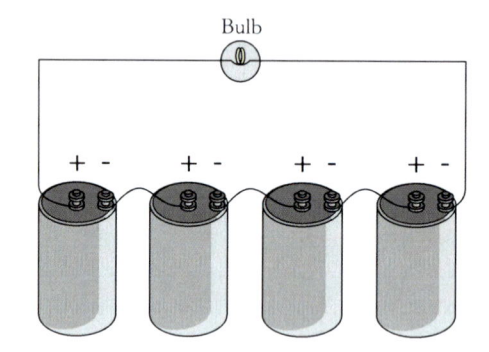

8. In the figure below, if each cell is 1.5 V and has a current capacity of 0.125A this arrangement will have an output of:

    a) 6V, 0.125 A

    b) 1.5 V, 0.5A

    c) 6V, 0.5 A

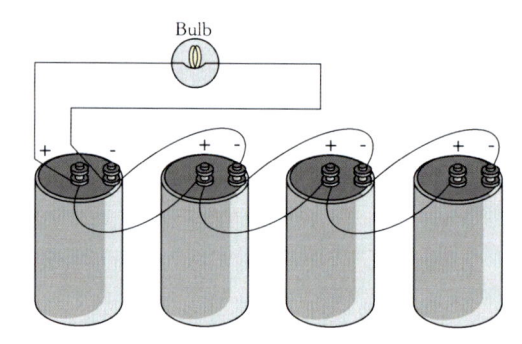

9. **A typical aircraft NiCad battery has:**

   a) 20 Cells

   b) 12 Cells

   c) 6 Cells

10. **A 50 ampere-hour battery will deliver:**

    a) 10 amperes for 5 hours

    b) 5 amperes for 20 hours

    c) 1 ampere for 100 hours

11. **Aircraft batteries are always rated according to:**

    a) A 10 hour rate of discharge

    b) A 20 hour rate of discharge

    c) A 1-hour rate of discharge

12. **To reduce a circuits internal resistance, batteries should be connected in:**

    a) Parallel

    b) Series/parallel

    c) Series

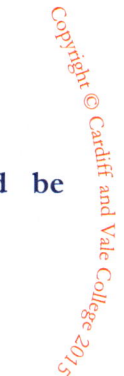

Revision

DC Sources of Electricity

Answers

1. A
2. C
3. B
4. A
5. A
6. B
7. A
8. B
9. A
10. A
11. C
12. A

DC Circuits

Introduction

Up to now in this Module, we have introduced several new terms and concepts without really studying in detail where these can be applied. Electricity can be considered from two (2) points of view:

1. The scientific view, which looks in detail at how things happen
2. The engineering view, which makes use of its properties

In chapters 1 to 5 of this Module we have looked at the 'how' of electricity but in this chapter we are going to put theory into practice and introduce some of the fundamental laws of electricity.

As far as electricity and electrical engineering are concern, we can summarise or split it down into one of four (4) main categories:

1. Electrical energy production
2. The transmission of electrical energy
3. The application of electrical energy
4. Electrical energy control

These four (4) categories cover all areas of the modern aircraft. However, before we start looking at complicated circuits it may be prudent to look at a very simple circuit that we all probably have experience of; the simple torch, figure 6.1.

CHAPTER SIX
DC CIRCUITS

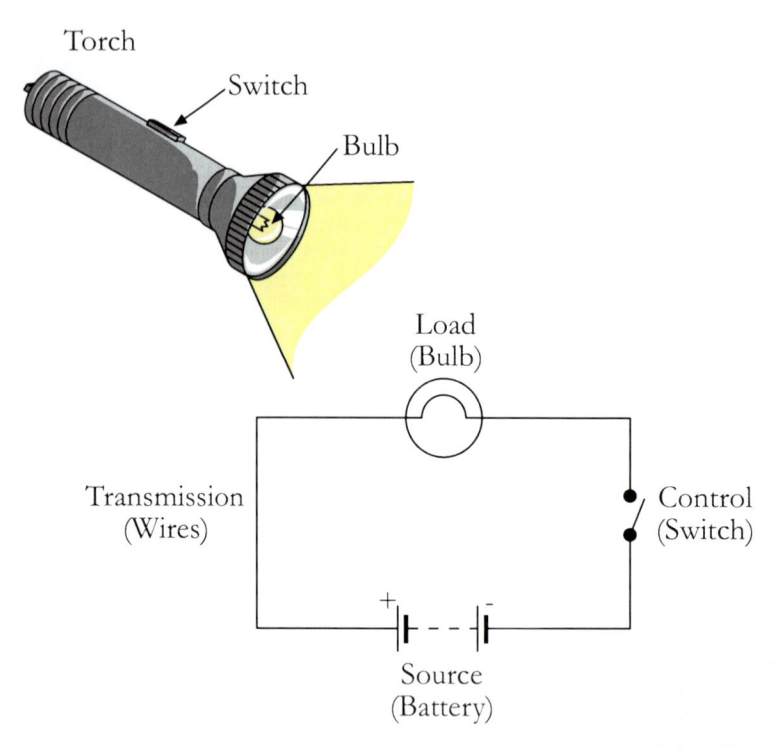

Figure 6.1 – Picture and Schematic Drawing of a Torch

The drawing in figure 6.1 is a basic electrical system that has four (4) basic elements:

1. Source
2. Load
3. Transmission system
4. Control

The Source

The source, as its name implies, is that part of the system that supplies energy, e.g. a generator, mains electric, battery etc.

The Load

The load's function is to absorb the source's energy and convert it into some other useful purpose; e.g. an electric light bulb is one type of load that takes the electrical energy and converts it into heat so that it can radiate light.

The Transmission System

The transmission system conducts the energy from the source to the load with as little loss between the two as possible. It is usually made up of wire, cables, bus-bars etc.

Control

As its name implies, the control's function is to manage the circuit and the simplest example of this is the straightforward on/off switch.

The simple circuit diagram in figure 6.1 illustrates two (2) of the most basic points concerning electrical systems:

1. It shows the basic function of any electrical system, i.e. to transport energy from the input source, the battery in this case, to the energy-converting load, i.e. the bulb.

2. Looking at the physical object, in this case the torch, as an electrical system is difficult to envisage how it works. Breaking down the circuit components as in the simple schematic line drawing makes it easy to understand and to work out exactly what is going on.

Although this is not your first introduction to a circuit drawing in this Module, it is the time to point out that there is a significant number of symbols, nomenclatures, abbreviations and acronyms used throughout the electrical, electronic and avionic systems found on an aircraft. In order to understand these circuits you should ensure you are familiar with the major ones in use today.

Ohm's Law

In the early part of the 19th century, a German physicist, **Georg Simon Ohm (1789-1854)** discovered that there was a direct relationship between current and voltage in a conductor. Ohm came to this conclusion by analysing the French Mathematician **Jean Fourier's (1768-1830)** work on heat flow along a metal rod.

In 1827, Ohm went on to put together his ideas into the law for which he is now best known, but received it little recognition for 20 years. He also discovered that the human ear is capable of breaking down complex musical sounds into their component frequencies, an important conclusion for the future development of Hi-Fi systems, but again one that was ignored at the time.

Ohm's Law was formulated after he proved by experimentation that a precise relationship exists between current, voltage, and resistance and it is stated as follows:

When current flows in a conductor, the difference in potential between the ends of the conductor, divided by the current flowing, is a constant, provided there is no change in the physical condition of the conductor, e.g. temperature.

CHAPTER SIX
DC CIRCUITS

This textbook definition does not really give a valid picture of something we can work with and so put into Basic English, Ohm's law states that the current flowing in an electrical circuit is directly proportional to the applied voltage and inversely proportional to the resistance. From this we get the formula:

$$I = \frac{V}{R} \text{ Amperes}$$

Where: 'I' is current in amperes

'V' is voltage in volts

'R' is resistance in ohms

Using simple transposition of formulae, we can then get:

$$V = IR$$

&

$$R = \frac{V}{I}$$

The simplest way to remember these formulae is to think of them as parts of a triangle as shown in figure 6.2.

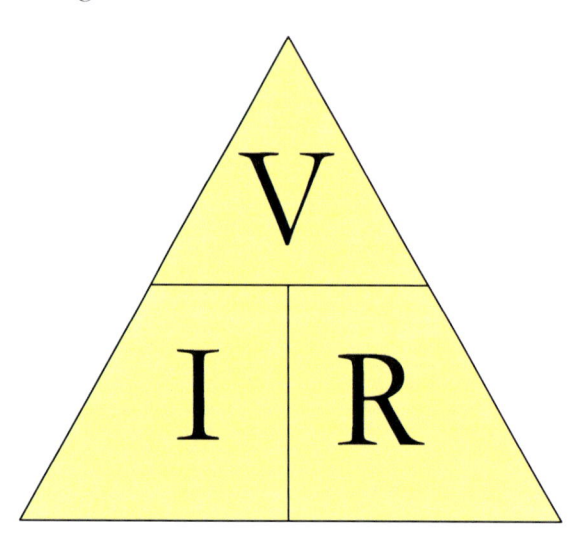

Figure 6.2 – Ohms Law Triangle

From this you can see that in a circuit in which the potential difference, or voltage, is constant, the current may be increased or decreased by adding or removing resistance respectively. Ohm's law may also be expressed in terms of the emf 'E', of the source of electric energy, such as a battery, e.g. $I = E/R$.

Application of Ohm's Law

By using Ohm's law, you will be are able to find the unknown in a circuit if you are given the values of the other two. For example, you will be able to discover the resistance of a circuit, knowing only the circuit's voltage and current. In any equation, if all the variables, sometimes called parameters, are known except one, that unknown can be found.

Without wanting to patronise you with regards to using transposition of formula, I have included here the steps taken when using Ohm's law, if the current (I) and voltage (V) in a circuit are known and the resistance (R) is not:

$$\text{Basic formula is: } I = \frac{V}{R}$$

We can now remove the divisor by multiplying both sides of the equation by R

$$R \times I = \frac{V}{R} \times \frac{R}{1}$$

The result of this step is that $R \times I = V$

To get R on it's own, we divide both sides by I to get:

$$\frac{RI}{I} = \frac{V}{I}$$

and the formula is now transposed for R to:

$$R = \frac{V}{I}$$

Let us take a look at a simple circuit, figure 6.3, where the source voltage V = 20 V and the current I = 2A and the resistance R is unknown.

CHAPTER SIX
DC CIRCUITS

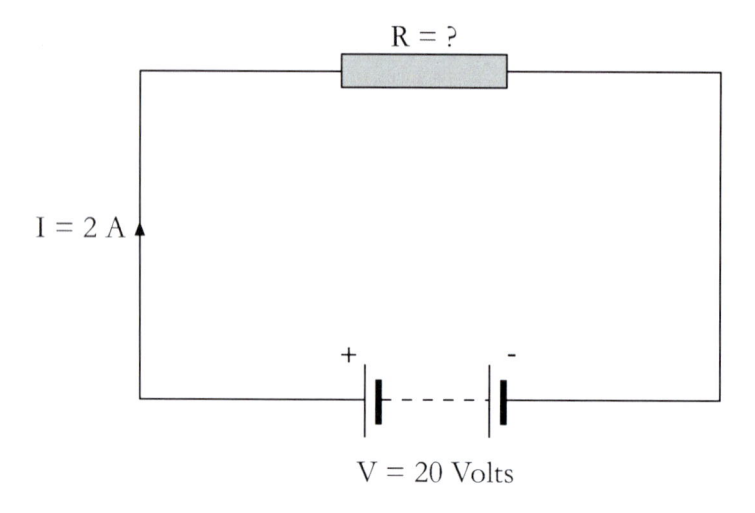

Figure 6.3 – Simple dc Circuit with one unknown

Solution:

$$R = \frac{V}{I} = \frac{20}{2}$$

$$\therefore R = 10\,\Omega$$

This simple example demonstrates the principle of discovering an unknown.

With modifications, Ohm's law also applies to alternating-current (ac) circuits, in which the relation between the voltage and the current is more complicated than for dc and it has also been extended to the constant ratio of the magneto-motive force (mmf) in a magnetic circuit; more on these in later chapters of this Module.

Kirchhoff's Laws

In 1845, the German Physicist **Gustav Robert Kirchhoff (1824 to 1887)** first announced he had extended the work of Georg Ohm to allow calculation of the currents, voltages, and resistances of electrical networks in three (3) dimensions. In further studies, he also demonstrated that current flows through a conductor at the speed of light and with the German chemist **Robert Bunsen (1811 to 1899)**, firmly established the theory of spectrum analysis. This is a technique for chemical analysis by analysing the light emitted by a heated material, which Kirchhoff applied to determine the composition of the Sun.

Kirchhoff's First Law

Kirchhoff's first law when formally stated sounds more complicated than it actually is:

The algebraic sum of the currents flowing through a junction is zero (0). Currents that are seen as approaching the junction are positive (+), while those leaving the junction are negative (-).

It plain English it means that the total current entering a junction must always equal the total current leaving the junction; as after all, no charges can simply disappear or get created, so current cannot disappear or be created either. A junction is defined as any place in a circuit where more than two paths come together, illustrated in figure 6.4.

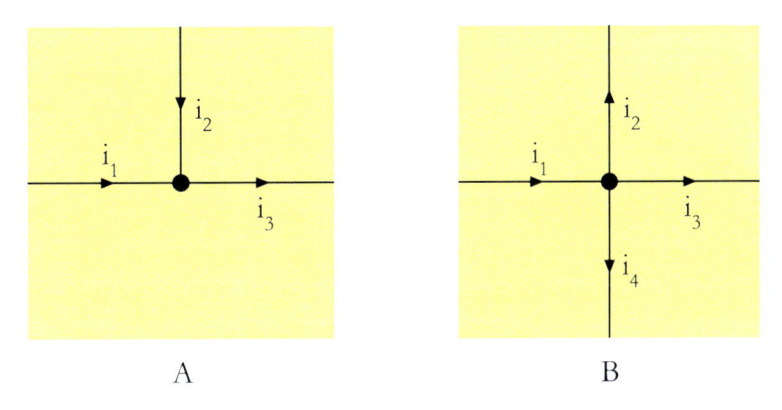

Figure 6.4 – Kirchhoff's First or Current Law

The first rule is also called the *Junction Theorem*, and again states that the sum of the currents into a specific junction in the circuit equals the sum of the currents out of the same junction.

Two simple relationships can be used to determine the value of currents in circuits and these are useful even in complicated situations such as circuits with multiple loops. The first relationship deals with currents at a junction of conductors. Figure 6.4 shows two such junctions, with the currents assumed to flow in the directions indicated on each illustration. For figure 6.4 drawing 'A', the sum is:

For figure 6.4 drawing A, the sum is:

$i_1 + i_2 = i_3$

For figure 6.4 drawing B, the sum is:

$i_1 = i_2 + i_3 + i_4$

Kirchhoff's Second Law

Kirchhoff's second law states:

The algebraic sum of the voltages in a loop is zero.

Again this complicated statement can be simplified as it just means that the sum of the emfs or source voltages in a circuit is equal to the sum of the potential drops in the circuit. In other words, all the energy passed on by the various sources is just equivalent to that lost in useful work and heat dissipation around each loop of the circuit.

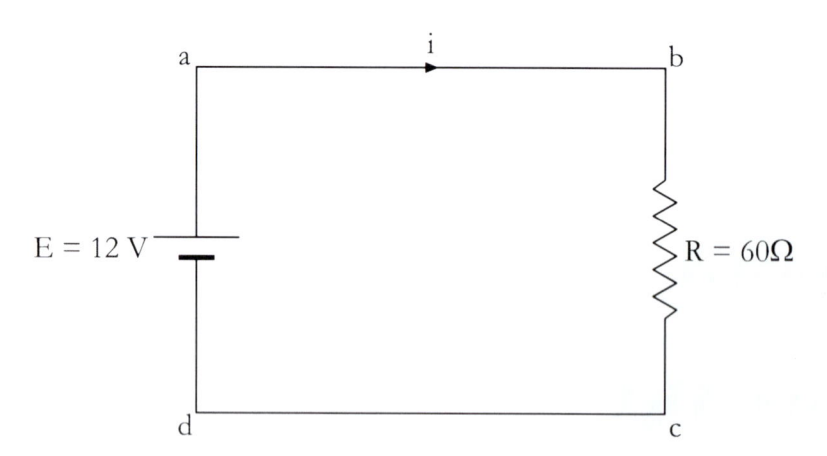

Figure 6.5 – Simple DC Circuit

With the circuit in figure 6.4, E = 12V and R = 60Ω and therefore the current flowing is 0.2A. If we now look at the voltage drop across R, this is simple the current flowing through the resistor, i.e. 0.2A, multiplied by the resistance of 60Ω, i.e. $V_R = IR, = 0.2 \times 60 = 12V$. This proves Kirchhoff's law that $E = V_R$ or $E - V_R = 0$.

When emfs in a circuit are symbolised as circuit components, as in figure 6.5, this law can be simplified further as:

The sum of the potential differences across all the components in a closed loop equals zero.

When applying Kirchhoff's laws to solve for circuit unknowns, the following definitions apply, Figure 6.6:

- An *element* is a resistance or impedance, emf as an ideal voltage source, or ideal current source

- A *node* is a point where three or more current-carrying elements are connected

- A *branch* is one element or several in series connecting two adjacent nodes

- An *interior loop* is a circuit loop not subdivided by a branch

Components are connected in *series* if they carry the same current.

Components are connected in *parallel* if the same voltage is across them.

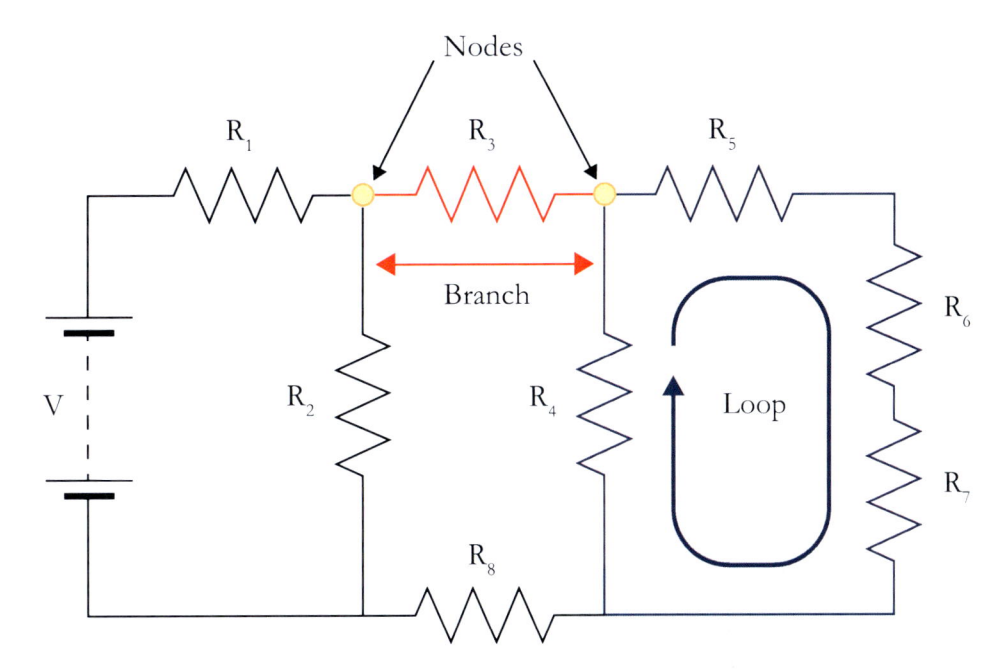

Figure 6.6 – Circuit Definitions

When applying the loop equation, i.e. $E_1 + E_2 + \ldots E_n = 0$, the first step is to choose a starting point on the loop and then walk around the loop, in either direction, and write down the change in potential when you go through an element. When the potential *increases*, the change is *positive*; when the potential *decreases*, the change is *negative*. When you get back to your starting point, add up all the potential changes and set this sum equal to zero, because the net change should be zero when you get back to where you started.

When you pass through a battery from *minus* to *plus*, that's a *positive* change in potential, equal to the emf of the battery. If you go through from *plus* to *minus*, the change in potential is equal to *minus* the emf of the battery.

Current flows from high to low potential through a resistor and if you pass through a resistor in the same direction as the current, the potential, given by IR, will decrease, so it will have a minus sign. If you go through a resistor opposite to the direction of the current, you are going from lower to higher potential, and the IR change in potential has a plus sign.

All this may seem a little confusing until you have gone through some examples.

Let us take a simple light circuit where there are three (3) light bulbs connected in series to a voltage source as in figure 6.7.

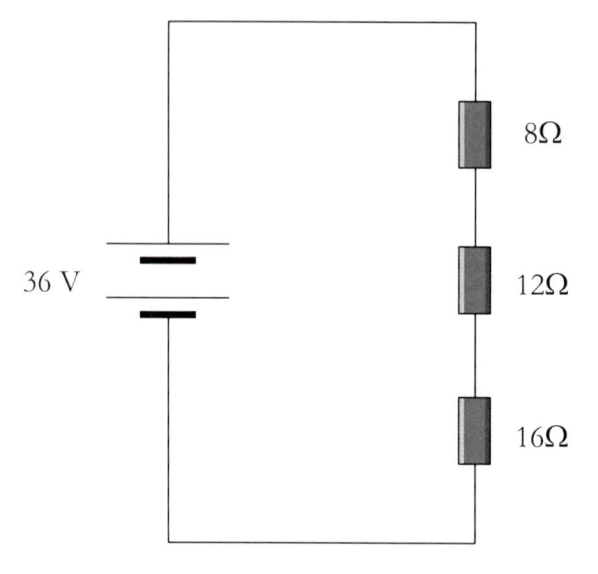

Figure 6.7 – Simple Series Light Circuit

We first need to find the current flowing, but although it is probably easier to do using Ohm's Law, let us do this using Kirchhoff's Voltage Law, which states that the sum of the voltages around a closed circuit equals 0.

$$(8 \times I) + (12 \times I) + (16 \times I) - 36 = 0$$
$$(36 \times I) - 36 = 0$$
$$36I = 36$$
$$\therefore I = 1A$$

Next, to find the voltage drops across each light bulb, i.e. resistance, we use Ohm's Law and so:

$$V_1 = 1 \times 8 = 8V$$
$$V_2 = 1 \times 12 = 12V$$
$$V_3 = 1 \times 16 = 16V$$

Now to check the answer using Kirchhoff's Voltage Law:

$$8 + 12 + 16 - 36 = 0$$

There are two well known methods of implementing Kirchhoff's Laws and these are briefly covered next.

Branch Current Method

The branch current method is a circuit analysis method using Kirchhoff's voltage and current laws to find the current in each branch of a circuit by generating simultaneous equations. Once you know the branch currents, you can determine voltages.

Loop Current Method

This method is also called the *mesh loop* method and the independent current variables are taken to be the circulating current in each of the interior loops. This method gets its name from the idea of these currents meshing together between loops like sets of spinning gears. Using this method requires three basic steps:

1. Label interior loop currents on a diagram
2. Obtain expressions for the voltage changes around each interior loop
3. Solve the system of algebraic equations

Depending on the problem, it may ultimately be necessary to algebraically sum two loop currents in order to obtain the needed interior branch current for the final answer, Figure 6.8.

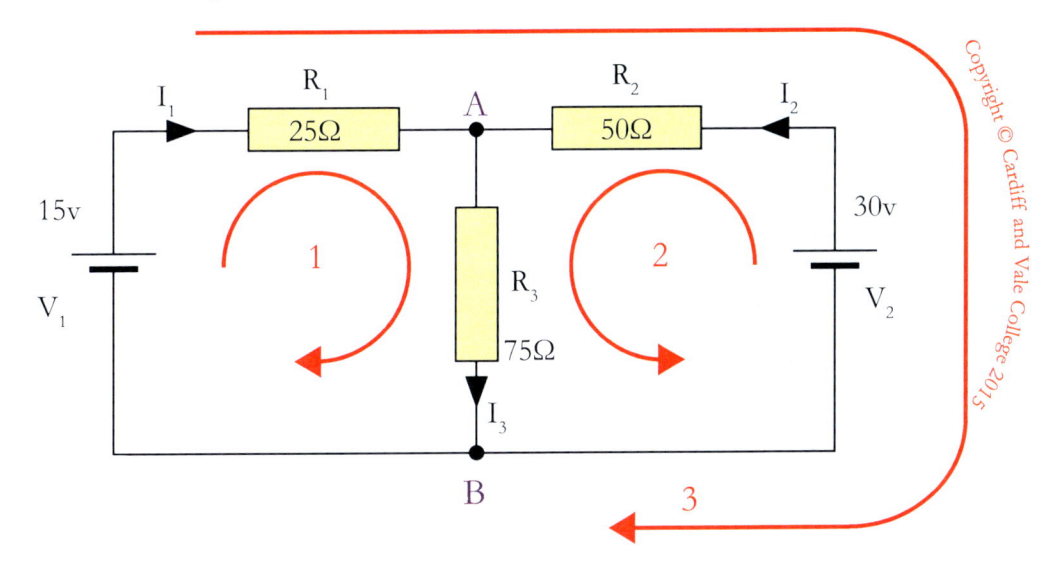

Figure 6.8 – Loop Current Circuit

The mathematics involved can be quite complicated and is beyond the scope of these notes.

You can use this method of circuit analysis to solve more things than just the current. If one or more of the currents are known then an unknown source emf or an unknown resistance could be found instead.

Significance of a Supply's Internal Resistance

A meter connected across the terminals of a good 12-volt battery not connected in a circuit would probably read approximately 12.25 Volts. However, when the same battery is then inserted into a complete circuit, the meter reading across the terminals with the circuit drawing current would decrease to something less than 12 volts. The battery's *Internal Resistance* causes this difference in terminal voltage. Indeed, all sources of emf will have some form of internal resistance that causes a drop in terminal voltage as current flows through the source.

This principle is illustrated in figure 6.9, where the internal resistance of a battery is shown by an additional resistor in series with the battery labelled as '*r*', although sometimes you will see internal resistance labelled as '*Ri*'.

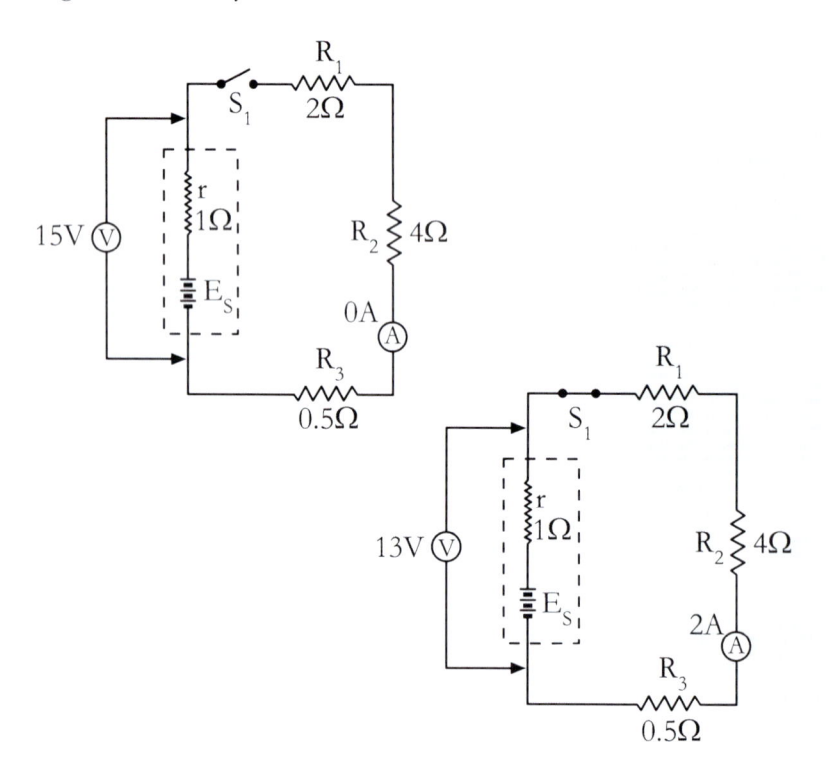

Figure 6.9 – Effect of Internal Resistance

The battery, with its internal resistance, is enclosed within the dotted lines of the schematic diagram, illustrating it is a component it its own right. With switch S_1 open, the voltage across the battery terminals reads 15 volts, but when the switch is closed, current flow causes voltage drops around the circuit in accordance with Ohm's and Kirchhoff's Laws. Taking all the circuit resistance into account, we have a circuit current of 2A and this causes a voltage drop of 2V across *r* and so drops the battery terminal voltage to 13 volts. Internal resistance cannot be measured directly with a meter, as an attempt to do so would damage the meter.

The result of a source's internal resistance is plain to see and so circuit designers must take this factor into account whenever developing new ideas. A simple way of analysing its affect over a range of loads on the output of a dc voltage source may be shown by looking at the circuit illustrated in figure 6.10.

CHAPTER SIX
DC CIRCUITS

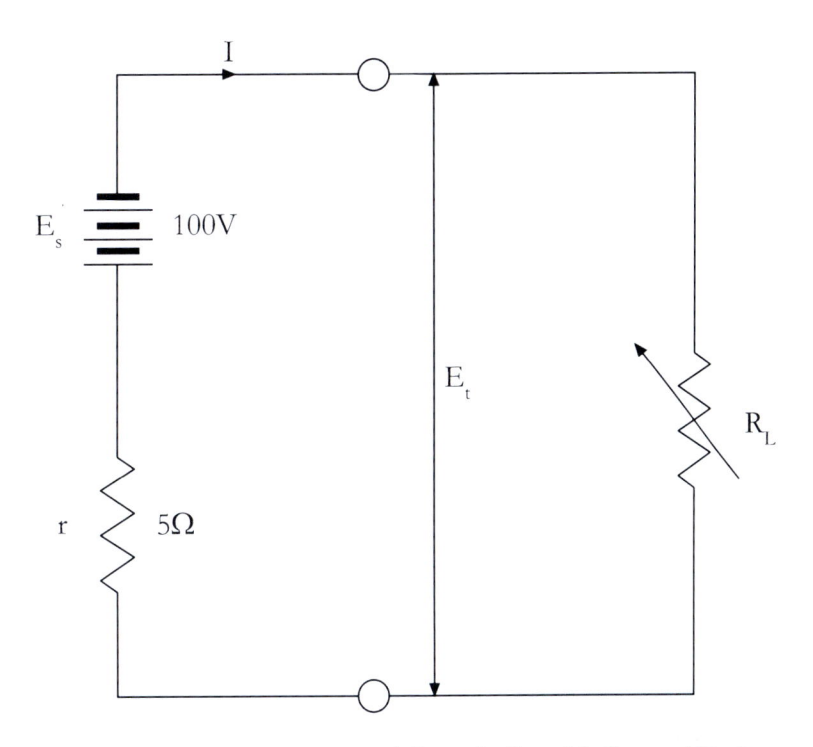

Figure 6.10 – Circuit to Analyse the Affect of a Supply's Internal Resistance

The idea of this circuit is to vary the load resistor R_L from its 0Ω position, i.e. equivalent to a short circuit, to its maximum of 50Ω, and then calculate the current I and terminal voltage E_t, for each value using Ohm's Law. The results of this are shown in table 6.1.

Load Resistance R_L (Ω)	Current I (A)	Terminal Voltage E_t (V)
0	20	0
1	16.7	16.7
2	14.3	28.6
3	12.5	37.5
4	11.1	44.4
5	10	50
10	6.7	66.7
20	4	80
30	2.9	85.7
40	2.2	88.9
50	1.9	90.9

Table 6.1 – Results of E_t versus R_L for an Internal Resistance of 5Ω

CHAPTER SIX
DC CIRCUITS

For instance when $R_L = 0$, the only opposition to current flow in the circuit is from the internal resistance $r = 5\Omega$. Using Ohms Law:

$$I = \frac{V}{R} \text{ Amperes}$$

Therefore:

$$I = 100 \div 5$$
$$I = 20 \text{ Amps}$$

In this scenario all of the supply voltage is dropped across the internal resistance. If we now increase R_L to 5Ω, then we also increase the circuit resistance. In the series circuit shown in Figure 6.10, total circuit resistance can be found by adding the load resistance and internal resistance, ie $5\Omega + 5\Omega = 10\Omega$. We can now substitute this value into Ohms Law:

Where we find:

$$I = 100 \div 10$$
$$I = 10 \text{ Amps}$$

We can also find the voltage drop across each resistor by using Ohms Law. For example when $R_L = 5\Omega$ we have already calculated that current flow in the circuit will be 10 Amps. Using these values in Ohms Law we find:

$$V_L = IR_L$$
$$V_L = 10 \times 5$$
$$V_L = 50 \text{ Volts}$$

50 volts will also be dropped across the internal resistance it is the same value as the load resistance, once again proving Kirchoff's Second Law

From table 6.1 you can see that as the load resistance R_L increases with the internal resistance remaining the same, the current drawn from the source also decreases and consequently, the voltage drop across the internal resistance would decrease. At the same time, the terminal voltage applied across the load increases and approaches a maximum as the current approaches 0A. Therefore, with any voltage source, if the internal resistance is in series with the load it should be as small as possible with respect to the load to have as little influence as possible on the overall circuit function.

Revision

DC Circuits

Questions

1. A basic electrical system that has four (4) elements:

    a) Power, resistance, voltage, current

    b) Source, load, transmission media & control

    c) Control, power, source & resistance

2. If a circuit has an EMF of 36 V and current of 3 A, its resistance is:

    a) 12Ω

    b) 102Ω

    c) 33Ω

3. In the figure below, if i1=5A, i3=7A, i2 is:

    a) 12 A

    b) 2 A

    c) 35 A

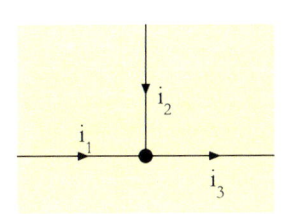

4. In a simple series circuit with 2 voltage sources and 3 resistances, by Kirchhoff's 2nd Law:

    a) E1 + E2 = IR1 + IR2 - IR3

    b) E1 + E2 + IR1 + IR2 + IR3 = 0

    c) E1 - E2 = IR1+IR2 + IR3

5. A series circuit's internal resistance:

    a) Has no affect on output voltage

    b) Reduces the output voltage

    c) Increases the output voltage

Revision

DC Circuits

Answers

1. **B**
2. **A**
3. **B**
4. **B**
5. **B**

Resistance & Resistors

Introduction

Resistance is defined as the opposition to current flow in an electric or electronic circuit. The amount of opposition to current flow produced by a material depends upon the amount of available free electrons it contains and the types of obstacles the electrons encounter as they attempt to move through the material. Resistance is often considered as localised in that we define it as discrete components such as light bulbs, radios, air conditioners, televisions and resistors. However, resistance is a characteristic of every part of a circuit, including the interconnecting wires, source internal resistance, etc. Even the best conductors have some resistance, although it is a relatively small amount and as resistance opposes the flow of electrons, if a current flows through a resistance, work is done. As discussed in a previous chapter, we know that the SI unit of resistance measurement is the ***Ohm (Ω)*** and its SI symbol is '***R***'.

One ohm (1Ω) is defined as a function of Ohm's Law, as discussed in the last chapter, which is put in a slightly different way as:

> *A conductor has a resistance of one ohm (1Ω) if it limits the current flowing in the conductor to one (1) ampere when the potential difference or voltage applied to the conductor is one (1) volt.*

By analysing this relationship we know that if a voltage is applied across a conductor, current flows. The amount of current flow depends upon the resistance of the conductor; i.e. the lower the resistance, the higher the current flow for a given amount of voltage or conversely, the higher the resistance, the lower the current flow.

A ***Resistor*** is a discrete electrical component that opposes the flow of either direct or alternating current, and is used to protect, operate, or control electrical and electronic circuits. They are useful devices that can be used to divide voltages, reduce current and in combination with other components, can be used to make and manipulate electrical waves into shapes that meet a designer's specific requirements. Resistors can have a fixed value of resistance, or they can be made variable or adjustable, within a certain range, when they are then called ***Rheostats***, or ***Potentiometers***. However, before we go on to look closely at resistors; let us take a look at the factors that influence resistance.

CHAPTER SEVEN
RESISTANCE & RESISTORS

What Affects Resistance?

As we have already discussed, electrical resistance may be thought of as an interference with the free flow of electrons. Therefore, the physical structure and type of material has a direct influence on its overall resistance. As such a conductor that has obstructions or barriers within it to the free flow of electrons will have resistance and the longer the wire, the more obstacles there are and so its resistance increases with length.

For example, if we consider a conductor made of a wire that has a resistance of 1Ω for each 20 cm of length and then double its length, we are effectively putting two sections of 20 cm each in series with each other. This being the case, the 40 cm conductor's resistance will have also doubled to 2Ω, i.e. if you double the length of the wire you have twice as many obstacles to overcome. This argument can be carried on so that 60 cm of wire will have a resistance of 3Ω and so on. This relationship is one of linear growth, i.e. if you plotted the conductor's length against its resistance it would be a straight line. Therefore, we can say that a conductor's resistance is directly proportional to its length; i.e.:

$$R \propto l$$

If we look again at our 20 cm conductor and now add another of exactly the same length but now put it in parallel with the first one, it will increase the overall cross-sectional area of conductor available to the passage of electrons. Therefore, in this case, as the cross-sectional area is increased the resistance will decrease, but this characteristic is not linear. For this Module you do not need to understand the mathematical reasons behind this but just have to accept that the resistance is inversely proportional to the cross-sectional area; i.e.:

$$R \propto \frac{1}{a}$$

The third factor that affects resistance is the type of material used in its construction. All materials have their own characteristic properties and one is their ability to conduct electrical current called its **Resistivity**. Resistivity is useful in comparing various materials on the basis of their ability to conduct electric currents, as it is totally independent of the size of the object. Resistivity is measured in **ohm-metres**, more on this later, and high Resistivity designates poor conductors. Resistivity is symbolised by the Greek letter 'rho' (ρ) and resistance is directly proportional to it, i.e.:

$$R \propto \rho$$

The value of resistivity depends also on the temperature of the material; tabulations of resistivity values are usually presented at 20°C. Resistivity of metallic conductors generally increases with a rise in temperature; but resistivity of battery electrolytes, and semiconductors, such as carbon and silicon, generally decreases with temperature rise.

The resistivity of an exceedingly good electrical conductor, such as hard-drawn copper, at 20°C is 1.77×10^{-8} ohm-metre and at the other extreme, electrical insulators have resistivity values in the range 10^{12} to 10^{20} ohm-metres. Figure 7.1 shows a broad spectrum of materials with their resistivity values.

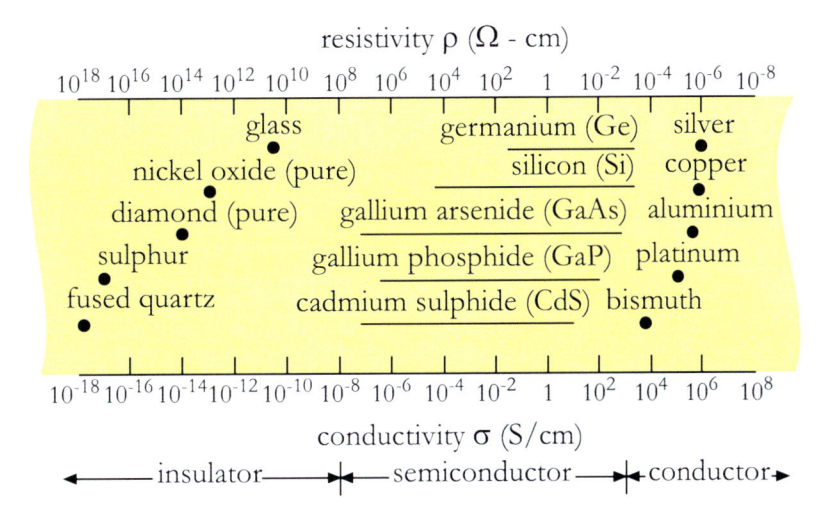

Figure 7.1 – Resistivity of Various Materials

Putting all these factors together, we arrive at an equation for the resistance of any conductor, which is:

$$R = \frac{\rho l}{A} \; \Omega$$

Specific Resistance

The term '*specific*' in any kind of property indicates that the value has been quoted for a unit amount of material, e.g. volume, area or mass, and some examples of specific and non-specific quantities are illustrated in table 7.1.

Non-specific		Specific	
Quantity	Units	Quantity	Units
Mass	kg	Density	kg/m³
Current	A	Current density	A/m²
Electrical resistance	Ω	Electrical resistivity	Ω/m

Table 7.1 – Non-Specific and Specific Quantities

CHAPTER SEVEN
RESISTANCE & RESISTORS

The key point regarding specific quantities is that they carry information about the material involved wherever it is fitted. Consider for example, electrical resistance versus resistivity. Suppose that a designer was selecting a resistance-heating element for an aircraft oven used to heat airline meals, i.e. an element that gets hot by virtue of its electrical resistance, when an electric current is passed through it. Some things the designer would need to consider include:

- How large are the heating elements and how many will be needed, i.e. will these fit into the restrictive space of an aircraft oven?

- How much will each element cost to buy?

- How long will these last, i.e. what is their reliability?

- How difficult will these elements be to change when they fail and how much will the resulting down time cost in terms of loss of use, i.e. what is the cost of ownership?

- How much electric power will be needed to run the heating elements?

This is quite a complex problem, as the answer to any one of these questions will influence the answer to the others; e.g. is it worth spending more on an element initially in order to enhance its reliability?

A full analysis of this problem is beyond the scope of our present discussion; but in order to make the point regarding specific resistance, let us look at one part of the problem, namely how much energy can be obtained from a given heating element in a given time?

We will be discussing power in the next chapter, so for now please accept that Power (P) is simply the amount of energy, measured in Joules, expended per second, i.e. J/s, and 1J/s = 1 Watt. For electrical energy, P is obtained using the formula:

$$P = V \times I \qquad \text{Equation (1)}$$

In words, power is calculated by multiplying the circuit voltage V by the circuit current I.

Now we already know from chapter 6 that the electrical resistance 'R', is given by Ohm's law, i.e.:

$$R = \frac{V}{I} \qquad \text{Equation (2)}$$

In other words, if the element's resistance is high, more source voltage is required for a given current.

Now substituting IR for the V of equation (1), we get equation (3):

$$P = I \times R \times I$$

$$\therefore P = I^2R \quad \text{Equation (3)}$$

Therefore, the amount of power that can be dissipated by a heating element with a certain resistance value is known, as long as the current is also known. When looking at this problem the designer must establish how high a resistance the heating elements will have. This is influenced by both the choice of material and the size of the heating element and it would be very helpful to have some knowledge of how these two variables interact.

The longer length (l) of the heating element the further electrons will have to travel through the resistor and the higher the measured resistance, but in contrast, the higher the cross sectional area of the wire used, the lower the measured resistance. Therefore, a *specific* measure of resistance, its resistivity, is preferred and by rearranging the formula for resistance

$$R = \frac{\rho l}{A} \; \Omega$$

we get:

$$\rho = \frac{RA}{l} \quad \text{Equation (4)}$$

(Remember - Resistance is measured in ohms, Area in metres^2 and Length in metres).

Now from this rearranged formula, we can see that the units of resistivity are the *ohm-metres (Ωm)*, as already mentioned earlier.

Resistivity is a fundamental parameter of the material making up the wire and it describes how easily the wire can convey an electrical current. High values of resistivity imply that the material is very resistant to the flow of electricity while low values imply that it allows electrical current to flow very easily.

Note: Unfortunately, the symbol ρ is used throughout geophysical literature to represent both density and resistivity. Although one would suspect that this could lead to some confusion, it rarely does, because the context in which ρ is used will usually define whether it is representing density or resistivity unambiguously. In these notes, we will follow standard geophysical practice and use ρ to represent both physical properties.

You should have already noticed how much the value of ρ differs from one material to another from figure 7.1, but you should also bear in mind the effect of water in reducing resistivity, especially to human skin. Wet hands and electrical equipment really are an unsafe combination!

Just to finish off the subject of specific versus non-specific properties, consider the following:

1. 1 kW electric heater
2. 1 kW microwave oven
3. 1 kW CO_2 laser

All three of these devices operate at the same power, i.e. 1 kW, but the power is concentrated into a much smaller volume in 'c' than in 'b' than in 'a'. A 1 kW heater will raise the temperature of a meal by a few °C, whereas the microwave can boil water and the laser is able to burn through steel! Therefore, it is important to know the specific resistance of the material used in constructing the element, especially in 'c', as the mathematics in producing a CO_2 laser are critical.

Resistor Rating Colour Code

To make them easy to identify, resistors are colour coded to indicate their resistance values or ratings. The code consists of various colour bands that indicate the resistance values in ohms, as well as its tolerance rating. The Resistor Colour Codes are listed in table 7.2 below.

Colour	1st Band	2nd Band	Multiplier	Tolerance
Black	0	0	1	0
Brown	1	1	10	
Red	2	2	100	2
Orange	3	3	1,000	
Yellow	4	4	10,000	
Green	5	5	100,000	
Blue	6	6	1,000,000	
Violet	7	7	10,000,000	
Grey	8	8	100,000,000	
White	9	9	1,000,000,000	
Tolerance Bands				
Gold			0.1	5%
Silver			0.01	10%
None				20%

Table 7.2 - Resistor Rating Colour Code

Resistors generally have four colour bands that are read as follows:

- First, look up the number values of the first two bands and combine them
- Multiply this two digit number by the value of the 3rd band, i.e. the multiplier band
- The resulting number is the resistor's value in ohms
- The fourth band gives the resistor's tolerance; i.e. if the 4th band is gold, the resistor is guaranteed to be within 5% of its rated value; if silver, it is guaranteed to be within 10%. If there is no 4th band, the resistor is guaranteed to be within 20% of the rated value.

As an example, let us look at a resistor with colour bands Brown, Black, Red and Silver as illustrated in figure 7.2.

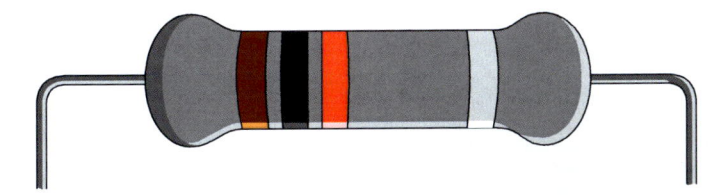

Figure 7.2 – Typical Resistor

With the example in figure 7.2, the 1st band, which is always the one closest to the end of the resistor, is brown and its value from table 7.2 is '1'. The 2nd band is black and has the value '0', and so combining the two numbers becomes 10. The 3rd band, i.e. the multiplier, is red and so has the multiplier value of 100. Multiplying the combined 10 of the first two bands by the multiplier results in the answer of 1000 and so we can say that this resistor has a value of 1000Ω, usually written as 1kΩ.

The 4th, or tolerance, band is silver, which means the resistor is guaranteed to have a resistance value within ±10% of 1kΩ, i.e. it could range from 900Ω to 1100Ω. This may seem an unexpected wide range for a given resistor value, but it reflects the problems associated with manufacturing specific resistances. The greater the tolerance requirement, the more expensive the resistor is to make. Fortunately, for most electrical and electronic circuits this is not a problem and there are other factors designers can use to compensate for this with little or no detriment to the circuit's performance.

Rarely, some resistors also have a 5th band, which is used to indicate its reliability factor, given as the percentage failure per 1000 hours of use; e.g. a 1% failure rate suggests that one in every 100 resistors would not remain within tolerance after 1000 hours of use. A typical example is shown in figure 7.3.

CHAPTER SEVEN
RESISTANCE & RESISTORS

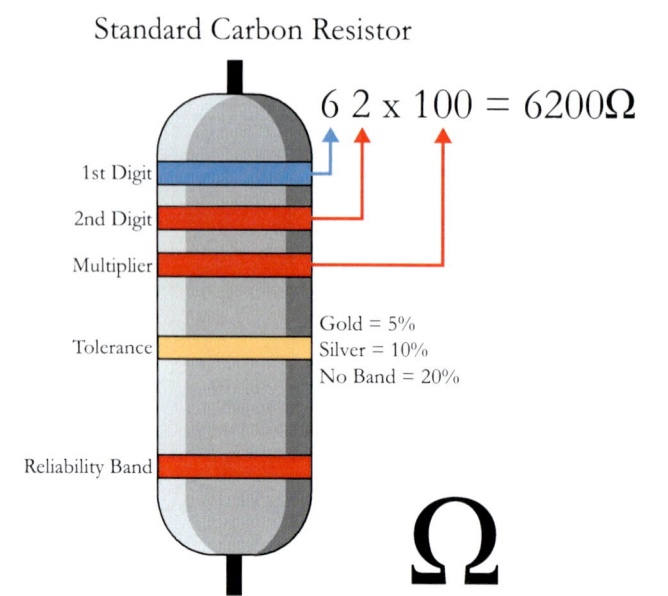

Standard Carbon Resistor

6 2 x 100 = 6200Ω

- 1st Digit
- 2nd Digit
- Multiplier
- Tolerance — Gold = 5%, Silver = 10%, No Band = 20%
- Reliability Band

Ω

Figure 7.3 – Resistor Reliability Band

The reliability band will always be positioned after the tolerance band.

The colours used for this 5th band are shown in table 7.3.

Colour	Percentage
Brown	1
Red	0.1
Orange	0.01
Yellow	0.001

Table 7.3 – 5th Band Colours

Resistors come in many different sizes, as you would expect, but there is not an infinite number. With the amount of resistors used every day in the various electrical and electronic industries, it would be impractical to make resistors of all denominations and so they come in *preferred values*, listed in table 7.4.

Ohms (Ω)					Kilo-Ohms (kΩ)		Megohms (MΩ)	
0.10	1.0	10	100	1000	10	100	1.0	10.0
0.11	1.1	11	110	1100	11	110	1.1	11.0
0.12	1.2	12	120	1200	12	120	1.2	12.0
0.13	1.3	13	130	1300	13	130	1.3	13.0
0.15	1.5	15	150	1500	15	150	1.5	15.0
0.16	1.6	16	160	1600	16	160	1.6	16.0
0.18	1.8	18	180	1800	18	180	1.8	18.0
0.20	2.0	20	200	2000	20	200	2.0	20.0
0.22	2.2	22	220	2200	22	220	2.2	22.0
0.24	2.4	24	240	2400	24	240	2.4	
0.27	2.7	27	270	2700	27	270	2.7	
0.30	3.0	30	300	3000	30	300	3.0	
0.33	3.3	33	330	3300	33	330	3.3	
0.36	3.6	36	360	3600	36	360	3.6	
0.39	3.9	39	390	3900	39	390	3.9	
0.43	4.3	43	430	4300	43	430	4.3	
0.47	4.7	47	470	4700	47	470	4.7	
0.51	5.1	51	510	5100	51	510	5.1	
0.56	5.6	56	560	5600	56	560	5.6	
0.62	6.2	62	620	6200	62	620	6.2	
0.68	6.8	68	680	6800	68	680	6.8	
0.75	7.5	75	750	7500	75	750	7.5	
0.82	8.2	82	820	8200	82	820	8.2	
0.91	9.1	91	910	9100	91	910	9.1	

Table 7.4 – Preferred Resistor Values

Many resistors, especially those made of metal-oxide, can be so small or of an odd shape that makes colour coding difficult to apply. To overcome this, a letter code is used to indicate the resistor's value and this is shown in table 7.5, using the 0.68Ω to 6.8MΩ range as an example.

Resistance	Code Marking
0.68Ω	R68
6.8Ω	6R8
68Ω	68R
680Ω	680R
6.8kΩ	6K8
68kΩ	68K
680kΩ	680K
6.8MΩ	6M8

Table 7.5 - Resistor Marking Code

With these markings a 4th letter indicates the tolerance and these are:

F = ±1%, G = ±2%, H = ±2.5%, J = ±5%, K = ±10%, M = ±20%

CHAPTER SEVEN
RESISTANCE & RESISTORS

In the example shown in Figure 7.4, a resistor marked 4K7J has a value of 4.7 kΩ ± 5%.

4K5 J

Figure 7.4 – Resistor Letter Code

Some new high value resistors use a 5-Band colour code, shown in Table 7.6.

Colour	1st Digit	2nd Digit	3rd Digit	Multiplier	Tolerance
Black	0	0	0	1	
Brown	1	1	1	10	1%
Red	2	2	2	100	2%
Orange	3	3	3	1000	
Yellow	4	4	4	10000	
Green	5	5	5	100000	
Blue	6	6	6	1000000	
Violet	7	7	7		
Grey	8	8	8	0.1 Gold	5% Gold
White	9	9	9	0.01 Silver	10% Silver

Table 7.6 - 5-Band Resistor Colour Codes

A sixth band can also used to show the *temperature coefficient* - a measure of how much the resistance of a resistor is changed by changes in temperature. Figure 7.5

114

CHAPTER SEVEN
RESISTANCE & RESISTORS

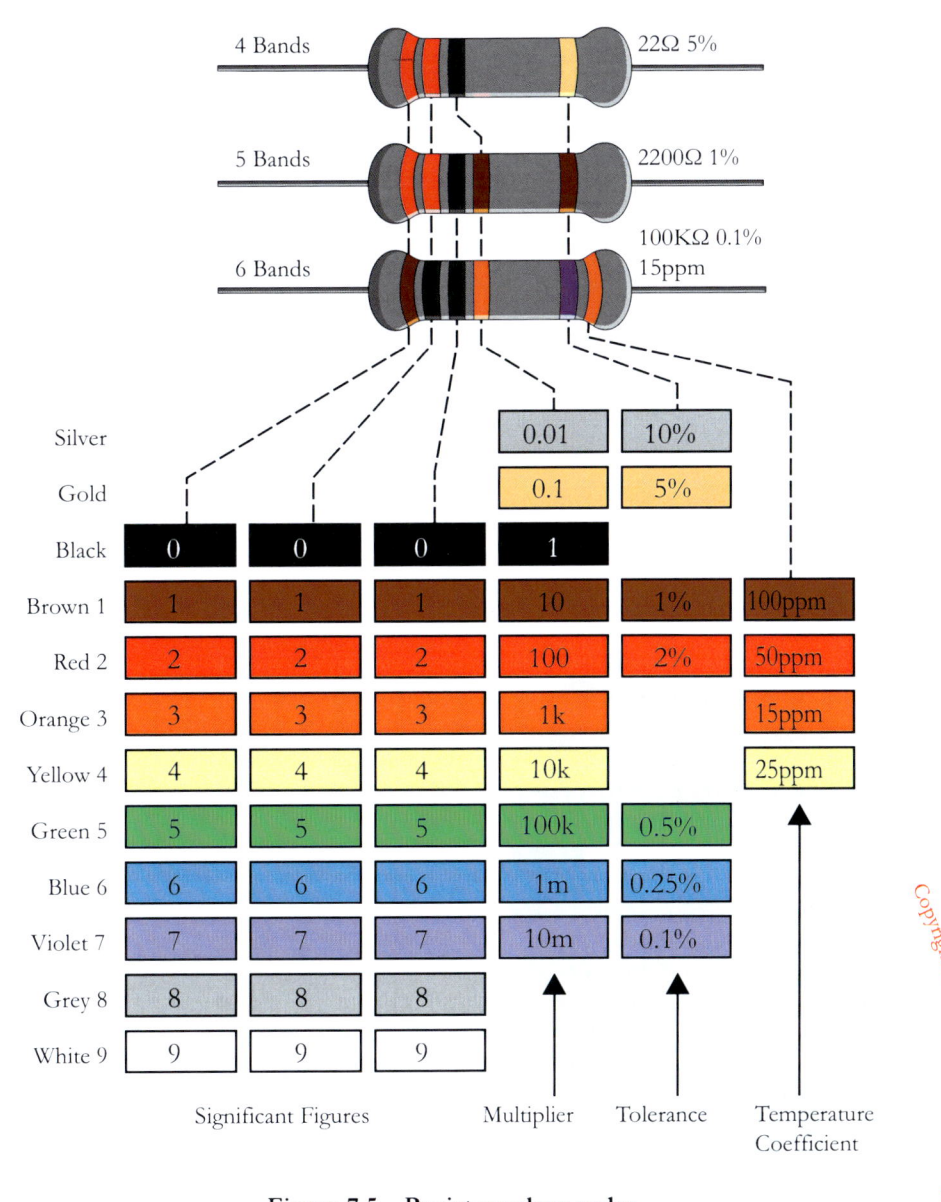

Figure 7.5 – Resistor colour codes

The temperature coefficient is expressed in parts per million resistance change per one degree celsius temperature change (ppm/C or more commonly just ppm). Ideally the value should be zero; however this is never attained in practice. Standard resistors typically have a temperature coefficient of 100-300 ppm, precision resistors 5-20 ppm and ultra high stability resistors <2 ppm. The very best resistors can achieve about a 0.2 ppm temperature coefficient. Resistors without a temperature coefficient band will generally have a high temperature coefficient.

CHAPTER SEVEN
RESISTANCE & RESISTORS

Wattage Ratings

Electrical components are often given a power rating, measured in **Watts**, which indicates the rate at which the device converts electrical energy into another form of energy, such as light, heat, or motion. An example of such a rating is the common household bulb when comparing a 100-watt lamp to a 60-watt lamp. The higher wattage rating of the 100-watt lamp indicates it is capable of converting more electrical energy into light than the lamp of the lower rating. In some electrical devices the wattage rating indicates the maximum power the device is designed to use rather than the normal operating power. A 100-watt lamp, for example, uses 100 watts when operated at the specified voltage printed on the bulb.

A component such as a resistor is not normally given a voltage or a current rating but is given a power rating in watts that indicates it can be operated at any combination of voltage and current as long as the power rating is not exceeded. In most circuits, the actual power used by a resistor is considerably less than its power rating because designers usually build in a 50% safety factor; e.g. if a circuit resistor normally used 2 watts of power, a resistor with a power rating of 3 watts would usually be used.

Resistors of the same resistance level are available in different wattage values. Carbon resistors, for example, are commonly made in wattage ratings of $1/8$, $1/4$, $1/2$, 1 and 2 watts. The larger the physical size of a carbon resistor the higher the wattage rating as a larger surface area of material can radiate a greater amount of heat.

When resistors with wattage ratings greater than five (5) watts are required, wire-wound resistors are used. Simple wire-wound resistors are made in values between 5 to 200 watts but special types can be constructed for power ratings in excess of 200 watts; figure 7.6 shows some typical examples of different wattage resistors.

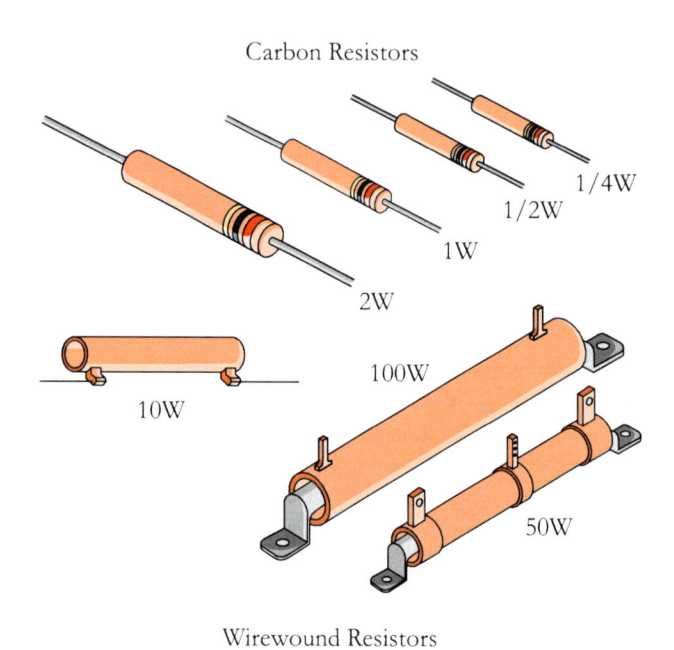

Figure 7.6 – Different Wattage Resistors

As with other electrical quantities, prefixes may be attached to the word watt when expressing very large or very small amounts of power. Some of the more common of these are the kilowatt (1000 watts), the megawatt (1,000,000 watts), and the milliwatt (0.001 watt).

Resistor Construction

There are many types of fixed value resistor in production including the following:

- Solid Carbon
- Wirewound
- Film Type

Solid Carbon Resistors

Carbon composition is the oldest design and usually the cheapest of the resistors. Carbon granules are mixed with a filler material and inserted into a tubular casing. In earlier types vulcanised rubber was used but in modern designs the carbon is mixed with a ceramic filler, Figure 7.7. The same process is used to make pencil lead which also measures like a resistor, whereby different hardness yields different resistance. The value of resistance is determined by the amount of carbon added to the filler mixture.

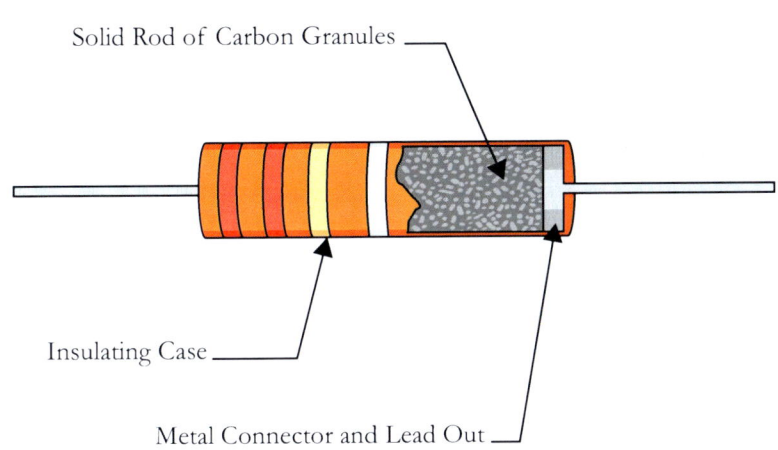

Figure 7.7 – Solid Carbon Resistor

The big advantage of carbon composition resistors is their ability to withstand high energy pulses. When current flows through the resistor, the entire carbon composition body conducts the energy.

The disadvantage is that precision is difficult to attain as it has to do with the exact chemical mix and dimensions. Because of the low stability of the resistance value, this type of resistor is not suitable for any modern high precision application.

However their ability to withstand high energy pulses, while having a relatively small size they are still used today. Applications include the protection of circuits (surge or discharge protection), current limiting, high voltage power supplies, high power or strobe lighting, and welding.

Wirewound Resistors

As the name implies wire wound resistors have a cylindrical core which is wound with metal resistance wire and because of this, they can be manufactured to precise values. Their resistive elements are commonly lengths of wire, usually an alloy such as Nickel/Chromium (Nichrome) or Manganin (Copper/Nickel/Manganese) wrapped around a small ceramic or glass fibre rod and coated in an insulating flameproof cement film, Figure 7.8.

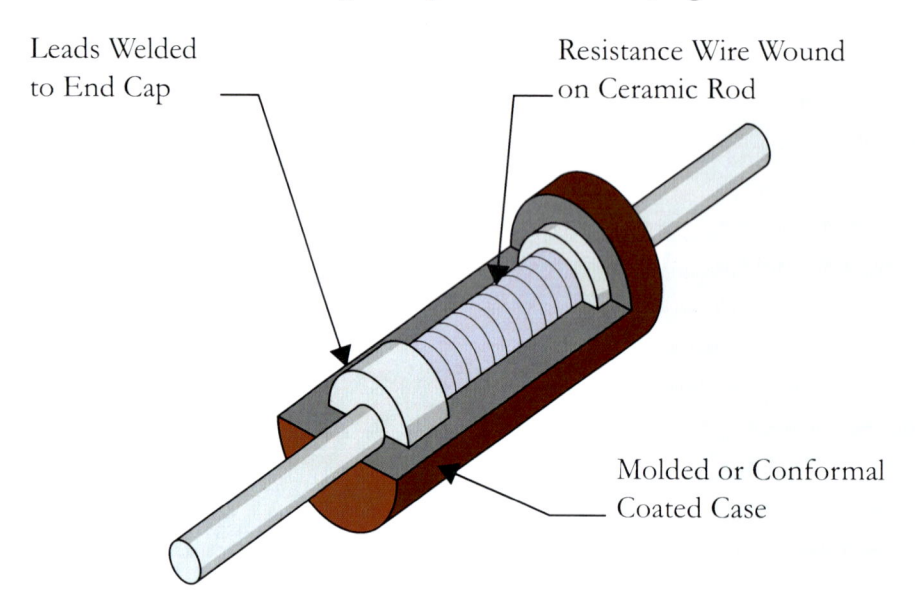

Figure 7.8 – Wirewound Resistor

They have very high power ratings, from 1 or 2 watts to dozens of watts. These resistors can become extremely hot when used for high power applications, and this must be taken into account when designing the circuit.

High power wirewound resistors may be housed in a finned metal case that can be bolted to a metal chassis to dissipate the heat generated as effectively as possible, Figure 7.9 shows a cutaway image of a high power wirewound resistor.

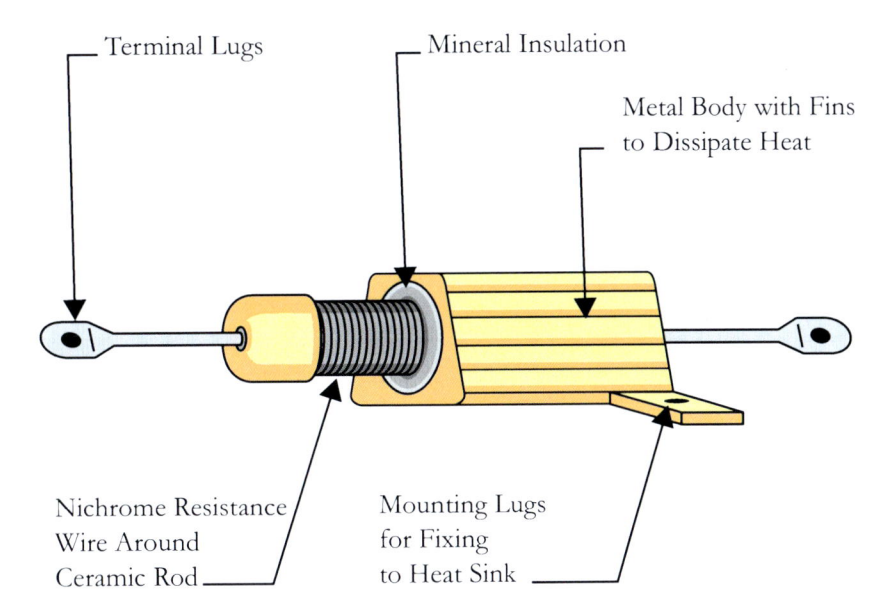

Terminal Lugs — Mineral Insulation — Metal Body with Fins to Dissipate Heat

Nichrome Resistance Wire Around Ceramic Rod — Mounting Lugs for Fixing to Heat Sink

Figure 7.9 – High Power Wirewound Resistor

Because wirewound resistors are coils they have more undesirable inductance than other types of resistor, because of this they **cannot** be used for high-frequency circuits.

The only available resistor types until the 1960s, were wirewound and carbon composition. In the 1960s and 70s there was a shift in use of carbon composition resistors to other types like the Film Type resistors.

Film Type Resistors

"Film Resistor" is a generic term to describe a range of resistor type's including Metal Film and Metal Oxide Film. These are generally made by depositing pure metals, such as nickel, or a metal oxide film, such as tin-oxide, onto an insulating ceramic rod or substrate, Figure 7.10.

Metal Oxide Film — Helical Cut to Reach the Desired Resistance Value

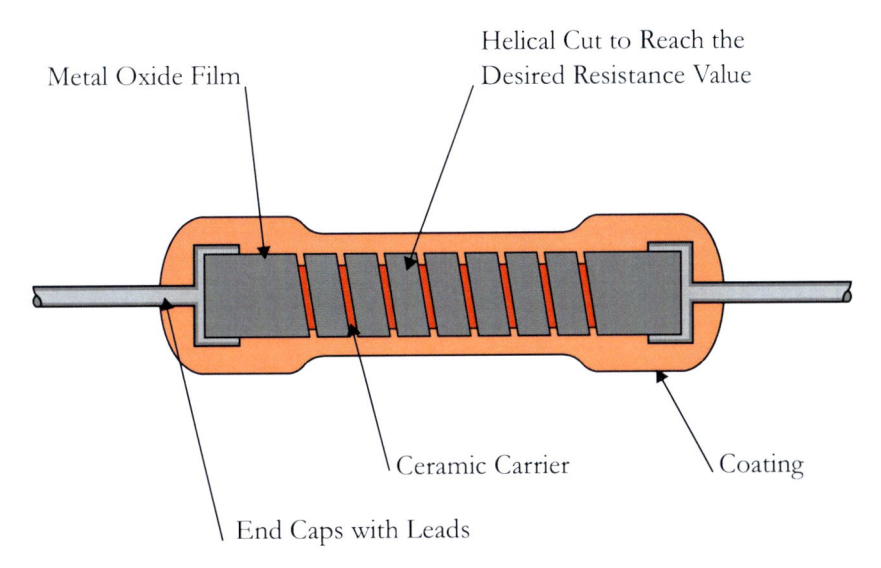

Ceramic Carrier — Coating — End Caps with Leads

Figure 7.10 – Film Type Resistor

The resistive value of the resistor can be controlled by increasing the desired thickness of the deposited film giving them the names of either "thick-film resistors" or "thin-film resistors". Once deposited, a fine spiral groove is cut along the rod using a laser or diamond cutter to cut the carbon or metal coating effectively into a long (spiral) strip. This has the effect of increasing the conductive or resistive path, a bit like taking a long length of straight wire and forming it into a coil.

This precise method of manufacture allows for much closer tolerance resistors (1% or less) when compared to carbon composition types.

Metal Oxide Resistors have better high surge current capability with a much higher temperature rating than the equivalent metal film resistors. Film resistors can also be used on Higher frequency circuits, when compared with wirewound resistors.

Resistors in Series

A series circuit is one in which the current has only one path and all the current passes through each circuit component. The circuit shown in figure 7.11 below has three resistors in series and the current from the battery flows through each of the components.

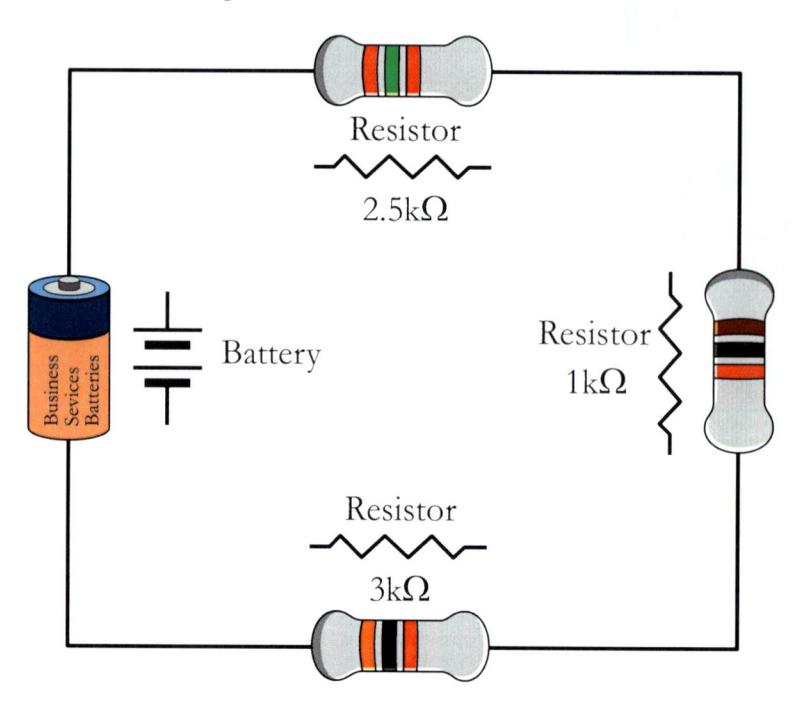

Figure 7.11 – Resistors in Series

When resistors are connected in series the total circuit resistance (Rt) is equal to the sum of the individual resistances, i.e.:

$$R_t = R_1 + R_2 + R_3 \ldots R_n$$

In the example in figure 7.11, the total resistance, Rt, of the circuit is:

$$R_t = R_1 + R_2 + R_3$$
$$= 2.5k\Omega + 1k\Omega + 3k\Omega$$
$$= 6.5k\Omega$$

In some circuit applications, the total resistance is known and the value of one of the circuit resistors has to be determined and the equation $R_t = R_1 + R_2 + R_3 \ldots R_n$ can then be transposed to solve for the value of the unknown resistance.

In a circuit, once two (2) of three elements, i.e. voltage, current and resistance, are known, the third can be calculated using Ohm's law. In addition, by Ohm's Law, the value of the voltage dropped by a resistor is determined by the applied voltage and is in proportion to the circuit resistances. Therefore, in a series circuit with several resistors, the voltage drops across each one are in direct proportion to their individual resistances, which is the result of having the same current flow through each resistor; i.e. the larger resistor value, the larger the voltage drop across it.

Each of the resistors in a series circuit consumes power that is dissipated in the form of heat. Since this power must come from the source, the total power must be equal to the power consumed by the circuit resistances. In a series circuit the total power is equal to the sum of the power dissipated by the individual resistors; i.e. total power (P_t) is equal to:

$$P_t = P_1 + P_2 + P_3 \ldots P_n$$

Rules for Series DC Circuits

To summarise, the following rules apply to all series circuits:

- The same current flows through each part of a series circuit

- The total resistance of a series circuit is equal to the sum of the individual resistances

- The total voltage across a series circuit is equal to the sum of the individual voltage drops

- The voltage drop across a resistor in a series circuit is proportional its value

- The total power in a series circuit is equal to the sum of the individual powers used by each circuit component

CHAPTER SEVEN
RESISTANCE & RESISTORS

Resistors in Parallel

Ohm's and Kirchhoff's Laws apply to all electrical circuits, but the characteristics of a parallel dc circuit are different than those of a series dc circuit. A parallel circuit is defined as one having more than one current path connected to a common voltage source. A parallel circuit therefore, is one in which components are arranged so that the current path divides, although not necessarily equally, figure 7.12 shows an example of a circuit with three resistors in parallel.

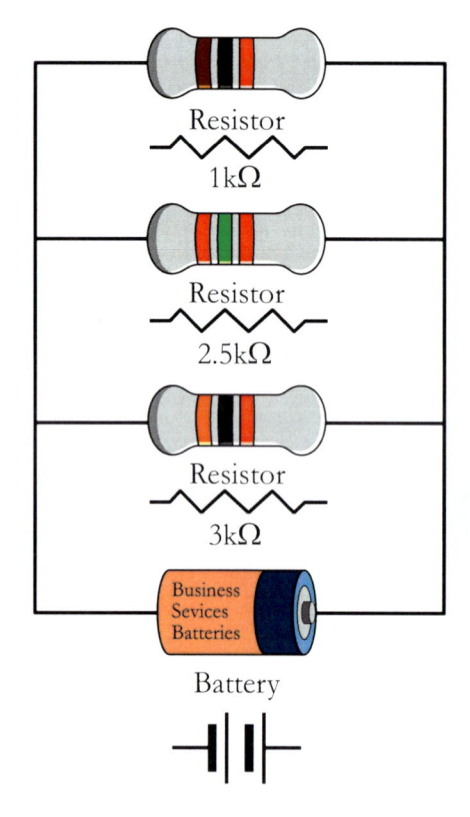

Figure 7.12 - Resistors in Parallel

You have seen that the source voltage in a series circuit divides proportionately across each resistor in the circuit. However, in a parallel circuit, the same voltage is present in each branch, i.e. a section of a circuit that has a complete path for current. In figure 7.12 this voltage is equal to the applied voltage from the battery. This is because there are only two sets of electrically common points in a parallel circuit, and voltage measured between sets of common points must always be the same at any given time. Therefore in Figure 7.12, the voltage across the 3kΩ resistor is equal to the voltage across the 2.5kΩ resistor which is equal to the voltage across the 1kΩ resistor which, in turn, is equal to the voltage across the battery.

As far as the current flow is concerned, Ohm's law states that the current in a circuit is inversely proportional to the circuit resistance and this is true in both series and parallel circuits. There is a single path for current in a series circuit and its amount is determined by the total resistance of the circuit and the applied voltage, but in a parallel circuit, the source current divides among the available paths. However, we can still use Ohm's Law to find the total resistance of a parallel circuit as the equation R = V/I still holds true.

Using this formula with any parallel circuit will result in a figure for R that is smaller than any of the individual resistances. Now since the total resistance of a parallel circuit is smaller than any of the individual resistors, it is not the sum of the individual resistor values as was the case in a series circuit. But why is this so?

The answer is that placing resistors in parallel always *decreases* the total resistance of the circuit because it is equivalent to placing them side-by-side; i.e. increasing the total cross sectional area available for current flow and so reducing the overall resistance. The total resistance of resistors in parallel is also referred to as the circuit's *Equivalent Resistance (R_{eq})* and the terms total resistance and equivalent resistance are interchangeable.

There are several methods used to determine the equivalent resistance of parallel circuits. The best method for a given circuit depends on the number and value of the resistors. For a circuit where *all* resistors have the *same* value, it is simply:

$$R_{eq} = \frac{R}{n}$$

Where R is the value of one (1) resistance and '*n*' is the total number of resistors. However, this equation is valid for any number of parallel resistors of *equal value*.

For example; if three (3) 27Ω resistors are connected in parallel, then the total resistance (R_t) equals:

$$R_t = 27Ω \div 3$$
$$R_t = 9Ω$$

Obviously this would not work if the resistors were of different values, so we need another method to calculate the total resistance. When resistors are connected in parallel, to calculate the total resistance of the circuit, the reciprocal of each resistor value is added as in the following formula:

$$\frac{1}{R_T} = \frac{1}{R_1} + \frac{1}{R_2} + \frac{1}{R_3} \ldots \frac{1}{R_n}$$

CHAPTER SEVEN
RESISTANCE & RESISTORS

In figure 7.12, as the three resistors are connected in parallel, the circuit current (I_T) divides between R_1 and R_2 and R_3, but not necessarily in equal parts, and the overall resistance is:

$$\frac{1}{R_T} = \frac{1}{1} + \frac{1}{2.5} + \frac{1}{3} = \frac{15 + 6 + 5}{15} = \frac{26}{15}$$

Remember this is $\dfrac{1}{R_T}$

$$\therefore R_T = \frac{15}{26} = 0.58 \text{k}\Omega$$

You should notice in the answer above that the final equivalent resistance R_t is less than the lowest value of individual resistor. This will **always** be the case with resistors in parallel and if your calculations show otherwise, it should give you a big hint that your calculations have gone wrong somewhere.

For finding the equivalent or total resistance of two parallel resistors, and **just** two resistors, there is another formula that you can use which is:

$$R_T = \frac{R_1 R_2}{R_1 + R_2} \; \Omega$$

This equation, called the product over the sum formula, is used so frequently it should be committed to memory for any future examination.

Power computations in a parallel circuit are essentially the same as those used for the series circuit. As resistor power dissipation consists of a heat loss, they affects are additive, regardless of how the resistors are connected in the circuit. The total power is equal to the sum of the power dissipated by the individual resistors. Like the series circuit, the total power consumed by the parallel circuit is:

$$P_t = P_1 + P_2 + P_3 + \ldots P_n$$

Rules for Parallel DC Circuits

To summarise, the following rules apply to all parallel circuits:

- The same voltage exists across each branch of a parallel circuit and is equal to the source voltage

- The current through a branch of a parallel network is inversely proportional to the amount of resistance of the branch

- The total current of a parallel circuit is equal to the sum of the individual branch currents of the circuit

- The total resistance of a parallel circuit is always less than the smallest resistance value in the circuit

- The total power consumed in a parallel circuit is equal to the sum of the power consumptions of the individual resistances

Series-Parallel Resistor Networks

In the discussions up to now, series and parallel dc circuits have been considered separately. However, you will usually encounter circuits consisting of both series and parallel elements. A circuit of this type is called a *Series-Parallel Network* or a *Combination Circuit* in the USA. Solving for the quantities and elements in a series-parallel circuit is simply a matter of applying the laws and rules you have learnt up to this point. The total resistance of series-parallel resistor networks is best calculated by completing the following steps:

1. Identify all branches of the network where resistors are in series and replace each with *one* single resistance equal to the total value of that series branch

2. Identify all branches of the network where resistors are in parallel and replace each with *one* single resistance equal to the total value of that parallel branch

3. Repeat steps 1 and 2 until the network is simplified down to one equivalent resistance

Example

Find the total resistance of the network shown in figure 7.13.

CHAPTER SEVEN
RESISTANCE & RESISTORS

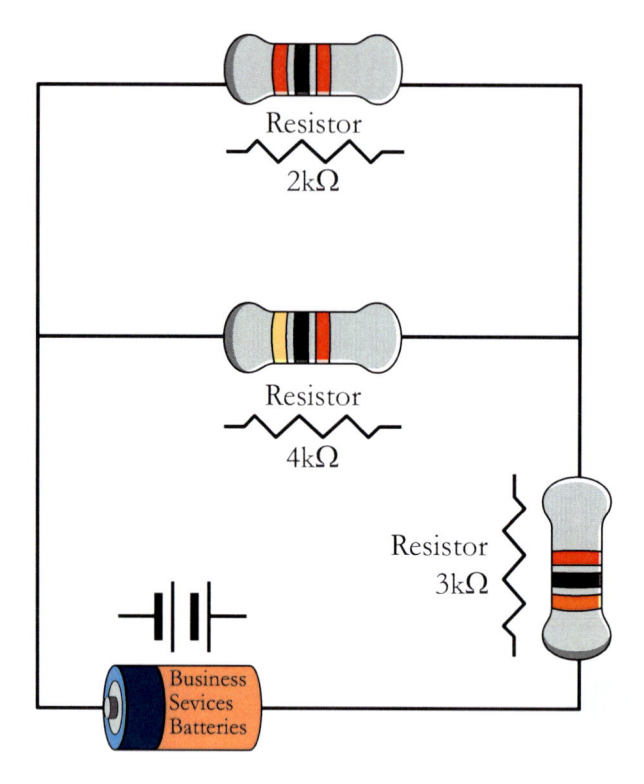

Figure 7.13 – Simple Series-Parallel Network

Using the steps mentioned above, the total or equivalent resistance of the above parallel circuit could be calculated as follows:

First calculate the equivalent resistance of the two resistors in parallel:

$$\frac{1}{R_T} = \frac{1}{2} + \frac{1}{4} = \frac{3}{4}$$

$$\therefore R_T = 1.33 k\Omega$$

At this point the circuit has been simplified to an equivalent series circuit consisting of an equivalent resistance of 1.33 kΩ and a resistance of 3 kΩ. Therefore, the total resistance of the compound circuit can be calculated as follows:

$$R_T = 1.33 k\Omega + 3 k\Omega$$

$$\therefore R_T = 4.33 k\Omega$$

Now let's have a go at a more complicated circuit. Calculate the total resistance of the circuit shown in Figure 7.14.

CHAPTER SEVEN
RESISTANCE & RESISTORS

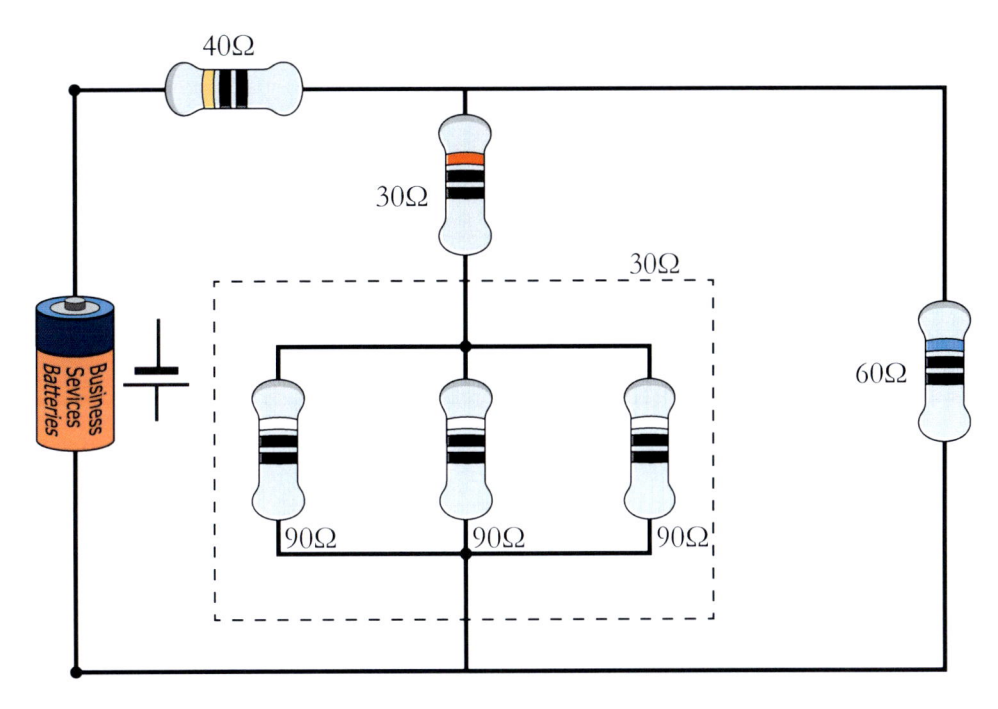

Figure new 7.14 – Series Parallel Circuit

Firstly we can replace the simple parallel group of three (3) 90Ω resistors with a single equivalent resistor. There are several ways to do this, but as we learnt earlier the simplest way, when the resistors are of the same value, is to divide the value of a single resistor by the number of resistors in the group.

In this case:

$$90Ω \div 3 = 30\ Ω$$

This now gives us an equivalent circuit as shown in Figure new 7.15:

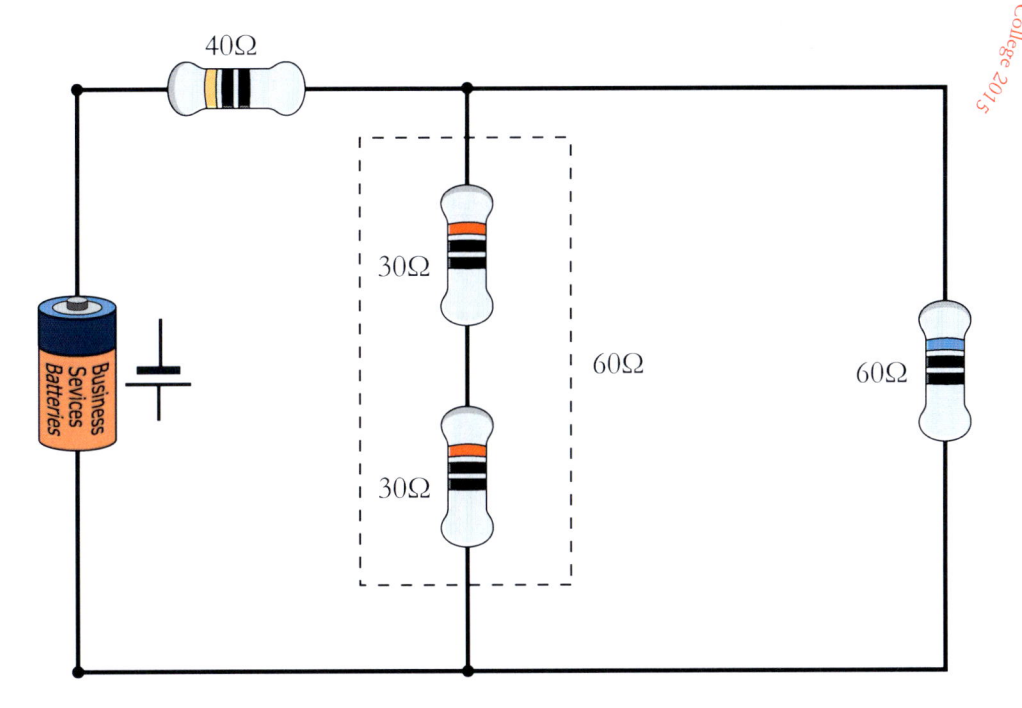

Figure 7.15 – Series Parallel Circuit

CHAPTER SEVEN
RESISTANCE & RESISTORS

We can now combine the two (2) 30Ω resistors, which are in series with each other, to produce a single equivalent resistor. In this case:

$$30\ \Omega + 30\ \Omega = 60\ \Omega$$

This now gives us an equivalent circuit as shown in Figure new 7.16:

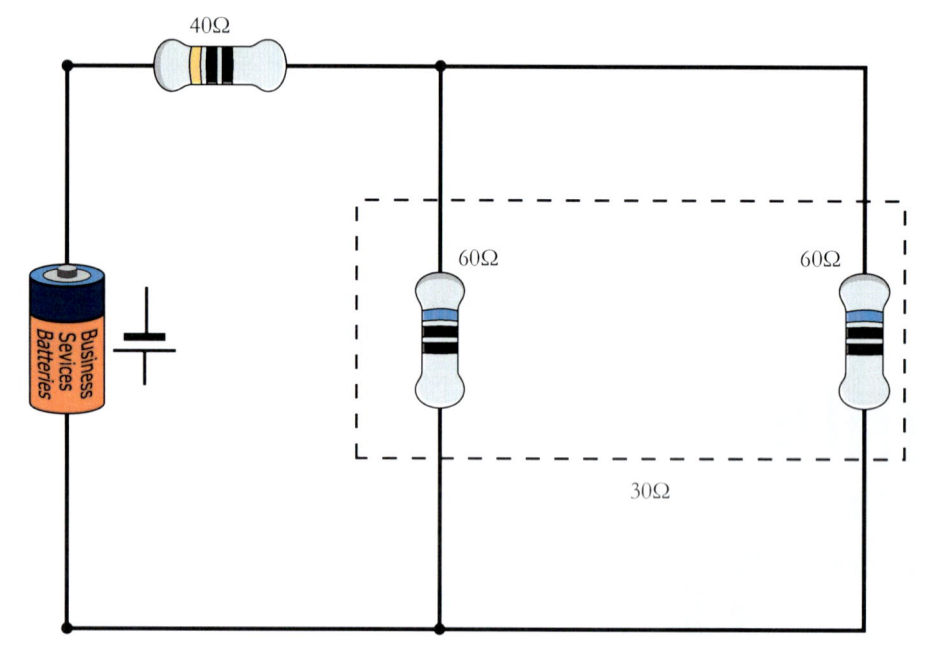

Figure new 7.16 – Series Parallel Circuit

Now we can replace the simple parallel group of two (2) 60Ω resistors with a single equivalent resistor:

$$60\Omega \div 2 = 30\ \Omega$$

This now gives us an equivalent circuit as shown in Figure new 7.17:

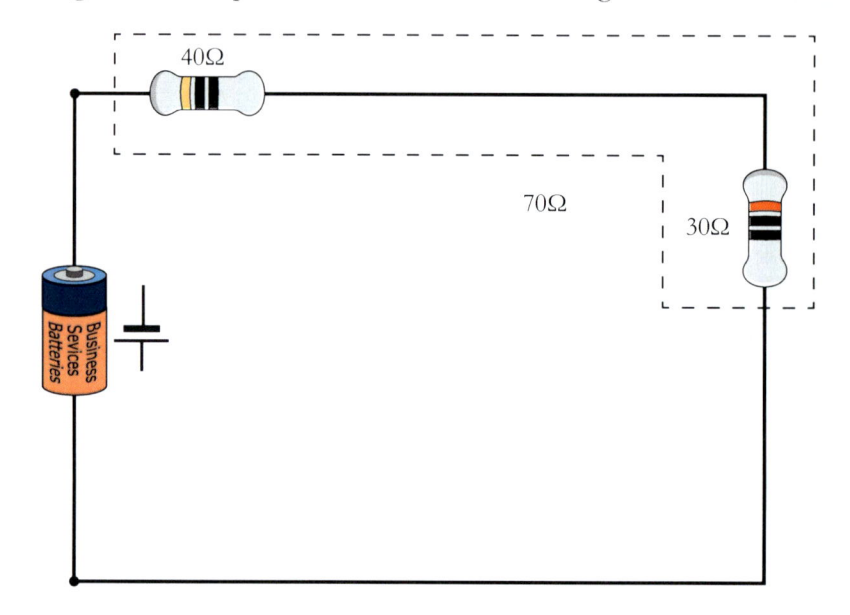

Figure 7.17 – Series Parallel Circuit

The two remaining resistors are connected in series, so it is a simple task to add the two together to find the total circuit resistance:

$$40 \, \Omega + 30 \, \Omega = 70 \, \Omega$$

We have now calculated the total circuit resistance in a Series/Parallel circuit. At this point the total circuit current (I_s) may be found if V_s is given, or V_s found if I_s is given. Having determined V_s or I_s, as appropriate, the current in any branch and the voltage drop across any resistor can be found by working backwards through the sequence, applying Ohm's law at each stage.

For example; if the supply Voltage is 140 Volts we can now calculate the total current flow in the circuit:

$$I = V \div R$$
$$\text{Therefore:} \quad I = 140 \div 70$$
$$= 2 \text{ Amps}$$

From this we can now calculate the voltage drop across the fixed 40Ω resistor and equivalent 30Ω resistor in Figure 7.17.

$$\text{Voltage drop across the 40}\Omega \text{ resistor:} \quad V = IR$$
$$\text{Therefore:} \quad V = 2 \times 40$$
$$V = 80 \text{ Volts}$$

This leaves us with 60 volts being dropped across the equivalent resistor. Hopefully you can see that in this scenario, the 40Ω fixed resistor will always have 80 volts dropped across it, as the total circuit current always has to flow through it.

We should note that the 30Ω resistor shown in Figure 7.17 is an equivalent resistor, determined by our earlier calculations. However as stated earlier the current in any branch and the voltage drop across any resistor can be found by working backwards through the sequence, applying Ohm's law at each stage.

With this new found knowledge, just think of the hours of endless fun that can be obtained from drawing your own circuits and calculating values of Resistance, Current and Voltage. Who would have thought that electrics could be so entertaining and interesting!

Open and Short Circuits

As with everything in life, electrical components and circuits can fail at times. The most common failures with electrical circuits are an *Open Circuit* or a *Short Circuit*.

Open Circuit

A circuit is said to be *Open* when a break exists in a complete conducting pathway. Although an open circuit occurs when a switch is used to de-energise a circuit, it may also develop accidentally. To restore a circuit to proper operation, the open circuit must be located, its cause determined, and repairs made.

Sometimes an open circuit can be located visually by a close inspection of the circuit components and any defective items, such as burned out resistors, can usually be revealed by this method. Others, such as a break in a wire covered by insulation or the melted element of an enclosed fuse, are not visible to the eye. Under such conditions, the understanding of the affect an open has on circuit conditions will enable you to make use of test equipment to locate the open component. In figure 7.18 is a series circuit that consists of two resistors and a fuse.

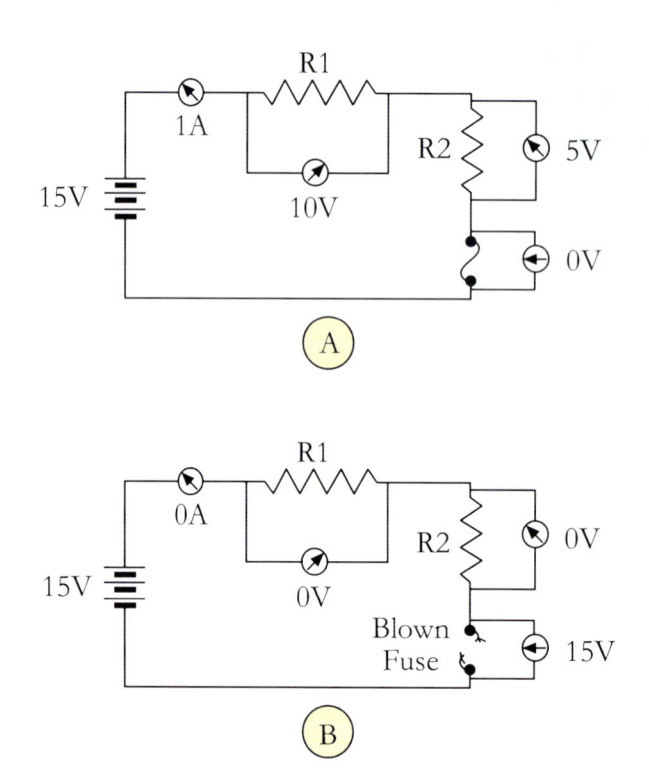

Figure 7.18 – Example of an Open Circuit and its Effect

In drawing 'A' of figure 7.18, current is flowing normally and if you follow the calculations through using Ohm's Law you will see that the voltage drop across each resistor is correct if $R_1 = 10\Omega$ and $R_2 = 5\Omega$. Notice what happens, however, in drawing 'B' when the fuse blows. In this case, current ceases to flow and so by Ohm's Law there is no longer a voltage drop across the resistors. Each end of the open conducting path now just becomes an extension of the battery terminals and the voltage felt across the open is equal to the applied voltage, i.e. 15V.

An open circuit has *infinite* resistance, which as you probably already know, represents a quantity so large it cannot be measured. The SI symbol for infinity is ∞ and if a circuit has an open circuit its total resistance $R_t = \infty$.

Short Circuit

A short circuit is an accidental path of zero (0) or low resistance that passes an abnormally high amount of current. A short circuit exists whenever the resistance of a circuit or part of a circuit drops in value to almost zero ohms (0Ω). A short circuit will often occur as a result of improper wiring or broken insulation. In figure 7.19 is a series circuit that consists of two resistors and a supply.

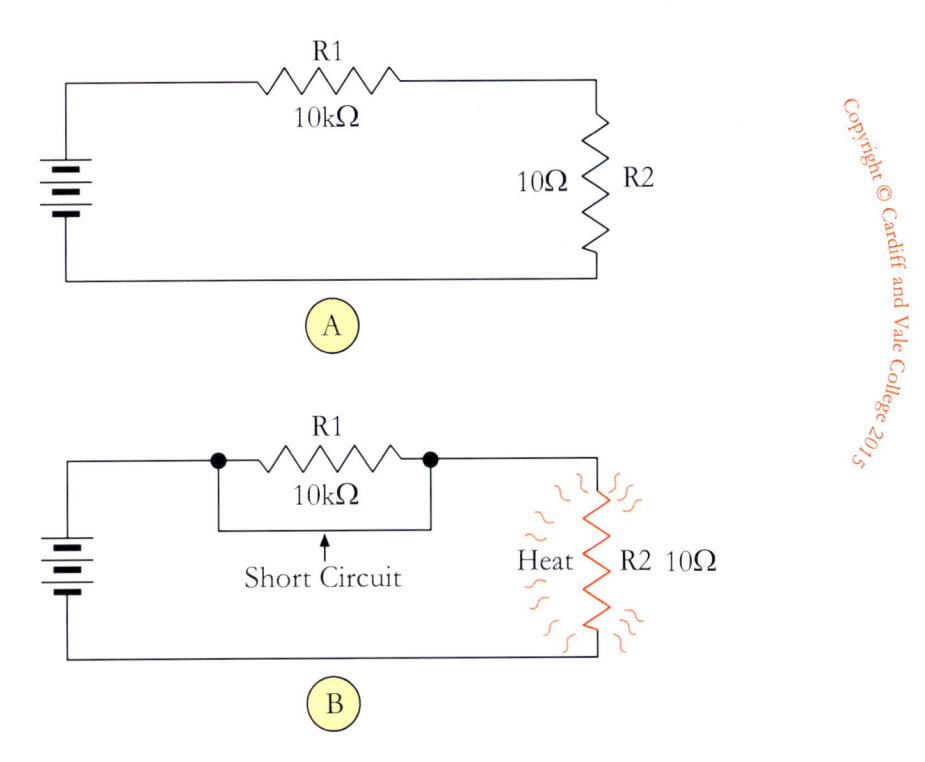

Figure 7.19 - Example of a Short Circuit and its Effect

In drawing 'A' of figure 7.19, the circuit is functioning normally but in drawing 'B' something has happened with the improper wiring causing a short circuit across R1, usually stated as just *a short*. With resistor R_1 now short-circuited in drawing 'B', it has been effectively replaced with a piece of wire. Now practically all the current flows through the short circuit and very little current flows through the resistor, i.e. electrons bypass R1 and flow through the remainder of the circuit by passing through resistor R2 and the battery.

The circuit's current flow now increases dramatically because its resistive path has decreased from 10010Ω to 10Ω. Due to this excessive current flow, the 10Ω resistor becomes heated and as it attempts to dissipate this heat, the resistor will probably be destroyed. Figure 7.20 shows a pictorial wiring diagram rather than a schematic one, to indicate how broken insulation might cause a short circuit.

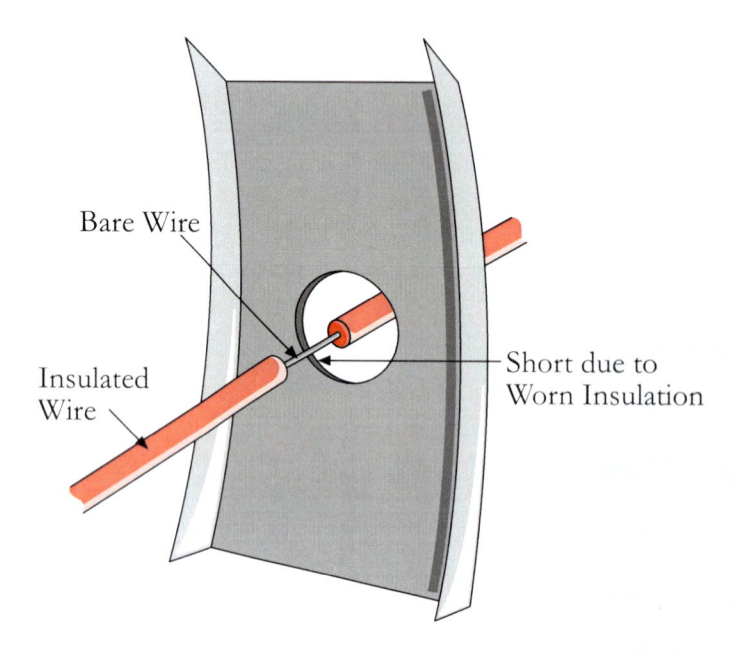

Figure 7.20 - Short Due to Broken Insulation

Operation & Construction of Potentiometers and Rheostats

A **Rheostat** is an adjustable or variable resistor. It is used in applications for introducing varying and known values of resistance into a circuit for controlling the amount of current flow. Its resistance element can be a metal wire or ribbon, carbon, or a conducting liquid, depending on its construction and proposed application. For average currents, the metallic type is most common; for very small currents, the carbon type is used; and for large currents the electrolytic type is used where electrodes are placed in a conducting fluid. This last type, for obvious reasons, is not used in the aviation industry.

A rheostat's resistance to electric current depends on the position of some mechanical element or control in the device. Typically, a rheostat consists of a resistance element equipped with two contacts, or terminals, by which it is attached to a circuit, i.e. a fixed contact at one end and a sliding contact that can be moved along the resistance element, figure 7.21.

CHAPTER SEVEN
RESISTANCE & RESISTORS

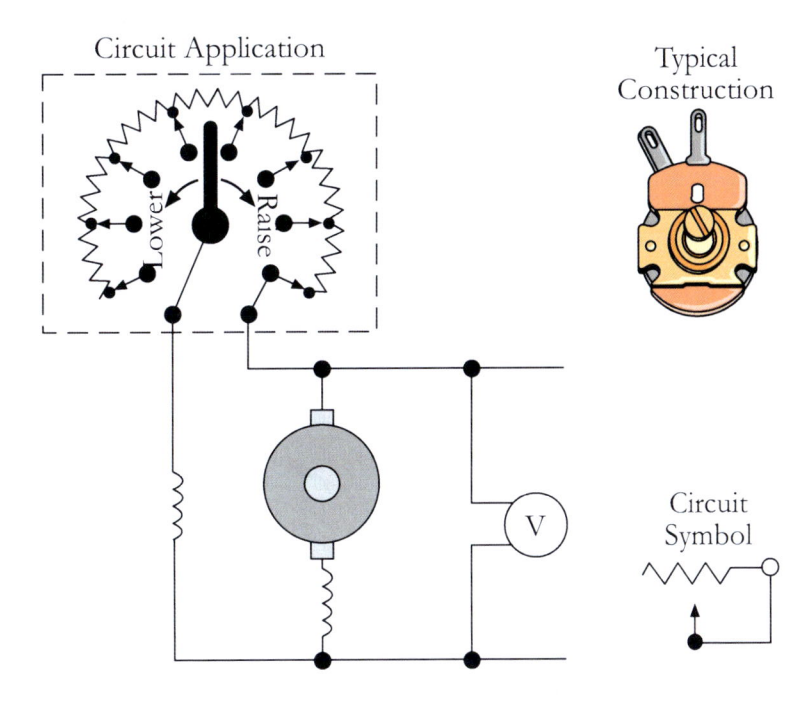

Figure 7.21 – Typical Rheostat

As illustrated in figure 7.21, electric current enters and leaves the resistance element through the dual contacts. By moving the sliding contact towards or away from the fixed contact, the length of the resistance element through which the current travels can be decreased or increased. In this way the current through the circuit can also be increased or decreased. Rheostats are widely used for dimming lights and controlling the speed of electric generators and motors, etc.

A *Potentiometer* is a special type of rheostat that is used for the accurate measurement or control of electrical potential. A common potentiometer is simply a resistance element that is attached to a circuit by three contacts, or terminals. The ends of the resistance element are attached to two input voltage conductors of the circuit, and the third contact, attached to its output, is usually a movable terminal that slides across the resistance element, effectively dividing it into two resistors. Since the position of the movable terminal determines what percentage of the input voltage will actually be applied to the circuit, the potentiometer can be used to vary its magnitude; for this reason it is sometimes called a *Voltage Divider*, figure 7.22.

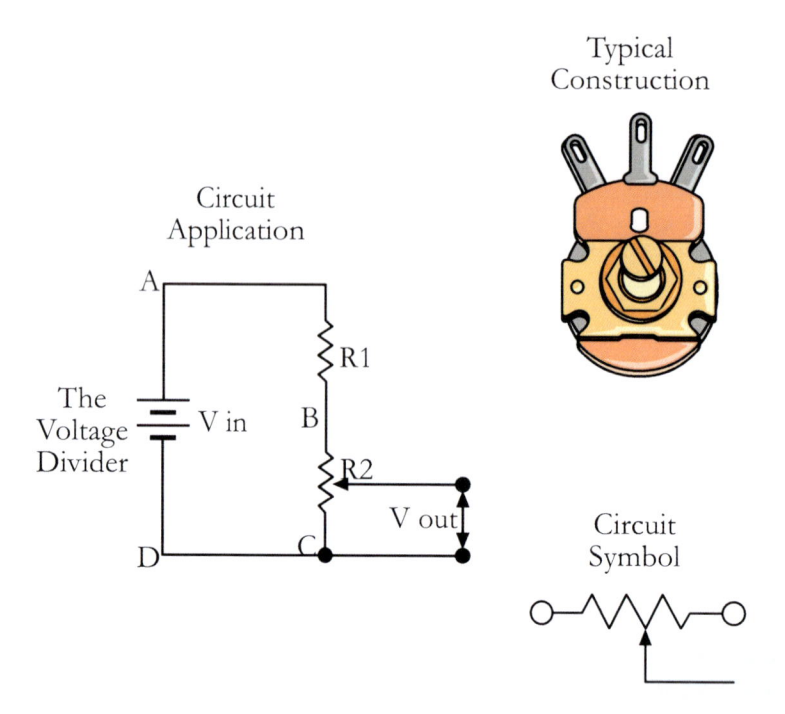

Figure 7.22 – Typical Potentiometer

Typical uses of potentiometers include radio volume controls, display brightness controls, voltage divider networks, etc.

The resistive element of inexpensive potentiometers is often made of graphite. Other materials used include resistance wire, carbon particles in plastic, and a ceramic/metal mixture called cermet. Conductive track potentiometers use conductive polymer resistor pastes that contain hard-wearing resins and polymers, solvents, and lubricant, in addition to the carbon that provides the conductive properties.

The tracks are made by screen-printing the paste onto a paper-based phenolic substrate and then curing it in an oven. The curing process removes all solvents and allows the conductive polymer to polymerize and cross-link. This produces a durable track with a stable electrical resistance.

Thermistors

Thermistors are temperature sensitive resistors which use a temperature-sensing element composed of a semiconductor material which exhibits a large change in resistance proportional to a small change in temperature. Although positive temperature coefficient units are available, most thermistors have a negative temperature coefficient (TC); which means the resistance of the thermistor decreases as the temperature increases.

Typical uses on aircraft include cabin temperature control. A probe type thermistor is shown in Figure 23.

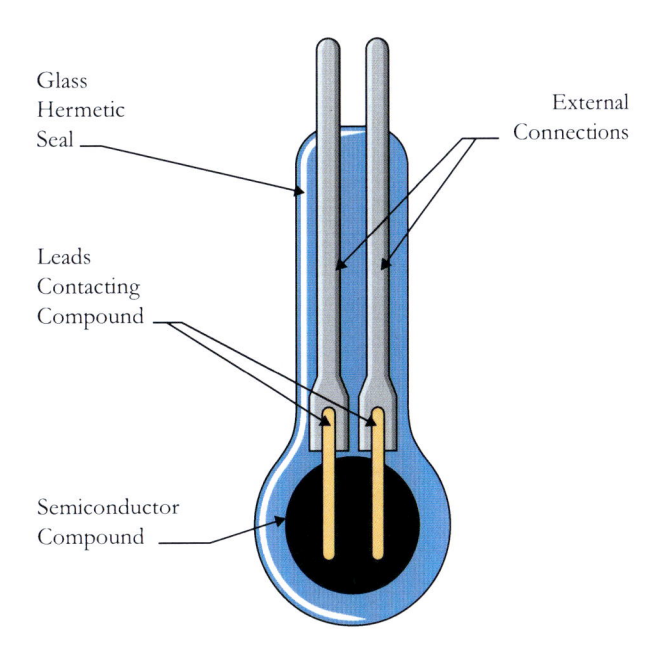

Figure 7.23 – Probe Type Thermistor

Voltage Dependent Resistors

Some components do not obey Ohm's law, in that the current flow through them does not vary linearly as the applied voltage is varied. These elements are known as non-linear resistors or non-linear conductors. Transistors, diodes and voltage dependent resistors all fall into this group.

A voltage dependent resistor, also known as a Varistor, has a nonlinear varying resistance, dependent on the voltage applied. Resistance is high under nominal load conditions, but will sharply decrease to a low value when a voltage threshold, known as the breakdown voltage, is exceeded. They are often used to protect circuits against excessive transient voltages. Mains 'spikes' or transients tend to be generated by switches like the thermostat controls that turn on and off inductive loads like fridge motors. These switching transients usually only last for a fraction of a second, but may generate large voltage spikes. Large voltage transients are in practice most likely to be a problem for equipment like home computers as the spikes could disrupt the digital electronics.

When the circuit is exposed to a high voltage transient, the varistor starts to conduct and clamps the transient voltage to a safe level. The energy of the incoming surge is partially conducted and partially absorbed, protecting the circuit. Figure 7.24.

CHAPTER SEVEN
RESISTANCE & RESISTORS

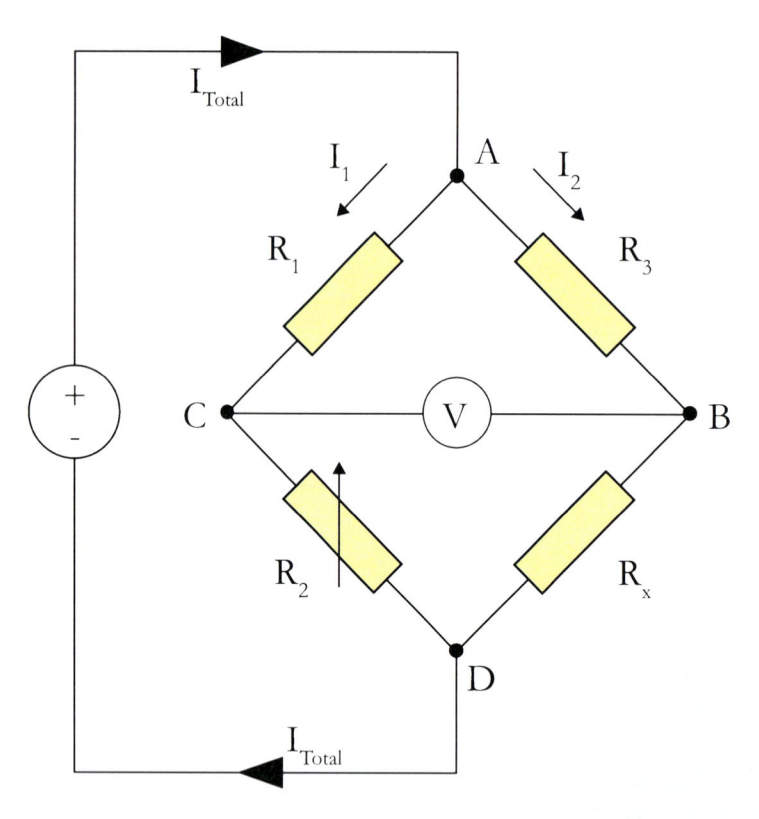

Figure 7.24 – Varistor Circuit Protection

Most voltage dependent resistors are made from a piece of Zinc Oxide or a similar material, usually doped with some other materials to adjust the electrical properties. Figure 7.25.

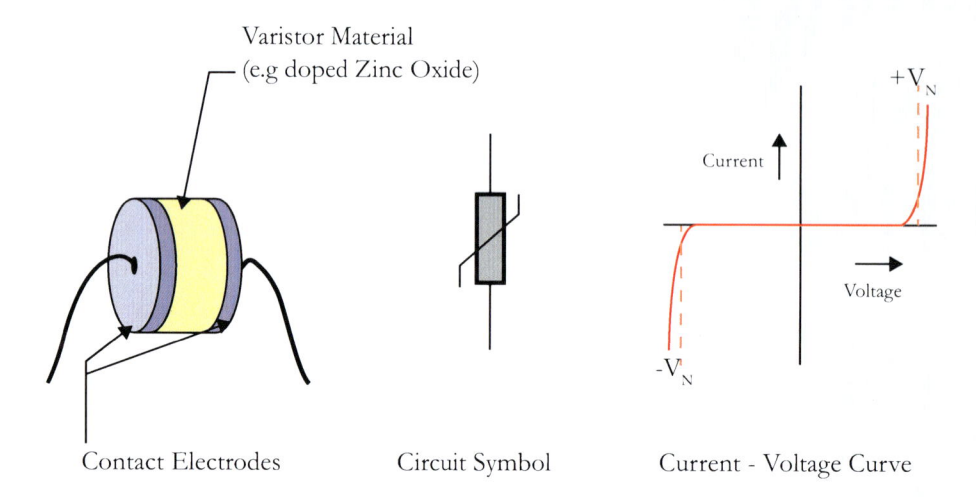

Figure 7.25 – Varistor Construction and Characteristics

As we can see in Figure 7.25, varistors exhibit a high resistance when the applied voltage is low. However they become considerably more conductive if the applied voltage is increased above a threshold value. Typically the device will have a resistance well above a million Ohms for small voltages, but the resistance will fall to a few Ohms or less when we try to apply high voltages between the leads.

Operation and Construction of a Wheatstone Bridge

The **Wheatstone Bridge** is named after the English physicist *Sir Charles Wheatstone (1802 to 1875)*, who never really laid claim to its invention, which was actually created by the British Mathematician Samuel Christie. Sir Charles Wheatstone popularised the Wheatstone bridge in 1843, and in 1837 with Sir William Fothergill Cooke, patented an early telegraph.

The Wheatstone bridge uses a galvanometer to provide the most accurate measurement of resistance available. This circuit consists of three known resistances and an unknown resistance connected in a diamond pattern, figure 7.26. The bridge circuit is really just a pair of voltage dividers connected in parallel. A voltmeter, ammeter or *galvanometer* (very sensitive ammeter) connects the two voltage divider chains together. It can be used to find the resistance of a resistor or it can be used with sensors, such as thermistors, to make temperature measurements. In our example we shall use a voltmeter to connect the voltage divider chains together.

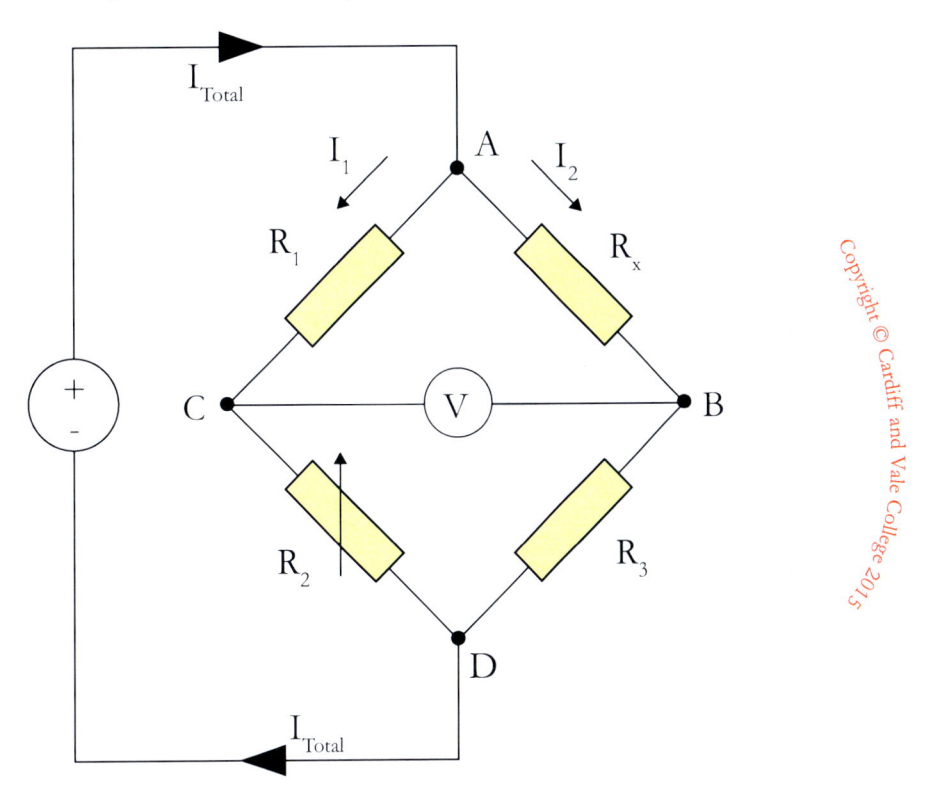

Figure 7.26 - Wheatstone Bridge

A DC voltage is connected across two opposite points of the diamond, and a voltmeter is bridged across the other two points. The variable resistor, R_2, is adjusted until the voltmeter reads zero volts. At this point we say that the bridge is balanced. When a Wheatstone bridge is balanced:

$$\frac{R1}{R2} = \frac{R3}{Rx}$$

CHAPTER SEVEN
RESISTANCE & RESISTORS

If we know the value of three of the resistors in a balanced Wheatstone bridge circuit we can calculate the value of the fourth resistor.

If a Wheatstone bridge is balanced the voltmeter will have a zero reading.

If the bridge is put out of balance by altering the resistance of one of the resistors, a reading will be obtained on the voltmeter. The reading on the voltmeter is proportional to the change in the resistance of the resistor.

Going back to the bridge in figure 7.26, resistors R1 and R3 are precision fixed value resistors, R2 is a variable resistor and the value of Rx is an unknown value of resistance that must be determined. The bridge works only when it has been properly balanced, i.e. the **Voltmeter**, reads zero. When the bridge is properly balanced, no difference in potential exists across terminals 'B' and 'C'; the voltmeter reading is zero. The bridge's operation can best be explained in a few logical steps:

1. **Conventional** current flows from the battery to point 'A' where the current divides as it would in any parallel circuit due to Kirchhoff's Law

2. Part of it passes through R1 and R2, i.e. I_1 and the remainder I_2 passes through R3 and Rx

3. The two currents, I_1 and I_2, unite at point 'D' and return to the negative terminal of the battery

4. By Ohm's Law, the value of I_1 depends on the sum of resistance R1 and R2, and the value of I_2 depends on the sum of resistance R3 and Rx; in each case, according to Ohm's law, the current is inversely proportional to the resistance

5. R2 can now be adjusted so that there is no difference of potential between points 'C' and 'B'

6. When this happens, all of I_1 follows the A to C to D path and all I_2 follows the A to B to D path

7. This means that a voltage drop V_1, across R1 between points A and C, is the same as voltage drop V_3, across R3 between points A and B

8. Similarly, the voltage drops across R2 and Rx, i.e. V_2 and V_x, are also equal

Expressed algebraically:

$$V_1 = V_3$$
$$\text{i.e. } I_1 R_1 = I_2 R_3$$
$$\& V_2 = V_x$$
$$\text{i.e. } I_1 R_2 = I_2 R_x$$

With this information, we can calculate the value of the unknown resistor Rx by dividing the voltage drops across R1 and R3 by their respective voltage drops across R2 and Rx as follows:

$$\frac{I_1 R1}{I_1 R2} = \frac{I_2 R3}{I_2 Rx}$$

We can then simplify this equation to:

$$\frac{R1}{R2} = \frac{R3}{Rx}$$

Now by multiplying both sides of the expression by Rx to separate it out, we get a formula for calculating the value of Rx, which is:

$$Rx = \frac{R2\,R3}{R1}$$

Although in context here we are talking about a resistance, Wheatstone bridges can measure other quantities depending upon the type of circuit elements used in their arms. These include inductance, capacitance, and frequency.

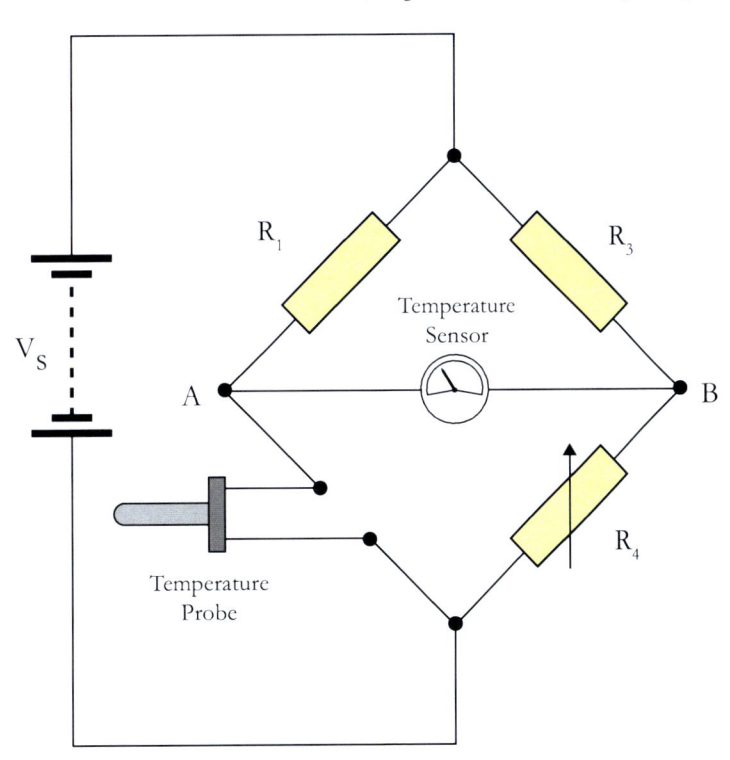

Figure 7.27 – Typical Wheatstone Bridge Circuit

A typical bridge circuit used to measure temperature is illustrated in Figure 7.27. When the bridge is in balance, the voltage drops across R1 and R3 are equal, and no potential difference exists between point A and B. A change in temperature will cause the temperature probes resistance to changes in value. This affects the voltage drop across R1 causing the bridge to become unbalanced. A potential difference now exists between points A and B, and current will flow through the gauge, proportional to the change in temperature. In this example variable resistor R4 is used to calibrate the circuit.

Temperature Coefficient of Resistance (α)

The resistance of any substance is usually affected by temperature; some increase in resistance, while others decrease. Temperature Coefficient of Resistance is the measure of change in electrical resistance of any substance per degree of temperature rise, and is represented by the Greek letter *Gamma (α)*.

Technically, a substance's temperature coefficient is the amount of increase or decrease in the resistance of a 1Ω sample of the substance per °C of temperature rise above 0°C. However, some countries use a temperature of 20°C as the norm starting point. Because of this difference, a substance's temperature of coefficient is usually identified as α_0 or α_{20} to signify the difference.

The resistance of pure metals *increases* with temperature, whereas those of carbon, electrolytes and insulating materials *decreases* with increasing temperature, but why is this so? This is because, as we have already discovered, a material with many free electrons that will allow current to flow is called a conductor, while a material that opposes current flow is called an insulator, and in a semiconductor there are a few free electrons and so a small current may or may not flow, depending on the circumstances.

As you would expect, conductors, insulators and semiconductors are all affected by changes in temperature. A conductor has a very large number of free electrons but still has some resistance because these free electrons, rather than passing unobstructed through the material, collide with other atomic particles that vibrate naturally. As temperature increases, these atomic particles will vibrate quicker, obstruct the path of the electrons more, so that collisions will occur more frequently. The result is that the resistance of conductors *increases* with temperature.

Insulators, due to the nature of their atomic bonding, have no free electrons, except for those due to thermal energy, that manage to break free from their fixed positions. However, as the temperature of an insulator increases, more electrons acquire sufficient thermal energy to break free, and so the number of free electrons increases. The result is that the resistance of insulators *decreases* as temperature increases.

As we know, semiconductors act somewhat like insulators and so their resistance also *decreases* with an increase in temperature but their resistance fall is more pronounced as illustrated in figure 7.28.

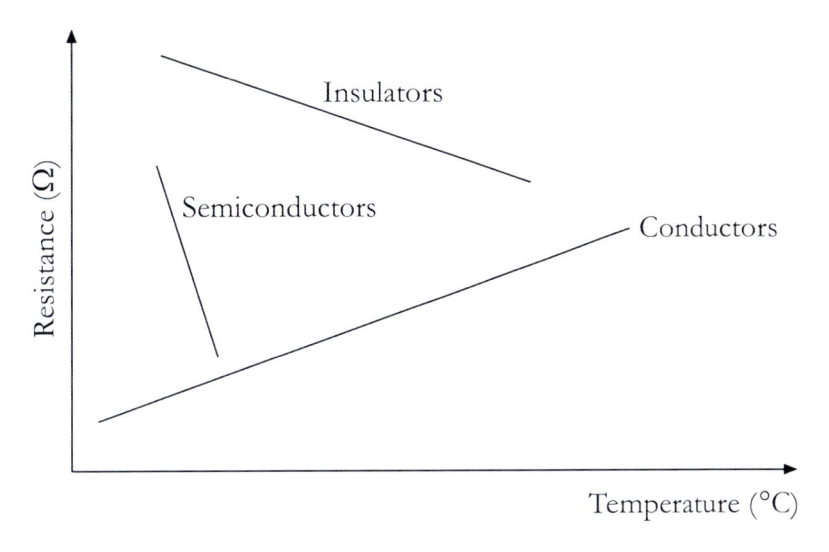

Figure 7.28 - Graph of Resistance Versus Temperature

Although the graph in figure 7.28 illustrates how temperature affects various substances, there are some modern alloys, e.g. **Manganin**, that show practically no change of resistance for a considerable variation of temperature. However, these alloys are expensive and are only used where a stable resistance is vital.

As an example of α, let us take a look at a very common conductor material, copper whose variation to resistance versus temperature is illustrated in figure 7.29.

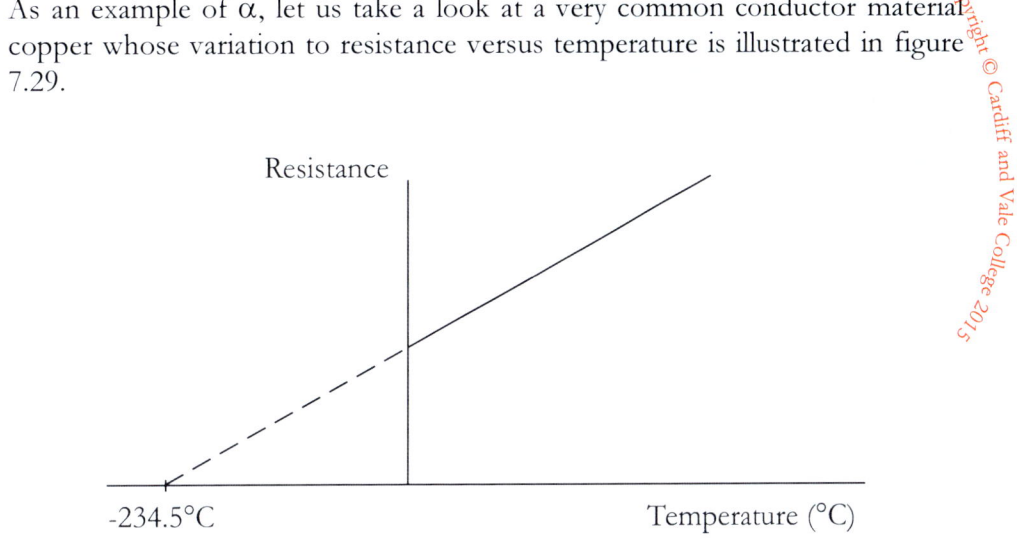

Figure 7.29 – Graph of a Copper Resistance Versus Temperature

The graph in figure 7.29 shows the copper having some resistance at 0°C, as expected, and if the graph is extended backwards it intersects the temperature axis at -234.5°C. Therefore, for a copper conductor having a resistance of 1Ω at 0°C, the change of resistance for 1°C temperature change is:

$$\frac{1}{234.5} = 0.004264 \Omega$$

Applying this to the ratio given for the temperature of coefficient of resistance we get:

$$\alpha_0 = \frac{0.004264 \Omega/°C}{1\Omega} = 0.004264/°C$$

As already mentioned, figures for α are usually given relative to 0°C and some typical values are given in table 7.7.

Material	α_0 at 0°C
Aluminium	0.00381
Brass	0.001
Carbon	-0.00048
Constantin (alloy)	0
Copper	0.004264
Manganin	0.00002
Nickel	0.00618
Silver	0.00408
Tin	0.0044
Zinc	0.00385

Table 7.7 – Typical α at 0°C

Revision

Resistance & Resistors

Questions

1. If a conductor is 10cm long, with a cross sectional area of 1 cm^2 and a resistivity of 5, its resistance is:

    a) 500Ω

    b) 5Ω

    c) 50Ω

2. The resistor in the figure below is:

    a) 1kΩ with 10% tolerance

    b) 1kΩ with 20% tolerance

    c) 1kΩ with 5% tolerance

3. A resistor marked with the letter number combination 4R7 is a:

    a) 4.7Ω resistor

    b) 47 Ω resistor

    c) 4.7 kΩ resistor

4. If 3 resistors of 1.5 kΩ are connected in series their total resistance is:

    a) 0.5 kΩ

    b) 4.5 kΩ

    c) 1.5 kΩ

5. If a resistor of 1.5 kΩ has a current of 2 A flowing through it dissipates:

   a) 3 kW of power

   b) 6 kW of power

   c) 0.75 W of power

6. If two (2) resistors in series are 10Ω each and the current flowing through them is 3 A, then the total power dissipated in the circuit is:

   a) 90 W

   b) 180 W

   c) 60 W

7. If three (3) resistors, each being 1.5 kΩ, are placed in parallel, their total equivalent resistance is:

   a) 500 Ω

   b) 4500 Ω

   c) 1500 Ω

8. If two (2) resistors of 2 kΩ & 4kΩ are placed in parallel, their equivalent resistance is:

   a) 6 kΩ

   b) 2 kΩ

   c) 1.33 kΩ

9. If two (2) resistors of 20 Ω & 30 Ω are placed in parallel, and the current going through each resistor is 4A & 2A respectively, the circuit's total power is:

   a) 140 W

   b) 72 W

   c) 440 W

10. In the figure below, what is the circuit's total resistance?

    a) 110 Ω

    b) 72 Ω

    c) 90 Ω

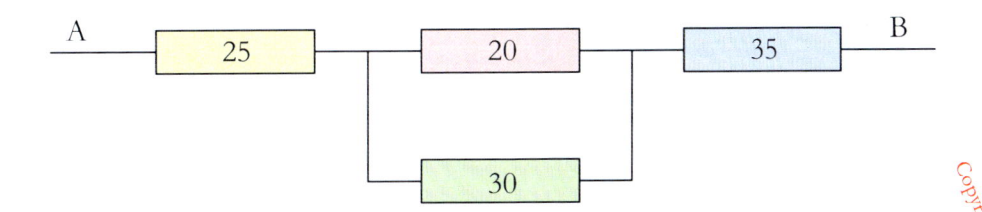

    A —[25]—+—[20]—+—[35]— B
              | |
              +—[30]—+

11. In the figure below, if the voltage across A & B is 36 V, what is the total circuit current?

    a) 500 mA

    b) 400 mA

    c) 330 mA

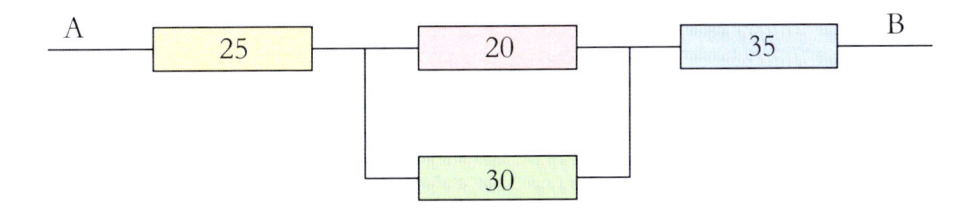

    A —[25]—+—[20]—+—[35]— B
              | |
              +—[30]—+

CHAPTER SEVEN
RESISTANCE & RESISTORS

12. In the figure below, if the voltage across A & B is 36 V, what is the current flowing through the 30Ω resistor?

    a) 200 mA

    b) 500 mA

    c) 300 mA

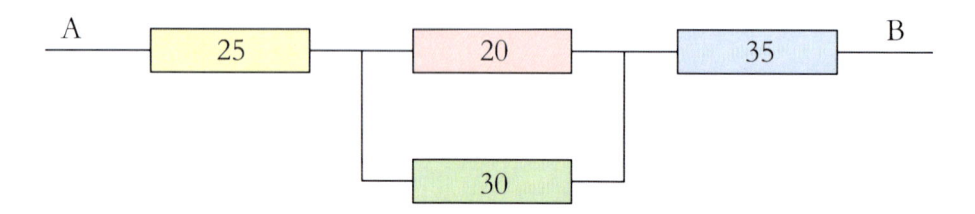

13. The symbol in the figure below is for a:

    a) Variable inductor

    b) Rheostat

    c) Potentiometer

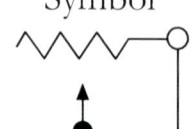

14. In the figure below the voltage between points b and c is 0 V. If R1 = 20Ω, R2 = 30 Ω & R3 = 40 Ω, Rx=:

    a) 60 Ω

    b) 10 Ω

    c) 30 Ω

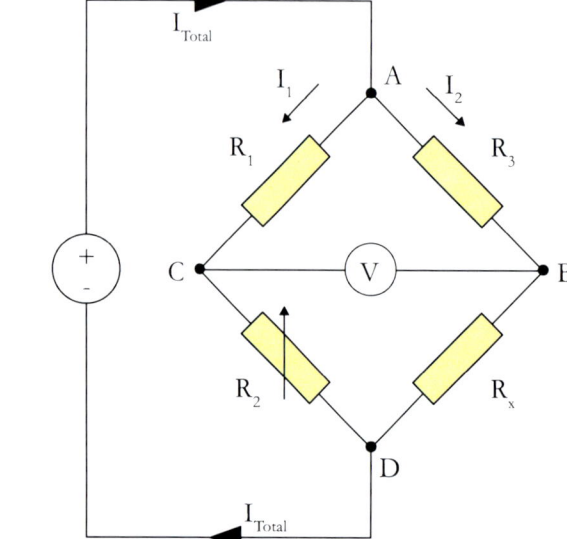

15. **As temperature increases, the resistance of conductors:**

    a) Decrease

    b) Increases

    c) Remains the same

16. **As temperature increases, the resistance of semiconductors:**

    a) Decrease

    b) Increases

    c) Remains the same

Revision

Resistance & Resistors

Answers

1. **C**   R = (5 × 10) ÷ 1 = 50Ω

2. **A**

3. **A**

4. **B**

5. **B**

6. **B**

7. **A**

8. **C**

9. **C**   16 × 20 = 320, 4 × 30 = 120, 320 + 120 = 440

10. **B**

11. **A**

12. **A**

13. **B**

14. **A**

15. **B**

16. **A**

Power

Introduction

Power, whether we are talking about electrical or mechanical, relates to the rate at which *Work* is being done or alternatively in physics is the process of transferring *Energy*. As with all forms of energy, the SI unit of work is the *Joule (J)*, named after *James Joule (1818 to 1889)*, a British physicist, who was the first to determine the mechanical equivalent of heat, i.e. the work required to produce a given amount of heat energy. Work is a scalar quantity calculated by multiplying the force and displacement vectors; i.e. the work done by a force is the product of that force and the distance over which it is applied in the direction in which it acts. The textbook definition of work is:

One (1) joule of work is done when one (1) Newton is applied over one (1) metre.

Power is also a scalar quantity and is measured in *Watts (W)*, where one (1) watt is equivalent to one joule per second (1 J/s). Common light bulbs are typically rated in terms of their wattage, as is the output from a stereo amplifier or electric fire, i.e. it is the rate at which an electrical device uses energy. When studying these concepts we have to make this distinction between power and energy, which is different from everyday usage, where they are almost synonyms.

Energy

When looking at things around us they can appear different in many ways, even though they are the same object; e.g.:

- A rock falling off a cliff at Beachy Head is different from the same rock when it is lying on the beach below

- The lively lecturer eager to begin lessons in the morning after a night's rest and a hearty breakfast is different from the lecturer who arrives at a IR Part-66 evening class after a full day's teaching

- A glowing light bulb is different from the same bulb when the electricity is switched off

CHAPTER EIGHT
POWER

In each case, an agent of change has acted upon the rock, lecturer and light bulb. It is the same rock, the same lecturer, the same light bulb; the difference is one of *Energy*.

Energy is one of the most basic ideas of science and all things in the universe can be explained in terms of energy and matter. However, the definition of energy is not simple since energy occurs in many different forms, e.g. chemical energy, thermal energy, electrical energy, nuclear energy. It is not always easy to tell how these forms are related to one another and what they have in common. One of the best-known definitions of energy is the classical one used in physics, which is:

Energy is the ability to do work.

However, physicists define work in a way that does not always agree with the average person's idea of work. In physics, work is done when a force applied to an object, moves it some distance in the direction of the force; i.e. no movement, no work is carried out. But what do we mean by this?

If we walk up a flight of stairs we regard it as work as we exert an effort to move our body to a higher level. In this instance, we are also doing work according to the definition accepted by physicists, i.e. we exert a force to lift ourselves over a distance, i.e. from the bottom to the top of the stairs. However, if a person stands without moving with a 10 kg weight in their outstretched arms, they are not doing any work as physicists define it. That person is exerting a force that keeps the 10 kg weight from falling to the floor, but the position of the weight remains unchanged, i.e. it is not moved any distance by the force. The person is of course, exerting considerable muscular effort to avoid dropping the weight, and would argue that they are working very hard, but they are not doing any work according to the definition accepted in physics.

Energy can be transferred from one body to another by work processes involving movement, e.g. the transfer of heat between objects that are at different temperatures; electromagnetic radiation, e.g. light and microwaves; and by electricity, i.e. the flow of electrical energy. Energy can also be converted from one form to another, e.g. the potential energy of water flowing quickly from a Welsh reservoir when passed over a turbine, converts it into electrical energy by a generator driven by the turbine.

One of the most useful facts about energy is that it can be changed from one form to another and these changes are happening all the time. Most machines' purpose is the conversion of energy from one form to another. However, energy cannot be created or destroyed and although it is conserved when converted from one form to another, some of the energy is converted into an unwanted form, usually heat and noise. This is especially true when thermal energy is converted into another form, as in your own car's internal-combustion engines.

Sources of energy used by the human race are usually directly or indirectly derived from sunlight, although nuclear energy must be excluded from this concept. The burning of fossil fuels like oil, coal and wood, and the use of water falling from a Welsh reservoir can be used directly to release energy or to drive various mechanical and electrical installations to produce electricity. Oil and coal were created over millions of years from living organisms whose lives depended upon the energy in sunlight to fuel the processes building their structures, while wood relies on the same process. Windmills also indirectly use solar energy, as a result of the Sun's affect on weather.

Kinetic Energy

The concept of energy only really emerged during the mid-19th century, when it was realised that moving bodies could be made to move against resisting forces, so doing work. *Kinetic energy* was the name given to this ability of a body to do work as a result of its motion, i.e. it is the work the body could do on coming to rest. The textbook definition of kinetic energy is:

> *For a body with a mass of 'm' kg and a velocity of 'v' m/s, its kinetic energy is equal to $½mv^2$, and in the SI system, like all energy, is measured in Joules (J).*

For a rotating body the kinetic energy is given by $½I\omega^2$, where 'I' is the moment of inertia and 'ω' the angular velocity.

Many energy changes involve the inter-conversion of kinetic energy and *Potential Energy*, which is the energy resulting from the position of a body. For example a pendulum has its greatest kinetic energy as it moves through the lowest point on its path as this is where it is moving fastest. As the pendulum rises it loses *kinetic energy* as it slows down, but gains *potential energy* as it rises to the top of its swing. The pendulum has its greatest potential energy where it has its least kinetic energy, i.e. at its highest point above the ground.

Potential Energy

Potential Energy, as mentioned above, is the ability of an object to do work because of its position. The textbook definition of potential energy is:

> *For a body with a mass of 'm' kg that is 'h' metres above the Earth's surface, its potential energy is equal to mgh, where 'g' is the acceleration of a free-falling object. In the SI system, like all energy, potential energy is also measured in Joules (J).*

The concept of potential energy can be applied to a wide range of situations in which there is a potential for work to be done; e.g. in a stretched spring awaiting release, an aircraft at full throttle at the end of a runway held by its brakes. With respect to the spring, its *potential energy* is converted to *kinetic energy* as the spring is released and returns rapidly to its original length. In astronomy, a planet in orbit about the Sun has potential energy *and* kinetic energy, continuously transferring one into the other as its distance from the Sun varies.

CHAPTER EIGHT
POWER

Electrical Energy

In an earlier chapter we discussed how voltage is an electrical force, and that voltage forces current to flow in a closed circuit and so when voltage causes electrons to move, work is done. However, when a circuit has an EMF but current is not flowing because the circuit is open, no work is done, which is similar to the spring under tension analogy that produces no motion as discussed earlier. Electrical energy is measured in joules, which has the technical definition of:

One (1) joule of electrical energy is used in moving one (1) coulomb of electrical charge (Q) through a potential difference of one (1) volt.

From this definition, therefore, we can say that Electrical energy = Charge × Voltage, i.e.:t is worth noting that an increase in voltage will also increase the current flow in a circuit, and therefore the amount of electrical charge. As an example 4 joules of work is done when a circuit has an electrical charge of 2 coulombs and a potential difference of 2 volts:

Therefore in this example:
Energy = Q x V
4 = 2 x 2

What would happen if we now doubled the voltage applied to the circuit to 4 volts? Firstly we must calculate how this increase in voltage affects the current flow in the circuit, using Ohms law:

$$V = IR$$

To achieve this we must assume that the resistor has a constant value. We could use any value we wanted, but to keep things simple I am going to use a 1Ω resistor. We can now calculate the current flow when the voltage is 2 volts:

I = V ÷ R
Therefore I = 2 ÷ 1

which equals 2 amps

If we now double the supply voltage to 4 volts, the resistance will remain constant, but the current will double.

I = 4 ÷ 1

which equals 4 amps

Since electrical charge (Q) is measures in Amps/second, then this must also have doubled to 4 coulombs, therefore:

$$\text{Energy} = Q \times V$$
$$\text{Energy} = 4 \text{ coulombs} \times 4 \text{ volts}$$
$$= 16 \text{ joules}$$

From this we can see that when the supply voltage is doubled, the work done is quadrupled.

Power is the rate of doing work and it is a measure of how much energy is used in one second. The SI unit of power is the **Watt (W)** and it has the SI symbol '**P**' and therefore:

$$\text{Power (P)} = \frac{\text{Energy (J)}}{\text{Time (t)}} = \frac{VIt}{t}$$

Simplifying this equation results in the formula:

$$\text{Power (P)} = VI$$

Now by using Ohm's law substitution, we can get the other formulae for power, which are:

$$P = I^2 R$$
$$\&$$
$$P = \frac{V^2}{R}$$

From the previous work we have done above, we can see that in a circuit with two or more resistors:

$$P_t = P_1 + P_2 + P_3 + \ldots P_n$$

The *instantaneous* rate at which this work is done is called the **Electric Power Rate**, and is also measured in **Watts**.

The total amount of work done in any given situation may be achieved over different time periods. For example, a given number of electrons may be moved from one point to another in 1 second or in 1 hour, depending on the *rate* at which they are moved; in both cases, the *total* work done is the same. However, when the work is done in a short time, the wattage, or

CHAPTER EIGHT
POWER

Instantaneous Power Rate, is greater than when the same amount of work is done over a longer period of time.

As already mentioned, the basic unit of power is the watt and power in watts is equal to the voltage across a circuit multiplied by current through the circuit, representing the rate at any given instant at which work is being done, which results in the formula P = V × I, as previously discussed. Consequently, the amount of power changes when either voltage or current, or both change.

In practice, the *only* factors that change are voltage and resistance. In explaining the different forms that formulas may take, current is sometimes presented as a quantity that is changed. Remember, if current is changed, it is because either voltage or resistance has been changed.

Up to this point in this module, we have discussed four (4) of the most important electrical quantities; i.e. voltage (V), current (I), resistance (R), and power (P). Over the past few chapters we have discussed the relationships that exist among these quantities because they are used throughout the study of electricity. In the preceding paragraphs of this chapter, power (P) has been expressed in terms of alternate pairs of the other three basic quantities V, I, and R. In practice, you should be able to express any one of these quantities in terms of any two of the others. Figure 8.1 shows a summary of 12 basic formulas you should know.

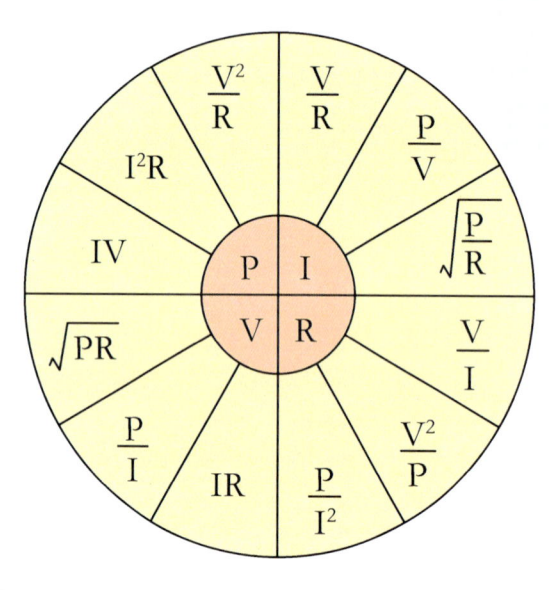

Figure 8.1 – Summary of Basic Electrical Formulae

The four quantities V, I, R, and P are at the centre of the figure and adjacent to each quantity there are three segments. Note that in each segment, the basic quantity is expressed in terms of two other basic quantities, and no two segments are alike. For example, the formula wheel in figure 8.1 could be used to find the formula to solve the following problem:

Worked Example

A circuit has a voltage source that delivers six (6) volts and the circuit uses 3 watts of power. What is the resistance of the load?

Since R is the quantity to find, look in the section of the wheel that has R in the centre. Now the segment V²/P has the quantities that have been given and therefore the formula you would use to solve the problem is:

$$R = \frac{V^2}{P} = \frac{6^2}{3} = 12\Omega$$

Power Rating

As already mentioned in chapter 7, electrical components are often given a power rating. The power rating, in watts, indicates the rate at which the device converts electrical energy into another form of energy, such as light, heat, or motion. For example, when comparing a 100-watt bulb to a 60-watt bulb, the higher wattage rating of the 100-watt bulb indicates it is capable of converting more electrical energy into light energy than the bulb of the lower rating. Other common examples of devices with power ratings are soldering irons and small electric motors.

In some electrical devices the wattage rating indicates the maximum power the device is designed to use rather than the normal operating power. The 100-watt bulb discussed above, for example, uses 100 watts when operated at the specified voltage printed on it. In contrast, a device such as a resistor is not normally given a voltage or a current rating. A resistor is given a power rating in watts and can be operated at any combination of voltage and current as long as the power rating is not exceeded. In most circuits, the actual power used by a resistor is considerably less than its power rating because a 50% safety factor built in to most circuits; e.g. if a resistor normally uses 2 watts of power, a resistor with a power rating of 3 watts is usually used.

As mentioned in chapter 7, resistors of the same resistance value are available in different wattage values; e.g. carbon resistors are commonly made in wattage ratings of 1/8, 1/4, 1/2, 1, and 2 watts and the larger the physical size of a carbon resistor the higher the wattage rating. This is true because a larger surface area of material radiates a greater amount of heat more easily.

CHAPTER EIGHT
POWER

Revision

Power

Questions

1. 120 Watts is equivalent to:

    a) 1200 Joules/second

    b) 120 Joules/second

    c) 12 Joules/second

2. If an object is moved over 8 metres with a force of 8 Newtons, the amount of work down is:

    a) 64 Joules

    b) 1 Joule

    c) 8 Joules

3. A body of 6 kg moving at a rate of 4m/s has kinetic energy of:

    a) 48 Joules

    b) 24 Joules

    c) 1.5 Joules

4. If a circuit has a voltage of 8 V and a current of 3A flows in the circuit for 12 seconds, the energy expended is:

    a) 24 Joules

    b) 2 Joules

    c) 288 Joules

5. If a circuit has a voltage source of 5 volts and the circuit uses 5 Watts of power, its resistance is:

   a) 25Ω

   b) 1Ω

   c) 5Ω

Revision

Power

Answers

1. **B**
2. **A**
3. **A**
4. **C**
5. **C**

Capacitance & Capacitors

Introduction

Capacitance was discovered independently by two scientists, *Ewald Georg von Kliest*, a Prussian, and *Pieter van Musschenbroek*, a Dutchman, in the middle of the 18th century. They were both studying electrostatics and as a result, found that an electric charge could be stored for a time. They were using a device now called a *Leyden jar*, figure 9.1, which consisted of a glass jar wrapped with a metal band that was filled with water and had a nail piercing the stopper, dipping into the water. They then connected the nail to an electrostatic charge, but after disconnecting the nail from the charge's source, they found that if they touched the nail they felt a shock, demonstrating that the device had stored the charge.

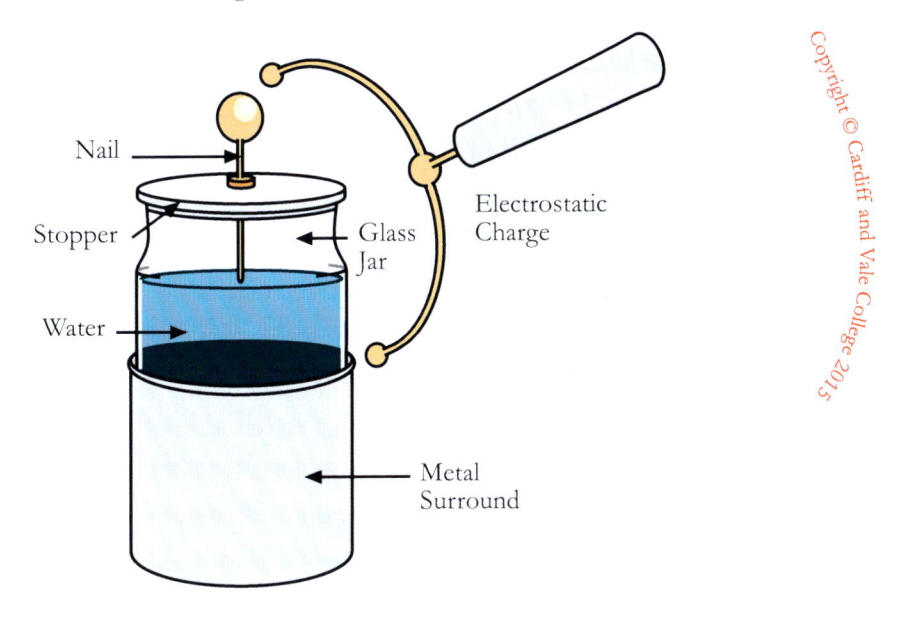

Figure 9.1 – Leyden Jar

A few years later, *John Bevis*, an English Physicist and amateur astronomer, refined the device further by replacing the water in the jar with metal foil. By lining both the inside and outside of the jar with the foil in this way, he created a capacitor with two conductors, i.e. the inside and outside metal foil layers, equally separated by the insulating glass. These design features are incorporated into the modern capacitor.

CHAPTER NINE
CAPACITANCE & CAPACITORS

As a matter of interest only, the Leyden jar was also used by *Benjamin Franklin* to store the charge from lightning and in other experiments.

Capacitors are used in many ways in electrical and electronic circuits, such as barriers to direct currents, storing memory in a computer chip, storing a charge for an electronic flash camera, or tuning a tuned circuit such as in a radio. How this is accomplished is the topic of this chapter.

What is Capacitance?

Capacitance is the property of an electrical system or circuit of conductors and insulators that enables it to store electric charge. *Capacitors* are electrical devices used to store electric charges in the same way as a sealed container is used to store liquid or gas. Looking at the sealed container as a storage device, we know that if the liquid or gas within it is put under pressure, the container can store more of that liquid or gas. The equivalent of pressure on a capacitor is the *electrical potential* across it. Therefore, using the same analogy as the sealed container, the *greater* the electrical potential across the capacitor, the *greater* the amounts of charge it can store.

However, before we look at this in more detail, now we have established a capacitor is a storage device for electric charge, let us look at typical applications. In aviation, capacitors are used for three (3) main purposes:

1. To store electrical charge in a dc circuit; i.e. they act as a charge reservoir.

2. To block dc but to allow ac to pass; i.e. they act as a *coupling* device.

3. To combine with an *Inductor*, more on this component later, to form a tuned circuit that resonates, i.e. oscillates, at one particular frequency.

As mentioned in (2) above, a capacitor presents an extremely high resistance to direct current (DC) and under normal circumstances, no current will flow in a dc circuit that includes a capacitor. However, alternating current (AC) is allowed to pass much more easily and the higher the frequency of the AC signal, the less opposition the capacitor presents. Because of this ability to separate AC and DC signals, capacitors are frequently used in both filters and power supplies.

There are several types of capacitor and each has their own characteristic. They are found in the majority of electrical and electronic circuits and are used for many purposes. One of the simplest examples you may have already come across is a small electrolytic capacitors used to eliminate noise from the generator on your car. We will be looking at more capacitor applications at the end of this chapter.

Description of a Capacitor

Typically, a capacitor in its simplest form consists of two (2) parallel conductors made of metal plates or electrodes, separated by an insulator or *dielectric*, figure 9.2.

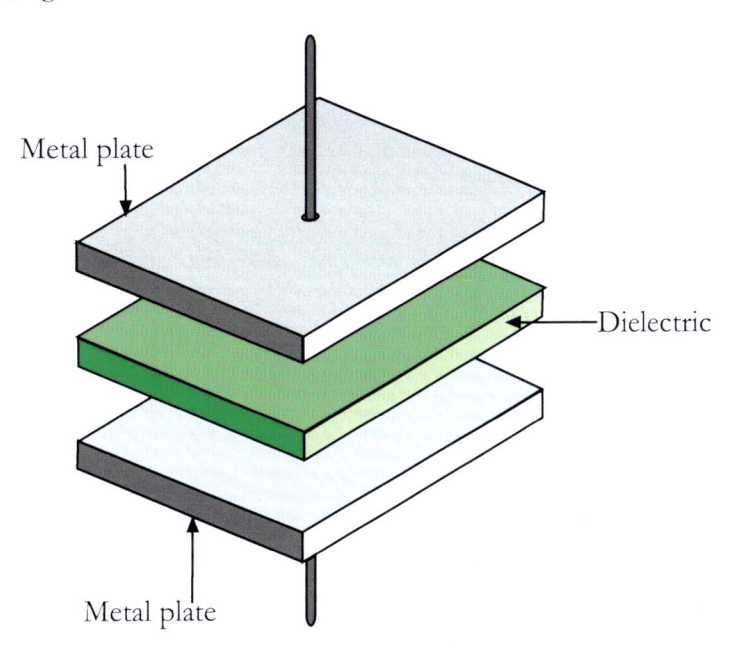

Figure 9.2 – Basic Capacitor Structure

If we have two metal plates close together, but separated by an insulator or dielectric (which could be air) and we apply a voltage across them, electrons are removed from one plate and applied to the other and each becomes charged.

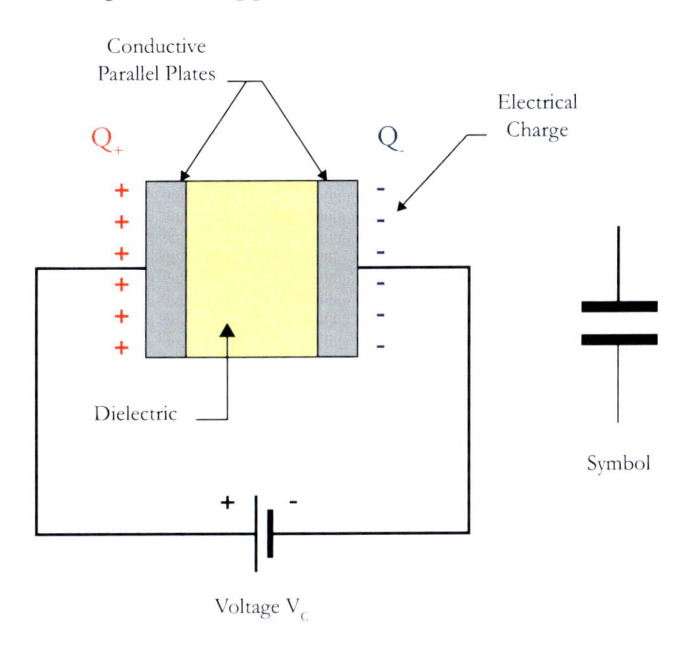

Figure 9.3 – Charged Capacitor

CHAPTER NINE
CAPACITANCE & CAPACITORS

Electrons are attracted from the left hand plate to the positive terminal (anode) of the battery, Figure 9.3. This leaves the left hand plate with a positive charge. Chemical action in the battery forces the electrons to the negative terminal (cathode) which then repels the electrons onto the right hand plate of the capacitor as shown in Figure 9.3.

This flow of electrons to the plates is known as the capacitors **Charging Current,** and will continue to flow until the voltage across both plates (and hence the capacitor) is equal to the applied voltage Vc. At this point the capacitor is said to be "fully charged" with electrons. The strength or rate of this charging current is at its maximum value when the plates are fully discharged (initial condition) and slowly reduces in value to zero as the plates charge up towards the supply voltage. From this we can see that capacitors effectively block DC current flow.

Although we have said that the charge is stored on the plates of a capacitor, it is more correct to say that the energy within the charge is stored in an "electrostatic field" between the two plates, Figure 9.4.

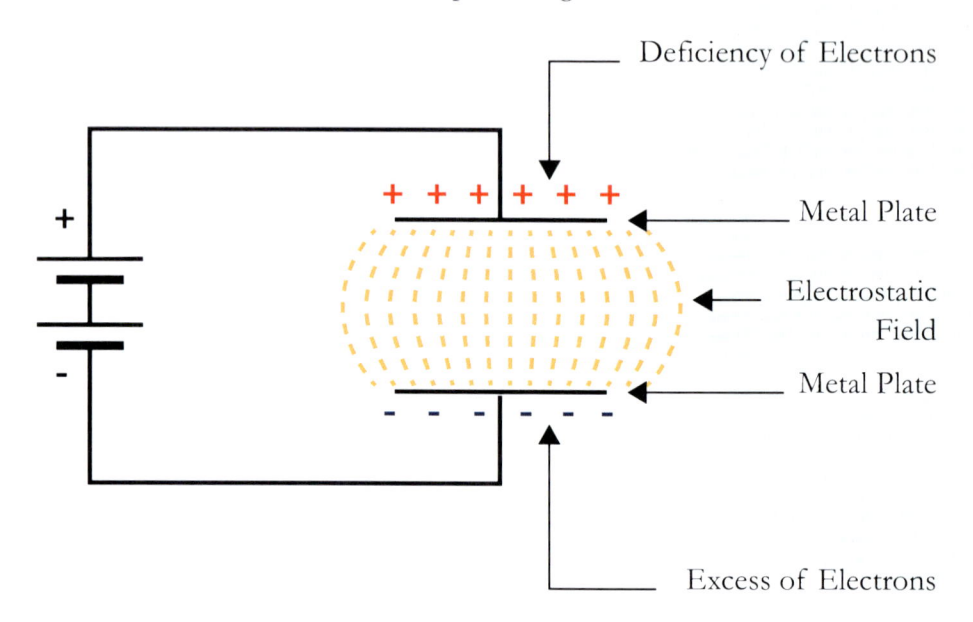

Figure 9.4 – Capacitor Electrostatic Field

When charging the capacitor up the electrostatic field becomes stronger as it stores more energy. Likewise, as the current flows out of the capacitor, discharging it, the potential difference between the two plates decreases and the electrostatic field decreases as the energy moves out of the plates. Theoretically capacitors could hold their charge indefinitely, even when the charging supply is removed. However all capacitors will eventually lose their stored charge due to internal leakage paths which allow electrons to flow from one plate to the other. Depending on the specific type of capacitor, the time it takes for a stored charge to self-dissipate can be incredibly long (several years for a charged capacitor sitting on a shelf!).

The charge held by the combination may be very large because of the concentration of the electrostatic field between the plates. This represents a basic capacitor.

However, capacitors differ in size and arrangement of the divided plates and the type of dielectric materials used; paper, ceramic, air, mica and electrolytic materials being the more common ones.

Dielectric Materials

As discussed earlier, the capacitor's dielectric material is an insulator and so prevents the flow of current between its plates and in addition, serves as a medium to support the electrostatic force of a charged capacitor. There are several types of material used for dielectrics and these are rated on their ability to support electrostatic forces in terms of a number called a *Dielectric Constant (K)*. The *higher* the dielectric constant, the *greater* its ability to support electrostatic forces. The most common dielectric materials are listed in table 9.1 below:

Material	Dielectric Constant
Vacuum	1.0
Air	1.00059
Paraffin	2.5
Polystyrene	2.5
Rubber	3.0
Paraffin coated paper	3.5
Paper	3.5
Pyrex	4.5
Mica	5.4
Slate	7.0
Flint Glass	9.9
Methyl Alcohol	35

Table 9.1 – Typical Dielectric Material

As shown above, all dielectrics are rated against a vacuum, which is used as the standard, and it has a dielectric constant of one (1). However, as shown in table 9.1, there is very little difference in the dielectric constant of a vacuum and air and so air is usually also referred to as having a dielectric constant of one (1).

CHAPTER NINE
CAPACITANCE & CAPACITORS

Capacitance Measurement

Capacitance is measured in *Farads*, named after *Michael Faraday (1791 to 1867)* and has the SI symbol '*F*'. The technical definition of a Farad is that:

If a charge of one (1) coulomb is placed on the plates of a capacitor and the potential difference between them is one (1) volt, the capacitance is then defined to be one (1) farad.

Note: From the earlier chapter on electrostatics, you should remember that one (1) coulomb is equivalent to a charge of 6.24×10^{18} electrons.

However, when looking at capacitance measurement, one Farad is an *extremely* large quantity and so micro-Farads (µF), i.e. 10^{-6} F, and Pico-Farads (pF), i.e. 10^{-12} F are more commonly used.

A capacitor's capacitance value is proportional to the quantity of charge that can be stored in it for each volt difference in potential between its plates. Mathematically this relationship is written as:

$$C = \frac{Q}{V} \text{ Farads}$$

Where:

C is capacitance in Farads

Q is the quantity of stored electrical charge in coulombs

V is the difference in potential in volts

Rearranging this formula, stored electric charge can be calculated by:

$$Q = CV$$

The difference in potential or voltage of the capacitor can be calculated by:

$$V = \frac{Q}{C} \text{ Volts}$$

Energy Stored in a Capacitor

Energy is stored in the electric field of a charged capacitor. If a dielectric is inserted, extra energy is stored above that stored in free space, due to the distortion of electron orbits in the atoms. A parallel plate capacitor can only store a finite amount of energy before *dielectric breakdown* occurs.

The energy stored in a capacitor is measured in Joules and can be calculated using the following equation:

$$\text{Energy} = \tfrac{1}{2}QV$$
$$= \tfrac{1}{2}CV^2$$

Since $Q = VC$

Now this may seem a little strange, as in Chapter 8 we determined that Energy could be calculated using the equation:

Energy = QV

This still holds true as the battery puts out energy QV in the process of charging the capacitor until it equals the battery voltage V. But half of that energy is dissipated in heat in the resistance of the charging circuit.

Now you may think if we lower the resistance of the charging circuit we will get more energy stored on the capacitor, but it doesn't quite work that way. The circuit current will increase and the capacitor will charge more rapidly, but the energy dissipated in the resistance (I2R) increases dramatically, therefore reducing the amount stored on the capacitor.

In a similar way if we increased the resistance in the circuit dramatically, then most of the supply voltage would be dropped across the resistance, and a lower amount of energy would be stored on the capacitor!

Factors Affecting Capacitance Value

There are three (3) main factors that influence a capacitor's capacitance value:

1. The area of the plates
2. The distance between the plates
3. The dielectric constant of the material between the plates

Plate Area

The *greater* the plate area the *greater* the capacitance, as larger plates with increased surface area have a greater capacity to store electric charge. Therefore, as the area of the plates *increase*, capacitance *increases*.

Plate Gap

Capacitance is directly proportional to the electrostatic force field between the plates and this field is stronger when the plates are closer together. Therefore, as the distance between the plates *decreases*, capacitance *increases* and conversely, as the distance between the plates *increases*, capacitance *decreases*.

Dielectric Constant

Dielectric materials are rated on their ability to support electrostatic forces in terms of dielectric constant (K), as shown in table 9.1. The **_higher_** the dielectric constant the **_greater_** the ability of the dielectric to support electrostatic forces and therefore, as the dielectric constant **_increases_**, capacitance **_increases_**.

Considering these factors, a capacitor's capacitance, with two parallel plates, can be calculated using the formula:

$$C = \frac{\varepsilon K A}{d}$$

Where:

- **C** is capacitance in Farads
- **ε** is the permittivity of the dielectric
- **K** is the dielectric constant
- **A** is the area of one plate in m^2
- **d** is the distance between plates in metres

In this formula we have introduced an additional factor Epsilon 'ε' where ε is the permittivity of a vacuum or air and has a value of 8.855 x 10-12. (The permittivity of a medium describes how much electric field is 'generated' per unit charge in that medium). For all practical purposes:

$$C = \frac{8.855 \times 10^{-12} K A}{d}$$

Capacitor Voltage Ratings

Capacitors come in all shapes and sizes, figure 9.5. In selecting an appropriate capacitor for any given application, designers must not only consider the value of capacitance, but also the voltage the capacitor will be subjected to during normal operating conditions.

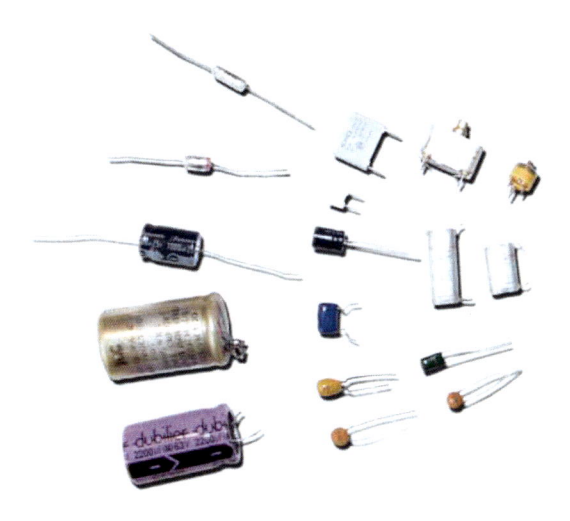

Figure 9.5 – Selection of Capacitors

Capacitors are designed to withstand, and so are rated, with a certain maximum working voltage. Exceeding this maximum voltage may result in current arcing through the dielectric and damaging the capacitor, usually beyond repair. The maximum voltage that a capacitor can withstand is called its ***working voltage***, sometimes also called its ***voltage rating***, which is usually indicated by the component manufacturer. However, when designing circuits, the standard margin of error in selecting a capacitor, is to choose one with a working voltage 50% higher than the maximum voltage it can be expected to encounter in any given application.

If the safe working voltage of a capacitor is exceeded, the electric field between the plates becomes strong enough to cause a '***flash-over***' from plate to plate. It follows then, to withstand a high working voltage the dielectric thickness is usually increased. However, this correspondingly reduces the capacitance value as the distance 'd' increases. To overcome this, the plates are usually enlarged and so capacitors with a high working voltage are usually physically bigger, as you might expect, than for the same capacitance at a lower working voltage. A capacitor's working voltage is usually written on the outside case, e.g. '***750 V DC WKG***'.

Note: A charged capacitor, especially large ones, can hold high capacity voltage for a considerable amount of time. Always ensure a capacitor is discharged before attempting to touch it.

Capacitor Symbols

Commonly used circuit symbols for capacitors are shown in figure 9.6.

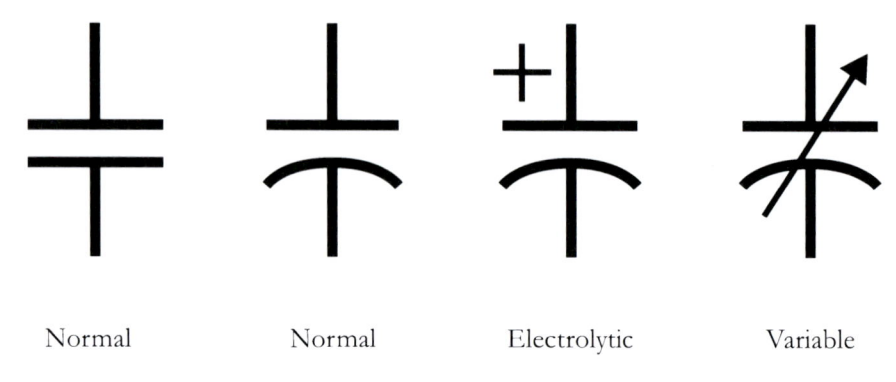

Normal   Normal   Electrolytic   Variable

Figure 9.6 – Capacitor Symbols

Note: you may see the normal capacitor symbol in circuit diagrams drawn with the straight line or curved line.

We shall now look at how capacitors are constructed; starting with the various types used as normal fixed value capacitors.

Capacitor Construction

The most common types of capacitor are paper, mica, ceramic, electrolytic, tantalum and polyester. Some of these are shown below.

Paper Construction

The paper capacitor is the cheapest and most common type used for routine circuits. The construction consists of two (2) aluminium foil strips separated by waxed paper acting as the dielectric, figure 9.7.

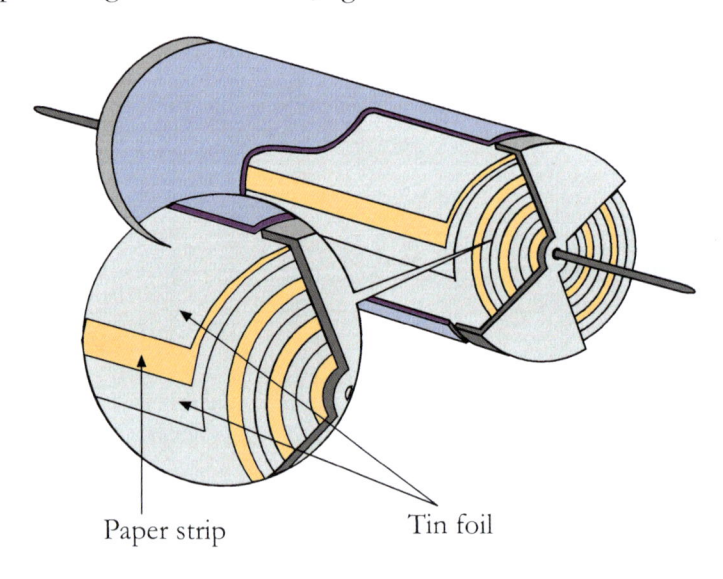

Paper strip   Tin foil

Figure 9.7 – Paper Capacitor

The foil strips and paper dielectric are rolled tightly into a tube, which is then sealed inside an outer container made of plastic, wood or metal.

The paper capacitor is usually made in the value range of 250pF to 15µF with a working voltage of up to 500 V DC. However, for special applications, they can be constructed with much higher working voltages, up to 150 kV DC, but these are very expensive and at these voltages, must be encased in an oil-filled metal case.

Mica Construction

Mica capacitors are used for high quality applications, and have several plates of metal foil, separated with sheets of Mica acting as the dielectric, figure 9.8.

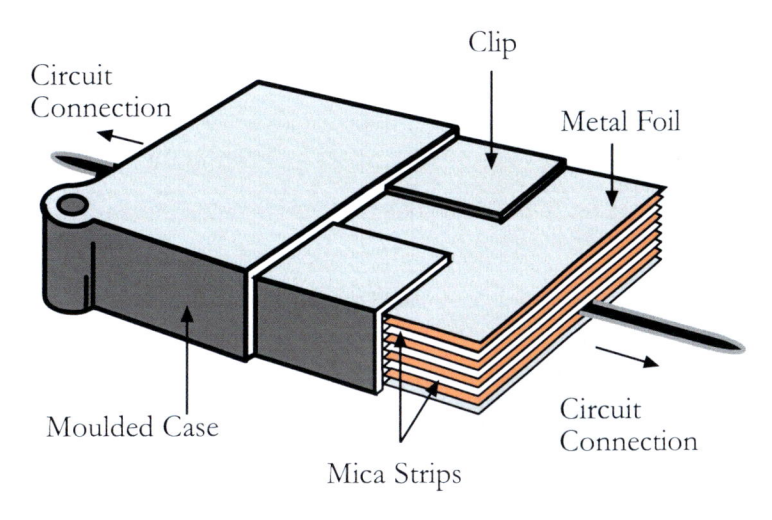

Figure 9.8 – Typical Mica Construction

Mica capacitors usually have low capacitance values, typically 1 pF to 0.1 µF, and are often used in *Radio Frequency (RF)* circuits.

Glass Capacitors

Glass capacitors are also used in high quality applications. They usually constructed with a capacitance range of 1pF to 0.05µF and have a greater capacitance to volume ratio than other construction methods.

Tantalum Capacitors

Tantalum capacitors can be constructed in two (2) ways. **_Dipped_** Tantalum capacitors are used where there is a requirement for high capacitance, high temperature, and reverse voltage with low leakage current. They provide excellent stability and are constructed in the range of 0.1μF to 2000μF.

Moulded Tantalum capacitors are used where there is a requirement for high frequency operation and high stability. They also provide excellent stability and are constructed in the range of 1μF to 3000μF.

Ceramic Capacitors

Ceramic capacitors offer a broad range of size versus performance and are easily the most popular in terms of numbers sold. They are available in capacitance values from less than 1pF to 1000s of μF and are constructed using a ceramic disc or rod as the dielectric with two (2) films of foil being bonded directly onto the ceramic to act as the plates, figure 9.9.

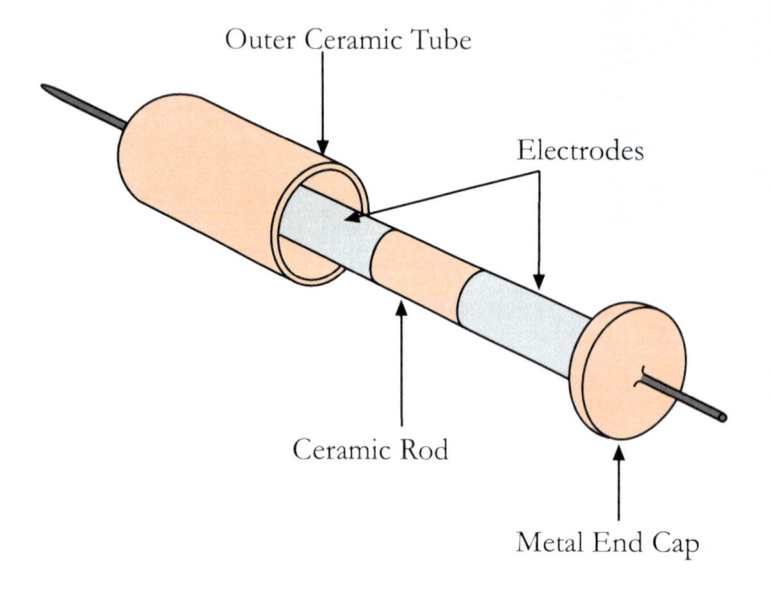

Figure 9.9 – Typical Ceramic Capacitor

Ceramic capacitors are especially suitable for surface mounting on modern PCBs due to their mechanical integrity, heat resistance and ability to make them in very small packages. However, they do suffer from low breakdown voltage and are usually found only in small values, typically 0.5pF to 0.01μF, in aviation applications.

Film Capacitors

Film capacitors include a variety of polymers, such as Polyester, Polycarbonate, Teflon, Polypropylene, and Polystyrene. However, traditionally film capacitors were only available in modest sizes, typically <10μF, but in recent years, with modern construction methods they can produce much larger values, even up to thousands of μF.

Film capacitors come in two broad categories, film-foil, and metalised film. Film-foil capacitors are made of alternating layers of plastic film dielectric and metal foil; while metalised film capacitors have the metal vacuum-deposited directly onto the film, figure 9.10.

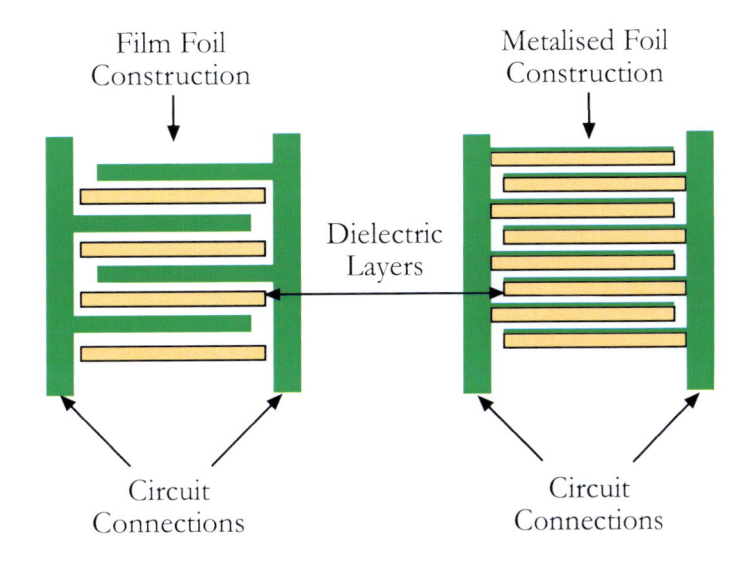

Figure 9.10 – Typical Film Capacitors

Capacitors of film-foil construction are better at handling high current, while metalised film ones are much better at '*self-healing*' (repair of broken down dielectric). To get the best of both worlds, manufacturer now produce various hybrid types.

On the plus side, film capacitors have reasonably well behaved electrical properties and have good characteristics regarding temperature, dissipation and dielectric absorption. They also have low leakage, and so good breakdown voltages, and low aging qualities.

On the minus side, they have low dielectric constants (K), which means that film capacitors are physically large for their capacitance values. While some film dielectrics are suitable for surface mounting, most cannot stand the heat of soldering and even polyester, the toughest of the traditional films, struggles to be good enough.

CHAPTER NINE
CAPACITANCE & CAPACITORS

Electrolytic Capacitors

The term '*Electrolytic*' means any capacitor that requires a **conductive** layer between the dielectric and one electrode. Electrolytic capacitors are best used when the circuit needs a lot of capacitance in a small space and at a reasonable price; e.g. power supply filtering, energy storage, etc. They are available in sizes far beyond that of other capacitors. Aluminium electrolytic capacitors are available from 0.1μF to several F, although it would be unusual to use a 0.1μF electrolytic capacitor. Tantalum electrolytic capacitors are available from 0.1μF to a few thousand μF.

In the original electrolytic capacitor, the layer was an actual electrolyte, i.e. a conductive salt in a solvent. Today, electrolytic capacitors are created by growing an oxide film, the dielectric, on a metal, the anode, by electrochemical means. The films are very thin with high 'K' values, typically 10 to 25, which makes for a lot of capacitance in a small package.

The majority, but not all, of electrolytic capacitors pass current much better in one direction than the other as the dielectric can usually only maintain charge if the capacitor has a component of dc applied to it in the correct sense. The two metals presently used are aluminium and tantalum, however, in theory, a number of other metals could be used.

Modern construction means the metal container of the capacitor is in direct contact with the electrolyte, which then acts as the negative plate. When a dc voltage is applied, a chemical reaction causes a thin oxide to form on the metal foil, which is connected to the positive terminal and becomes the positive plate with the oxide being the dielectric, figure 9.11.

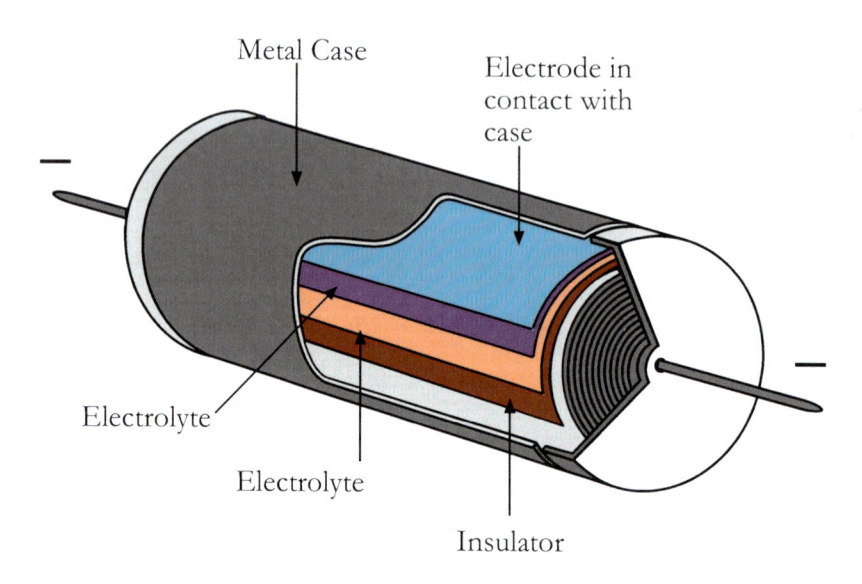

Figure 9.11 – Typical Electrolytic Capacitor Construction

The majority of electrolytic types of capacitors are **Polarized**, that is the voltage applied to the capacitor terminals must be of the correct polarity as an incorrect polarization will break down the insulating oxide layer and permanent damage may result. In extreme cases they are liable to explode.

Variable Capacitors

There are many occasions in electronics where the fixed capacitor is of no use and we need to vary the circuit's capacitance. Variable capacitors allows for a range of capacitance and are designed so that it can be changed using mechanical means, such as by adjusting a screw or turning a shaft, figure 9.12.

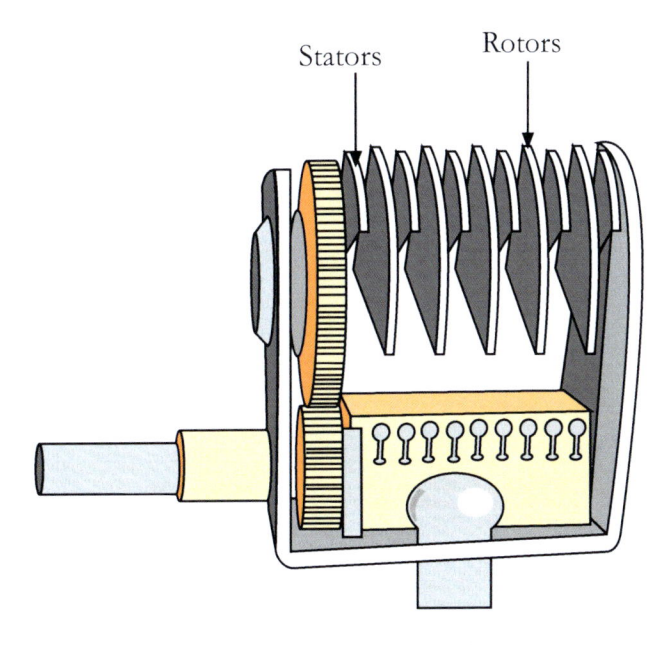

Figure 9.12 – Typical Variable Capacitor

The capacitor illustrated in figure 9.12 has two sets of plates; one set is called the *rotors* and the other the *stators*. The rotors are usually connected to an external knob and can move as it is turned. The two sets of plates are close together, but not touching and air is usually used as the dielectric. As the knob is rotated, the sets of plates become more or less meshed, increasing or decreasing the distance between the plates. As the plates become more meshed, capacitance increases; conversely, as the plates become less meshed, capacitance decreases.

As shown in figure 9.12, capacitors usually have several plates that make up its construction. This is true for variable as well as fixed value capacitors.

Variable Capacitor Symbol   Trimmer Capacitor Symbol

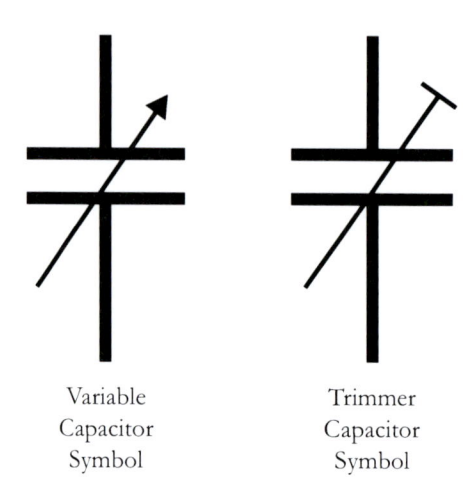

Figure 9.13 – Variable Capacitor Symbols

As well as the continuously variable type capacitors, preset types are also available known as Trimmers. These are generally small devices that can be adjusted or "pre-set" to a particular capacitance with the aid of a screwdriver They are available in very small capacitances of 100pF or less and are non-polarized.

As we have seen there are several methods used for manufacturing capacitors and the construction detail and sealing methods employed in finishing the product can have a significant impact on its performance and reliability. The majority of capacitors are constructed so that air and moisture are sealed out to prevent degradation or contamination of the dielectric and corrosion of the metal films. The best grade capacitors used in military and space applications may be hermetically sealed in metal and glass. Lesser grades may be moulded in plastic, dipped in an epoxy resin, or inserted in a plastic case and sealed with epoxy or urethane. In practice, the moisture absorption property of the dielectric dictates the method of encapsulation.

With modern microelectronics, **Service Mounted Devices (SMD)** are replacing the conventional '*through the hole*' construction of PCBs. However, this introduces new problems of miniaturisation as components get smaller and smaller. Some capacitor types, e.g. ceramics, are compatible with SMD but others like polystyrene and polypropylene, cannot stand up to the heat of soldering and so are rarely used for surface mounting.

Identification of Capacitors

Capacitors, like all electrical and electronic components, require identifying marks to indicate its value, working voltage, tolerances, etc. This can be in the form of colour coding or alphanumeric ident.

Capacitor Colour Coding

The capacitor colour coding is similar to resistance colour coding. Figure 9.14 shows a typical colour coded capacitor.

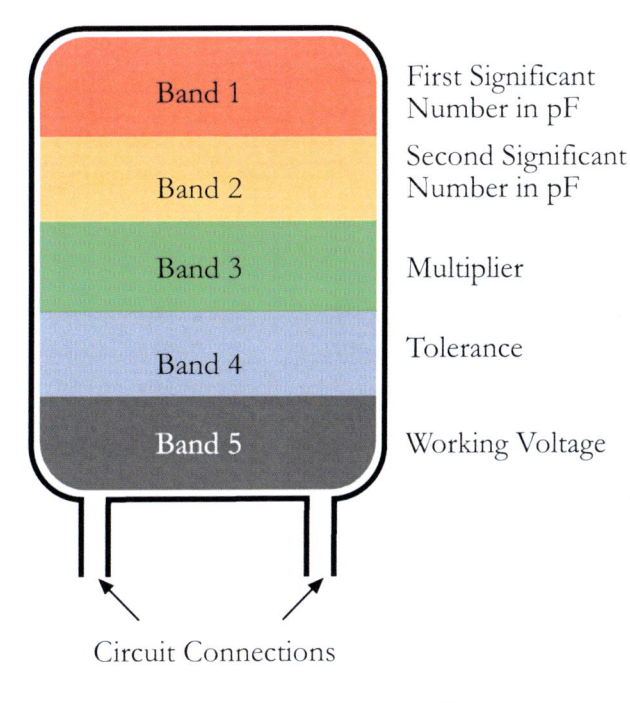

Band 1 — First Significant Number in pF
Band 2 — Second Significant Number in pF
Band 3 — Multiplier
Band 4 — Tolerance
Band 5 — Working Voltage

Circuit Connections

Figure 9.14 – Colour banded Capacitor

As shown in figure 9.14, several different colours band the capacitor; these are listed with their significances in table 9.2.

Colour	Significant Figure	Decimal Multiplier	Tolerance >10pF	Tolerance <10pF	Temperature Coefficient
Black	0	1	±20%	±2pF	0
Brown	1	10	±1%	-	-30
Red	2	100	±2%	-	-80
Orange	3	1000	-	-	-150
Yellow	4	-	-	-	-220
Green	5	-	±5%	±0.5pF	-330
Blue	6	-	-	-	-470
Violet	7	-	-	-	-750
Grey	8	0.001	-	±0.25pF	30
White	9	0.1	±10%	±1pF	500

Table 9.2 – Capacitor Colour Coding

CHAPTER NINE
CAPACITANCE & CAPACITORS

As shown in table 9.2, the 1st and 2nd significant figures as well as the multiplier, follow the standard colour code for capacitors. The working voltage and tolerance colour code follows that in table 9.3.

Working Voltage	Tolerances
Brown - 100V	Black ±20%
Red - 250V	White ±10%
Yellow - 400V	Green ±5%

Table 9.3 – Working Voltage Values and Tolerances

Some capacitor manufacturers use a numeric coding as illustrated in figure 9.15.

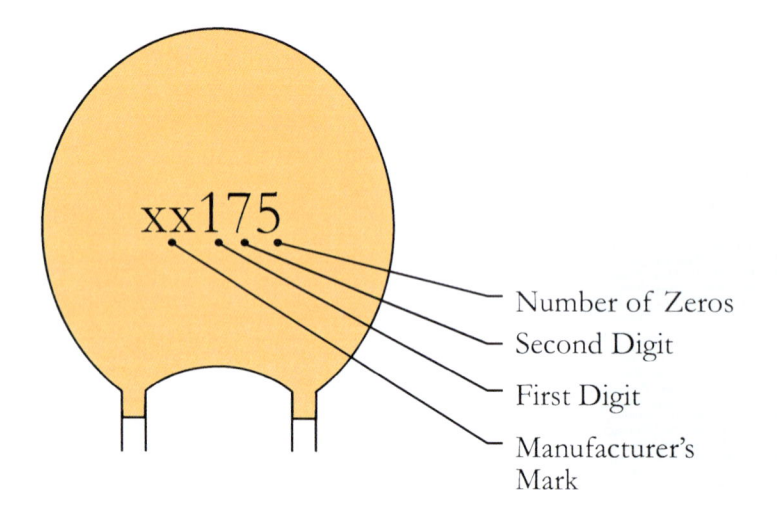

Figure 9.15 – Capacitor Numeric Coding

The value of some capacitors like aluminium and glass are usually written on the component, figure 9.16. Other information, e.g. operating voltage, is also stamped on the component.

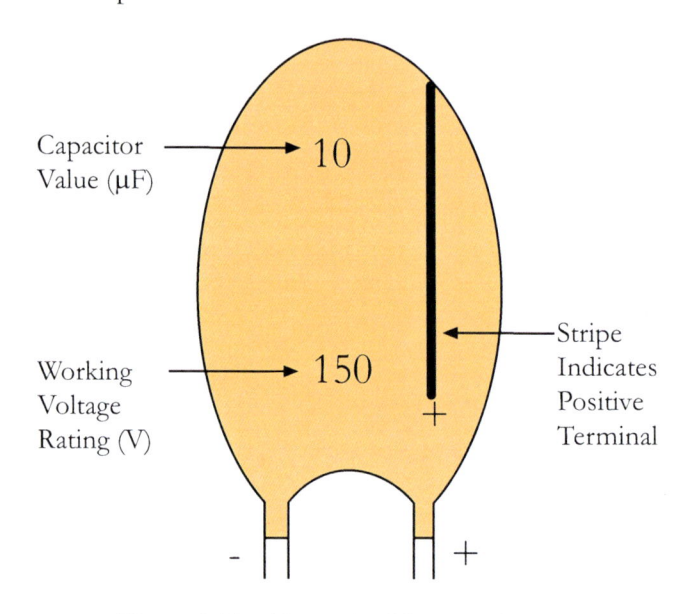

Figure 9.16 – Alternative Numeric Coding

When handling capacitors, especially those that rely on a numeric or stamped coding, it is important that the capacitor value, working voltage, etc is read accurately as they can look very similar as illustrated in figure 9.17.

Figure 9.17 – Typical Capacitors

Charging of a Capacitor

When a capacitor is connected in a circuit across a voltage source such as a battery, the voltage's polarity forces electrons onto the surface of one plate and pull electrons off the surface of the other plate. This results in a potential difference between them. As the dielectric between the plates is an insulator, figure 9.2, current cannot flow through it. However, a capacitor has a finite amount of capacity to store charges and when it reaches maximum it is *fully charged*.

The following sequence of diagrams, illustrates the charging of a capacitor step by step. Figure 9.18 shows a circuit containing a conductor connecting a battery, an open switch, and a capacitor. The capacitor in this circuit is not charged, as there is no potential difference between the plates.

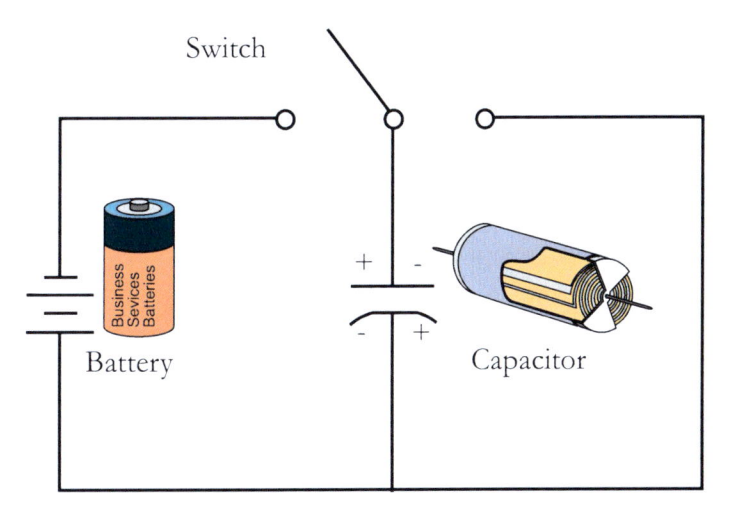

Figure 9.18 – Circuit with Open Switch

CHAPTER NINE
CAPACITANCE & CAPACITORS

When the switch is closed, as in figure 9.19, there is a momentary surge of current through the conductor to and from the plates of the capacitor. However, when the current reaches the negative plate of the capacitor, it is stopped by the dielectric.

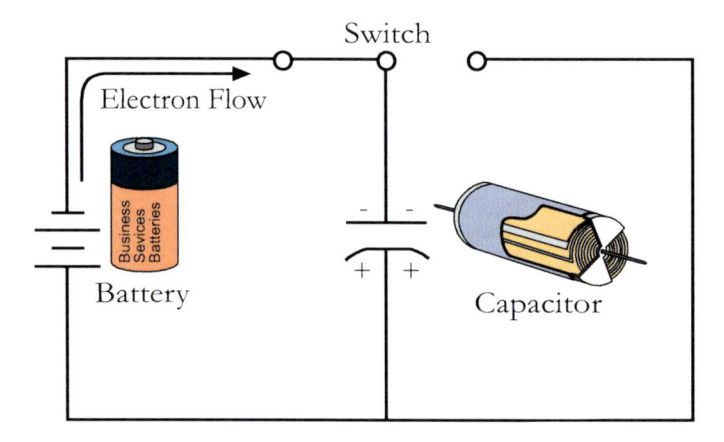

Figure 9.19 – Initial Current Surge

The surge of electric current to the capacitor induces a counter electromotive force in the conductor and the plates, which is called **reactance**. When reactance has reached a level equal to the voltage of the battery, the capacitor is fully charged and there is no further flow of current. In this state, i.e. when the capacitor is fully charged, the switch may be opened and it will retain its charge as in figure 9.20.

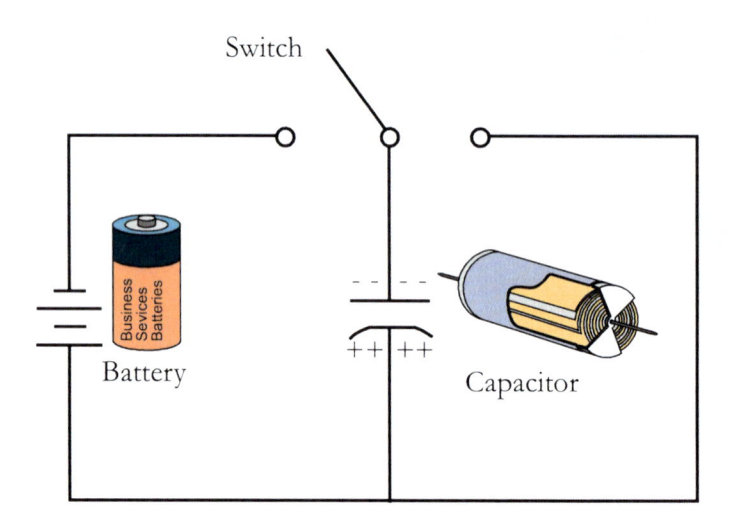

Figure 9.20 – Steady State Charge

Now there is a difference of charges on the plates, there is a source of potential energy in the capacitor; the energy stored being that which was required to charge the capacitor in the first place.

The lines of force between the plates of the capacitor, as shown in figures 9.19 and 9.20, can be said to represent an electric force field because of the unequal charges, positive and negative, on the inside surfaces of the plates. Current cannot flow through the *electrostatic* field because of the dielectric insulator. In other words, the difference in potential between the plates induces within the dielectric an *electrostatic field* that retains the charge.

Discharging of a Capacitor

The charged capacitor shown in figure 9.20 above is now a source of potential energy that is now available for its intended electronic application. If the switch is closed in the opposite direction, as in figure 9.21, current will immediately begin to flow through from the negative plate to the positive plate; i.e. the capacitor is discharging.

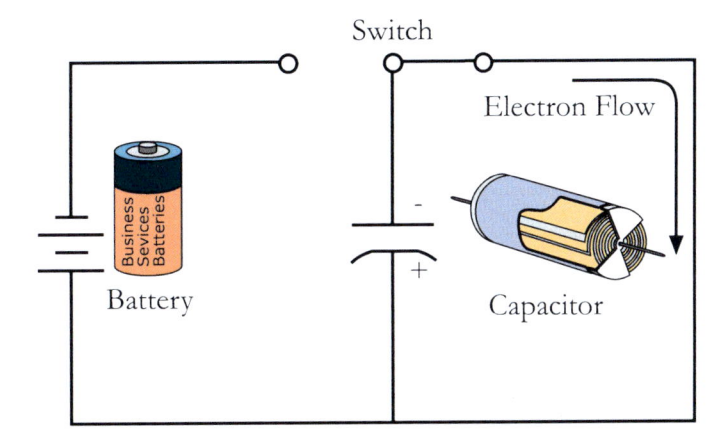

Figure 9.21 – Capacitor Discharging

As the battery is not in the circuit now, the charged capacitor is the source of voltage for the current flow. The current will cease flowing when the charges of the two plates are again equalised and the capacitor is completely discharged.

Resistive-Capacitive Series Circuit Time Constant

As a capacitor becomes charged in a dc circuit such as in figure 9.24, the current flow decreases as the voltage developed by the capacitor increases over time and *opposes* the source voltage. Therefore, the rate of charge of a capacitor reduces over time. The amount of time taken to charge and discharge a capacitor is a very important factor in the design of electronic circuits. Resistor/Capacitor combinations are often used to control the rate of charge and discharge of an intended application. This is represented in figure 9.22

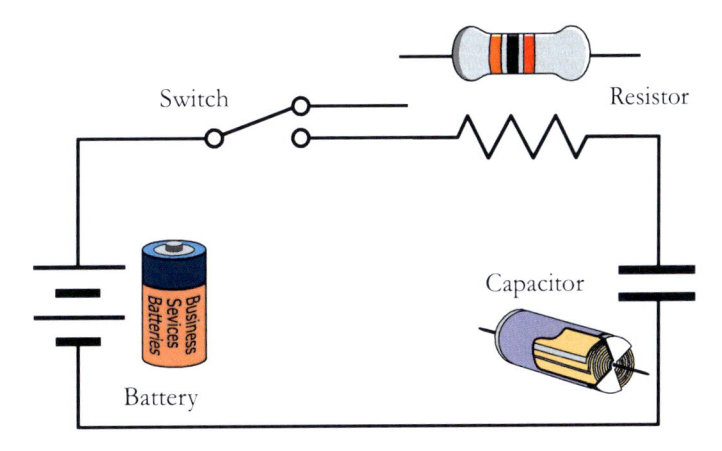

Figure 9.22 – Resistive-Capacitive Series Circuit

CHAPTER NINE
CAPACITANCE & CAPACITORS

Resistance *directly* affects the time required to charge a capacitor; in other words, as the circuit resistance increases, it takes more time to charge a capacitor. The amount of time for the capacitor in a resistor-capacitor (RC) circuit to become fully charged depends on the values of the capacitor and resistor. However, the rate of change is not constant nor is it linear. The graph in figure 9.23 illustrates this relationship.

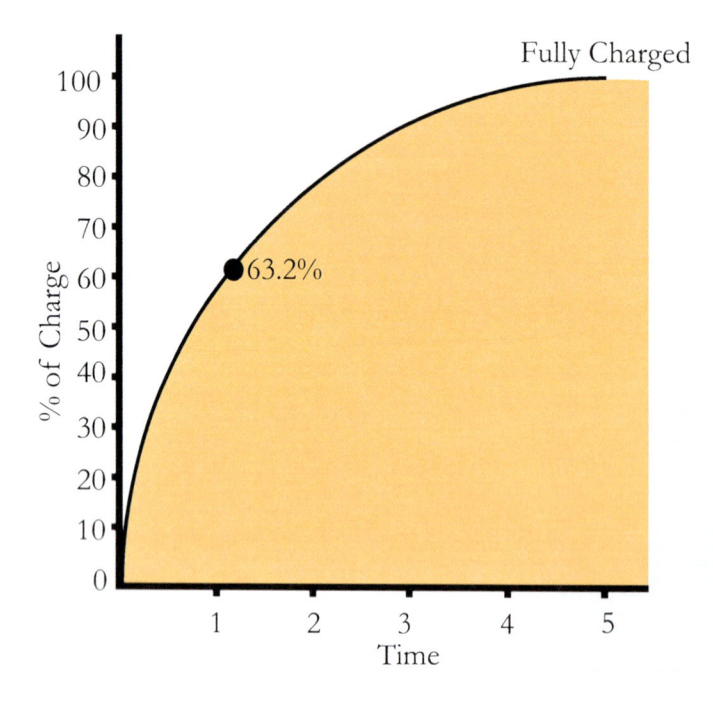

Figure 9.23 – CR Time Constant Graph

The graph in figure 9.23 shows the rate of charge of a capacitor in a RC circuit. Note that the rate of charge greatly decreases over time and that the latter stages of its charging time is many times longer than when the voltage is initially applied.

A capacitor reaches **63.2%** of its charge in approximately one fifth of the time it takes to become fully charged. Because of this, capacitors in actual applications are not necessarily fully charged. The time taken for a capacitor to charge to 63.2% of its full capacity is known as its resistive-capacitive, i.e. ***RC, Time Constant***.

The RC time constant of a circuit is calculated by using the formula:

$$t = C \times R$$

Where:

> **t** is time in seconds
>
> **C** is capacitance in farads
>
> **R** is resistance in ohms

180

Example:

If a circuit has a resistance of 20kΩ and a capacitor of 15μF, what is it RC time constant?

Answer:

$$t = 20 \times 10^3 \times 15 \times 10^{-6} = 0.3 \text{ seconds}$$

In a RC circuit, the capacitor becomes fully charged in approximately **5CR seconds**.

Circuit Discharging

As I am sure you would expect, a capacitor does not discharge in a linear manner either. When a capacitor is discharged through a RC circuit, the voltage does not immediately fall to zero but decreases exponentially as shown in figure 9.24.

Figure 9.24 – RC Discharge Graph

In this case, the voltage drops to approximately 36.8% from its initial value in the same CR time constant and reaches zero in approximately **5CR**.

Capacitors in a Parallel Circuits

Capacitance in a circuit, as opposed to resistance, can be *increased* by connecting capacitors in parallel as illustrated in figure 9.25.

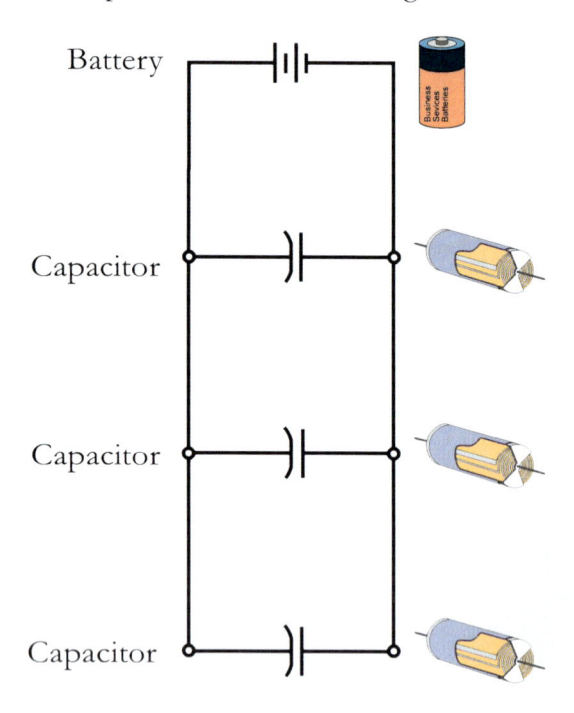

Figure 9.25 – Capacitors in Parallel

Obviously, this is opposite to what you would expect if talking about resistors. However, from the paragraphs already gone through above, you already know that the capacitance of a capacitor, and therefore a capacitive circuit, can be amplified by increasing the size of its plates. Connecting two or more capacitors in parallel in effect increases plate size and this makes it possible to store more charge and so create greater capacitance.

To determine the total capacitance of several parallel capacitors, we simply add their individual values together as in the following formula:

$$C_T = C_1 + C_2 + C_3 + ...C_n$$

Capacitors in Series Circuits

Capacitance in a circuit can be *decreased* by connecting capacitors in series as illustrated in figure 9.26.

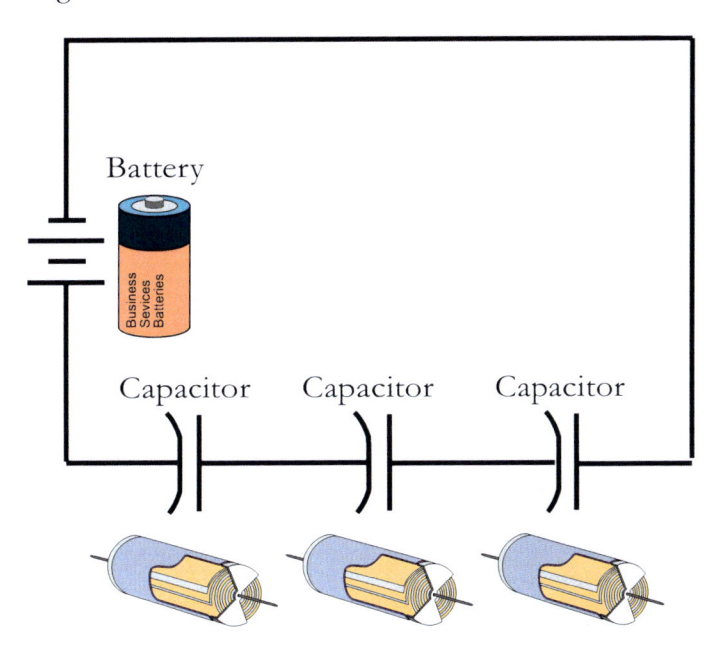

Figure 9.26 – Capacitors in Series

From our earlier discussions, you know that placing the plates further apart *decreases* the capacitance. Connecting two or more capacitors in series effectively increases the distance between the plates and in addition, the thickness of the dielectric, thereby *decreasing* the amount of capacitance.

To determine the total capacitance of several series capacitors, we simply apply the following formula:

$$\frac{1}{C_T} = \frac{1}{C_1} + \frac{1}{C_2} + \frac{1}{C_3} + \ldots \ldots \frac{1}{C_n}$$

Testing Capacitors

Due to their characteristics, it is not easy to test capacitors although some *Digital Multi-Meters (DMM)* have modes for capacitor testing. These work fairly well in determining the approximate µF rating. However, for most applications, they do not test at anywhere near the normal working voltage or test for leakage. However, an analogue meter and DMM without capacitance ranges can make certain types of tests.

For small capacitors, e.g. 0.01µF or less, about all you can really test for is shorts or leakage. However, on an analogue multimeter on the high ohms scale you may see a momentary deflection when you touch the probes to the capacitor or reverse them, but a DMM may not provide any indication at all. Any capacitor that measures a few ohms or less is bad as most should test infinite even on the highest resistance range.

For electrolytic capacitors in the 1µF range or above, you should be able to see the capacitor charge when you use a high ohms scale with the proper polarity; i.e. the resistance will increase until it goes to almost infinity. If the capacitor is shorted, then it will never charge; if it is open, the resistance will be infinite immediately and will not change. However, do check the polarity of the probes on your meter.

With any capacitor under test, if the resistance never goes very high, the capacitor is leaky.

Some Example Capacitor Applications

Modern computers make use of capacitors for memory storage. In most cases, the main memory of a computer is high-speed **Random Access Memory (RAM)**. To illustrate the point, there are two types of main memory with RAM circuits, **Static Random Access Memory (SRAM)** and **Dynamic Random Access Memory (DRAM)**. A single memory chip is usually made up of several million-memory cells. In a SRAM chip, each memory cell consists of a resistor circuit flip-flop, storing the binary bits at logic '1' or logic '0'. In a DRAM chip, each memory cell consists of a capacitor rather than a resistor circuit flip-flop. When a capacitor is electrically charged, it is said to store a logic '1' and when discharged a logic '0'.

Condenser Microphones

A microphone converts sound waves into an electric signal and all microphones have a diaphragm that vibrates as sound waves hit it. The vibrating diaphragm in turn causes an electrical component to create an output flow of current at a frequency proportional to the sound waves. A condenser microphone uses a capacitor to achieve this.

In a condenser microphone the diaphragm is the negatively charged plate of a charged capacitor. When a sound wave compresses the diaphragm, the diaphragm is moved closer to the positive plate. Decreasing the distance between the plates increases the electrostatic attraction between them. This results in a flow of current to the negative plate. As the diaphragm moves out in response to sound waves, the diaphragm moves further from the positive plate. Increasing the distance between the plates decreases the electrostatic attraction between them. This results in a flow of current back to the positive plate. These alternating flows of current provide weak electronic signals, which are usually fed to an amplifier before transmission.

Revision

Capacitance & Capacitors

Questions

1. **A capacitor presents an extremely high resistance to:**

    a) Direct current (DC)

    b) Alternating Current (AC)

    c) AC & DC

2. **Capacitance is the property of an electrical system or circuit of conductors and insulators that enables it to store:**

    a) Electrical Charge

    b) Current

    c) Resistance

3. **A Capacitor's dielectric is:**

    a) A conductor

    b) An insulator

    c) A semi-conductor

4. Types of dielectric materials include air, paper, ceramic and

   a) Mica
   b) Copper
   c) Ferrite

5. When the Dielectric Constant (K) increases, it means a material's ability to support an electrostatic charge:

   a) Decreases
   b) Remains unchanged
   c) Increases

6. If Q, the quantity of stored electrical charge in coulombs, on a capacitor is 40 mC and 500 V is the difference in potential in volts across the capacitor, what is its capacitance?

   a) 80μF
   b) 80ρF
   c) 12.5 μF

7. As the area of the plates *increase*, capacitance

   a) Decreases
   b) Increases
   c) Remains the same

8. As the distance between the plates *increases*, capacitance

   a) Decreases
   b) Increases
   c) Remains the same

9. If ε is 8 × 10⁻¹², K is 10, A is 0.2 m² and the distance between the plates is 4 mm, what is its capacitance?

   a) 6 ρF

   b) 4 ρF

   c) 4000 ρF

10. If the maximum voltage in a particular circuit is 500 V dc, the capacitor must be rated with a working voltage of at least:

    a) 750 V

    b) 1,000 V

    c) 500 V

11. Electrolytic capacitors are:

    a) Polarity sensitive

    b) Are always of paper construction

    c) The most selling type of capacitor

12. The figure below shows a simple capacitive circuit; If C1 = 10 μF and C2 = 15μF, what is their combined capacitance?

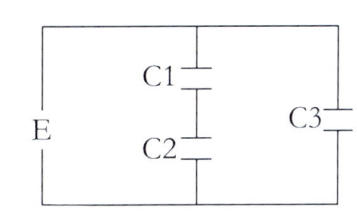

    a) 25 μF

    b) 6 μF

    c) 5 μF

13. If in the figure below, capacitors C1 = 10μF, C2 = 15μF and C3 = 10μF, what is the circuit's total capacitance?

    a) 20 μF

    b) 16 μF

    c) 35 μF

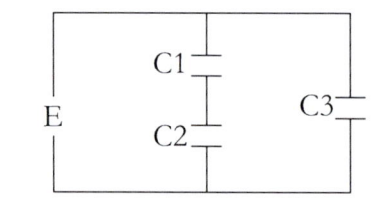

14. With a variable capacitor, to alter its capacitance usually only:

    a) The rotor moves

    b) The rotor and stator move

    c) The stator moves

15. If a capacitor has colour-coding bands of Yellow, Red, Red respectively, its capacitance value is:

    a) 42 pF

    b) 420 pF

    c) 4200 pF

16. If a circuit has a resistance of 40kΩ and a capacitor of 15μF, its RC time constant is:

    a) 0.6 Seconds

    b) 3 Seconds

    c) 0.3 Seconds

17. If a circuit has a resistance of 10kΩ and a capacitor of 15μF, the time taken for the capacitor to be fully charged is:

    a) 0.5 Seconds

    b) 0.15 Seconds

    c) 0.75 Seconds

18. If a circuit has a resistance of 30kΩ and a capacitor of 10μF, the time taken for the capacitor to be fully discharged is:

   a) 0.3 Seconds

   b) 1.5 Seconds

   c) 3.0 Seconds

Revision

Capacitance & Capacitors

Answers

1. A
2. A
3. B
4. A
5. C
6. A
7. B
8. A
9. B
10. A
11. A
12. B
13. B
14. A
15. C
16. A $40 \times 10^3 \times 15 \times 10^{-6} = 600 \times 10^{-3} = 0.6$
17. C $10 \times 10^3 \times 15 \times 10^{-6} = 150 \times 10^{-3} = 0.15 \times 5 = 0.75$
18. B $30 \times 10^3 \times 10 \times 10^{-6} = 300 \times 10^{-3} = 0.3 \times 5 = 1.5$

Magnetism

The Theory of Magnetism

Magnetism, what is it? In the context we are looking at, i.e. as part of electrical theory, magnetism is a set of phenomena associated with magnets and electric charge; both of which produce magnetic fields in the area surrounding them. As you probably already know, all matter is made up of *Atoms*, and an atom is the smallest particle that can usually be differentiated in chemical tests. In nature, *Hydrogen* is the simplest and lightest atom while those of *Uranium* are the heaviest.

All atoms consist of a central core called the *nucleus*. This possesses one or more *protons* that have a *positive* charge and is surrounded by a cloud of *electrons* that have a *negative* charge. The total negative charge of the electrons is exactly equal to the positive charge of the nucleus protons and so the atom as a whole is electrically neutral. Electricity and magnetism are associated with the properties of the electrons of an atom.

The magnetic properties of matter are largely determined by the behaviour of the negatively charged electrons that orbit the nuclei of atoms. The magnetic field of a single electron has two (2) components, one resulting from its spin about its own axis, and the other from its orbital motion about the nucleus. Both kinds of motion may be considered as tiny circular currents, ie moving charges, so connecting electrical and magnetic effects at the fundamental level.

The ancient Greeks, originally those near the city of *Magnesia*, and the early Chinese knew about strange and rare stones, now thought to be large pieces of iron ore struck by lightning, with the power to attract some metals. The Greeks and Chinese then went on to discover the magnetic properties of the mineral *Magnetite*, commonly called *Lodestone*, and magnetism was first studied seriously as far back as the 13th century. During continuous experimentation, they found that a steel needle stroked with such a lodestone became magnetic as well, and the Chinese also found that such a needle, when freely suspended, pointed between North and South.

The *Magnetic Compass*, as it was latterly called, soon spread to Europe and Columbus used it when crossing the Atlantic Ocean. He noted that the needle deviated slightly from exact geographic north, as indicated when comparing its position with the stars, and that the deviation also changed during the voyage.

CHAPTER TEN
MAGNETISM

Around 1600 *William Gilbert*, physician to Queen Elizabeth I, proposed an explanation. He stated that the Earth itself was a giant magnet, with its magnetic poles some distance away from its geographic ones that defined the axis around which the Earth turned.

On Earth, it requires a sensitive needle to detect magnetic forces, and they are usually much weaker as altitude increases. However, beyond our dense atmosphere, these magnetic forces have a much bigger role as they contain a mixture of electrically charged particles, and electric and magnetic phenomena determines its structure rather than gravity. This area is called the Earth's *Magnetosphere*, figure 10.1.

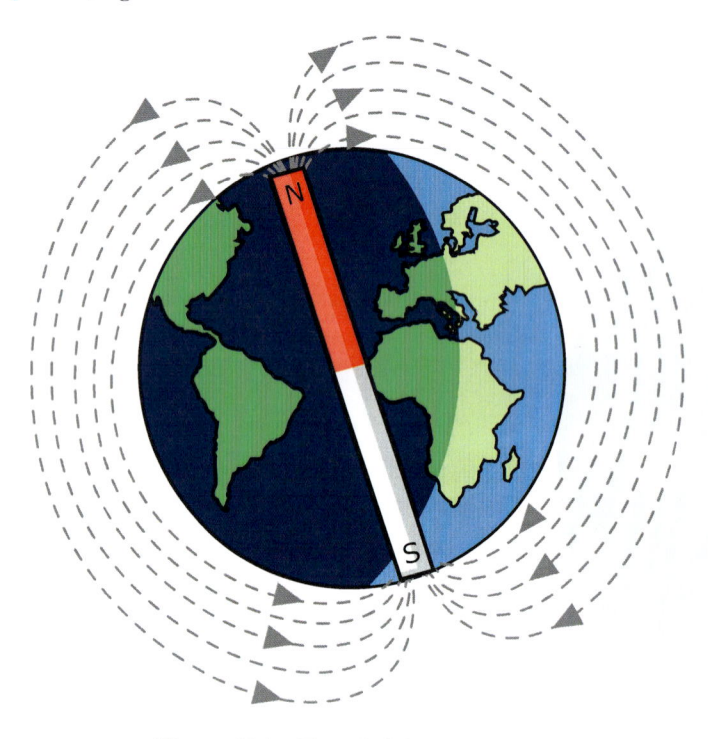

Figure 10.1 – Earth's Magnetosphere

However, only a few of the magnetic phenomena observed on the ground come from the magnetosphere, but it does manifests itself as the *Polar Aurora* or '*Northern Lights*' appearing in the night skies of Alaska and Norway.

It was in the 19th century that the connection between electricity and magnetism was first discovered. Experimenters found that charged particles are surrounded by electric fields that create magnetic fields when they move, or spin. At the same time, it was also discovered that an electric current flowing through a conductor produces a magnetic field surrounding the current, and the strength of the field was directly proportional to the amount of current flow.

Conversely, if the conductor was moved through a magnetic field, an electric current flowed in it. These are discussed in more detail on the following pages.

What is Magnetism?

As mentioned earlier, at the atomic level electric charge is a basic property of certain particles, which causes them to attract or repel each other, and there are two types, negative and positive. At the atomic level, *electrons* carry a *negative* charge and *protons* an equal quantity of *positive* charge. It is the attraction between protons and electrons that hold the latter in orbitals round the nucleus of each atom, and on a wider scale, it is a similar attraction that binds atoms and molecules together to form solids, gases, liquids etc, figure 10.2.

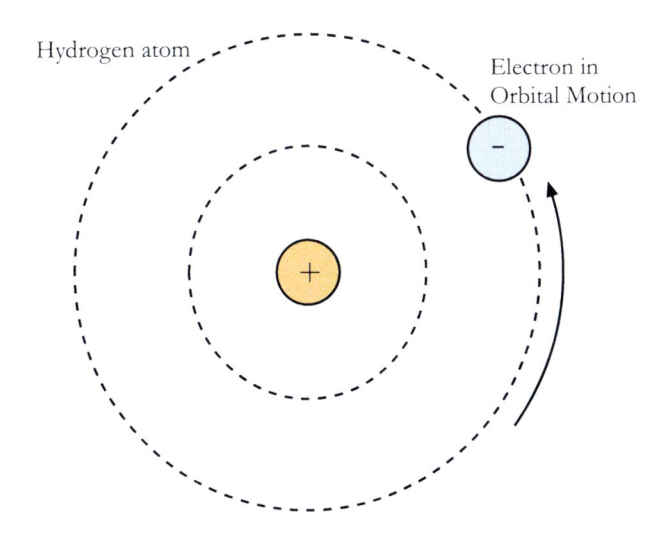

Figure 10.2 – Electron Motion

Magnetism is caused by '*eternal*' electric currents due to electron motion at microscopic scale in permanent magnets. However, looked at in this way then, why is not everything metal magnetic?

The answer is that *Ferromagnetic* materials, ie iron, cobalt, nickel, have electrons with '*unpaired*' spin directions and it is *these* spinning motions that give rise to electromagnetism in permanent magnets. However, not all iron, nickel, etc is magnetic so what is going on?

Magnetism is due to the orientation of the *magnetic domains* of metallic materials. For example, an iron bar made from liquid iron that solidifies naturally has domains that randomly orient so that the magnetism of the bar as a whole cancels, ① in figure 10.3. However, if the bar is put in a strong magnetic field during cooling, the domains align with the field and its collective effect is to cause permanent magnetism, ② in figure 10.3.

1. Domains randomly oriented
Not Magnetised

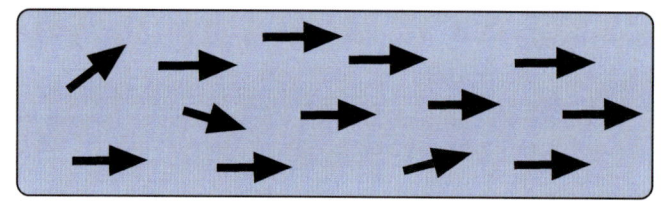

2. Domains prefer one direction and are greater size than non magnetised
Magnetised

Figure 10.3 – Non-Magnetised and Magnetised Iron Bar

The alignment of magnetic domains in a magnetic material can be destroyed by heat or shock and so they should be handled properly.

Once these effects were understood in the permanent magnet, the existence of magnetic poles and lines of flux were established and it was discovered that *like poles repel* and *unlike poles attract*. This can easily be illustrated by placing two (2) magnets in a test tube or cylinder with their like poles facing each other and then again with their unlike poles facing each other, figure 10.4.

Unlike Poles Attract

Like Poles Repel

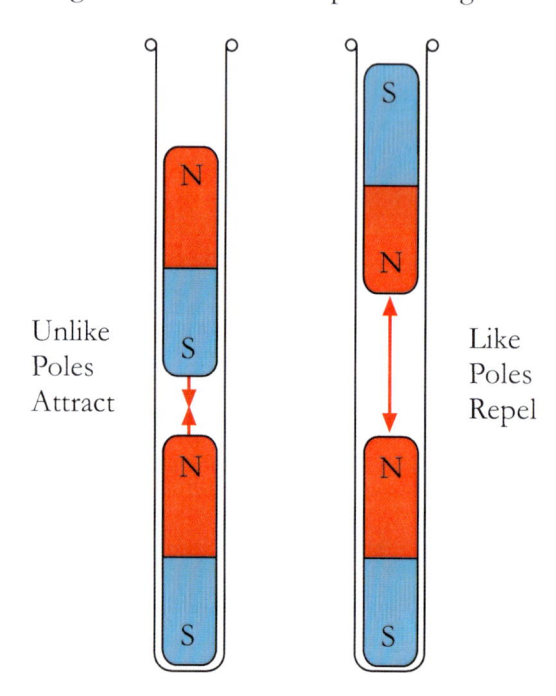

Figure 10.4 – Demonstration of Magnetic Pole Reaction

Properties of a Magnet

As demonstrated in figure 10.4, a magnetised bar can be seen as having its '*power*' concentrated at two ends. These are known as its poles; more commonly called its North (N) and South (S) poles. This is because if a bar magnet is suspended from its middle by string, its 'N' end tends to point North and its 'S' end south. This phenomena remains the same even when magnets break in pieces, figure 10.5.

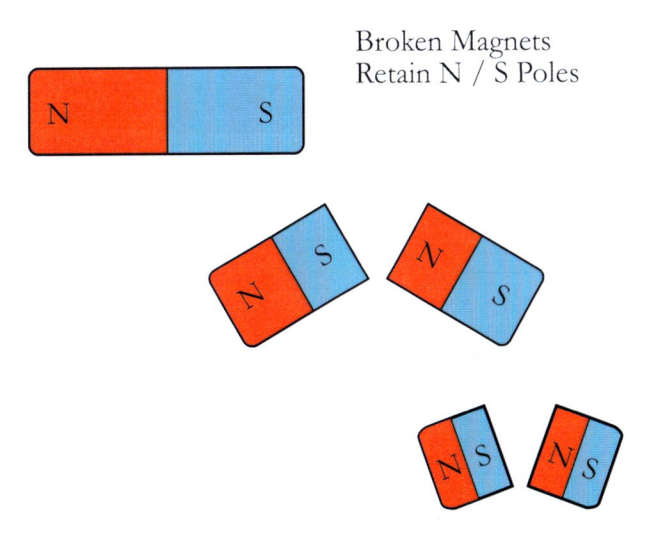

Figure 10.5 – Broken Magnets Retain N/S Arrangement

These days, magnets are made of steel for mechanical integrity, but their action remains the same.

Permanent magnets are so named as they always retain their magnetic properties. They do this as their magnetic fields at the individual atom level are aligned in one preferred direction, giving rise to a net *permanent* magnetic field.

Magnetic Field Lines

Michael Faraday is credited with discovering the fundamental properties of electricity and magnetism, and as we discovered in the previous chapter, an electric unit, the Farad, is named in his honour. He also proposed a method for visualising magnetic fields surrounding a magnet or electrical field, which is still widely used today, by introducing a compass needle near the magnet. He demonstrated that a compass needle, freely suspended in three dimensions, near a magnet or an electrical current could be used to trace in space the lines obtained when following the direction of the compass needle as it is moved around the magnet.

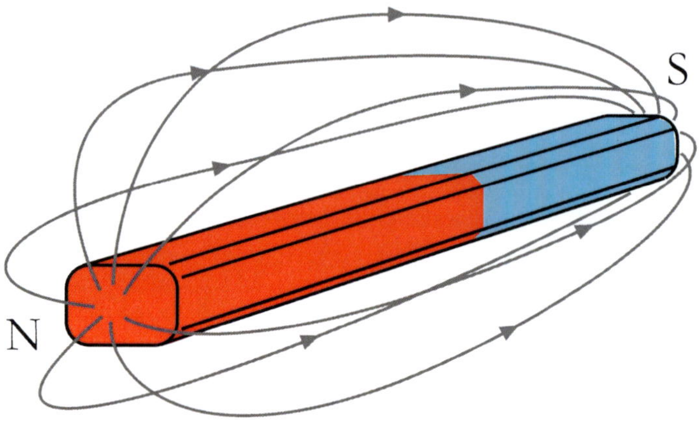

Figure 10.6 – Magnetic Field Lines

As you can see from figure 10.6, the shape and density of the chains allows a mental picture to form of the magnetic condition of the space or field surrounding the magnet.

Faraday called these *lines of force*, but the term *field lines* or *lines of magnetic flux* are now more commonly used. It should be noted though that these lines of magnetic flux are imaginary and have no physical existence but were introduced by Faraday as a means to visualise the distribution and density of a magnetic field. It is vitally important to realise that the magnetic flux occupies all of the space of the magnetic field.

The SI symbol for magnetic flux is Φ (phi) and its unit of measurement is the *Weber (Wb)*.

Magnetic Flux Density

Magnetic flux density, given the SI symbol 'B', is a measurement of the number of flux lines per square meter in a given circuit. This is calculated by the following formula:

$$B = \frac{\Phi}{A} \text{ Wb/m}^2$$

Note: The area should always be in square metres (m^2).

The unit of magnetic flux density (B) is the *Tesla (T)* and: *1 T = 1 Wb/m^2*

Magnetic Field Strength & Direction

When looking at figure 10.6, it is easy to see that the magnetic flux lines converge where the magnetic force is strong, and spread out where it is weak; i.e. they spread out from one pole and converge towards the other, and the magnetic force is strongest near the poles where they come together. The behaviour of flux lines in the Earth's magnetic field is very similar. By convention, the north pole of a magnet is the pole that aligns itself towards *geographic* north.

Magnetic Flux Line Characteristics

Although I have made the statement that magnetic flux lines do not have a physical existence, they do provide us with a convenient and useful method of explaining magnetic effects and allow us to make mathematical calculations regarding their strength. To understand this, we need to know the basic properties of magnetic flux lines and these are:

1. The direction of a line of magnetic flux, not influence by any other magnetic source, is that of a north-seeking pole of a compass needle placed at that point.

2. Lines of magnetic flux *never* intersect.

3. Each line of magnetic flux forms a closed loop north to south to north, figure 10.7.

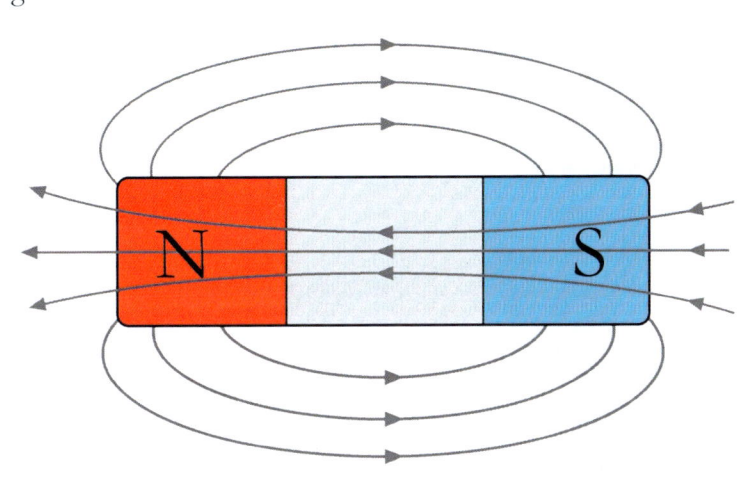

Figure 10.7 – Closed Loop Magnetic Lines of Flux

4. Lines of magnetic flux that are parallel and in the same direction repel each other.

5. Lines of magnetic flux are like stretched elastic bands, i.e. always trying to shorten as far as possible.

The phenomena of 4 and 5 above are illustrated in figure 10.8.

CHAPTER TEN
MAGNETISM

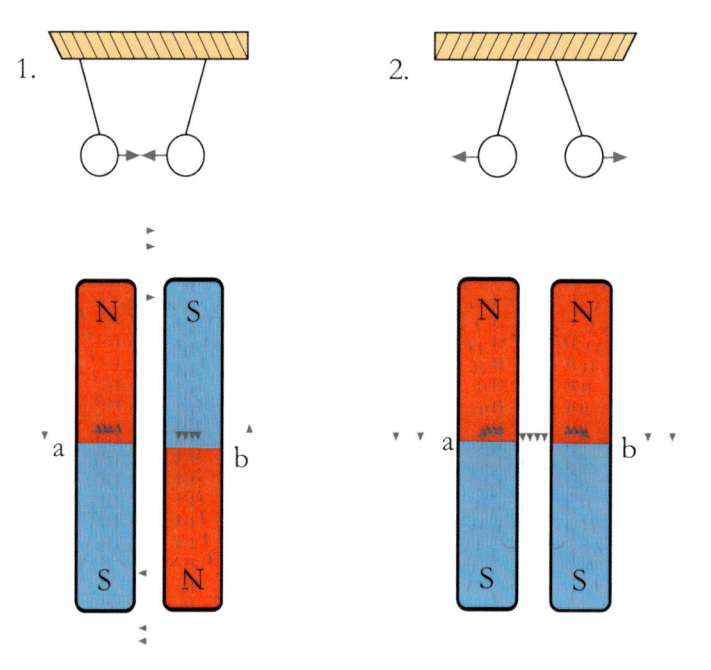

Figure 10.8 – Magnetic Reactions

In scenario ① of figure 10.8 with their magnetic poles in this arrangement, they are attracted to each other by the flux lines trying to take the shortest path. However, in scenario ②, the lines of flux of each magnet are parallel and in the same direction. These flux lines cannot intersect, and so exert a lateral force between them that drives the magnets apart.

Magnetic Fields due to Electric Current

Until 1821, only one kind of magnetism was known; ie that produced by iron magnets. Then a Danish scientist, *Hans Christian Oersted*, while demonstrating the flow of an electric current in a wire, noticed that the current caused a nearby compass needle to move, figure 10.9.

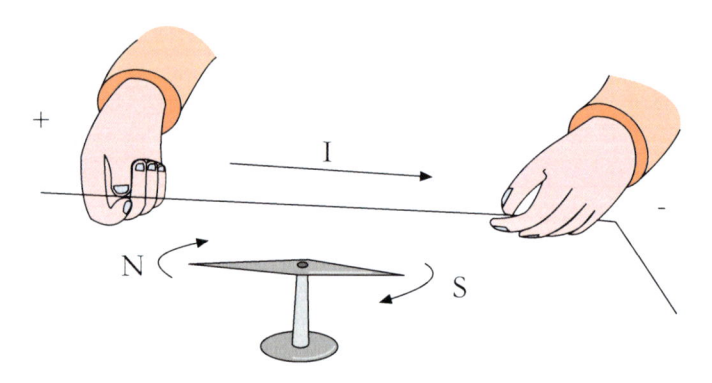

Figure 10.9 – What Oersted Saw

Experimentation by Oersted, found that when an electric current is fed through a conductor, a magnetic field is produced around the conductor.

This new phenomenon was further studied in France by *Andre-Marie Ampere*. He concluded that the nature of magnetism was quite different from what everyone had believed up until then. He guessed that each iron atom contained a circulating current, turning it into a small magnet and that in an iron magnet all these atomic magnets were lined up in the same direction, allowing their magnetic forces to combine. He went on to determine that magnetism was a *force between electric currents*, and that two parallel currents going in the same direction *attract*, and two going in opposite directions *repel,* Figure 10.10.

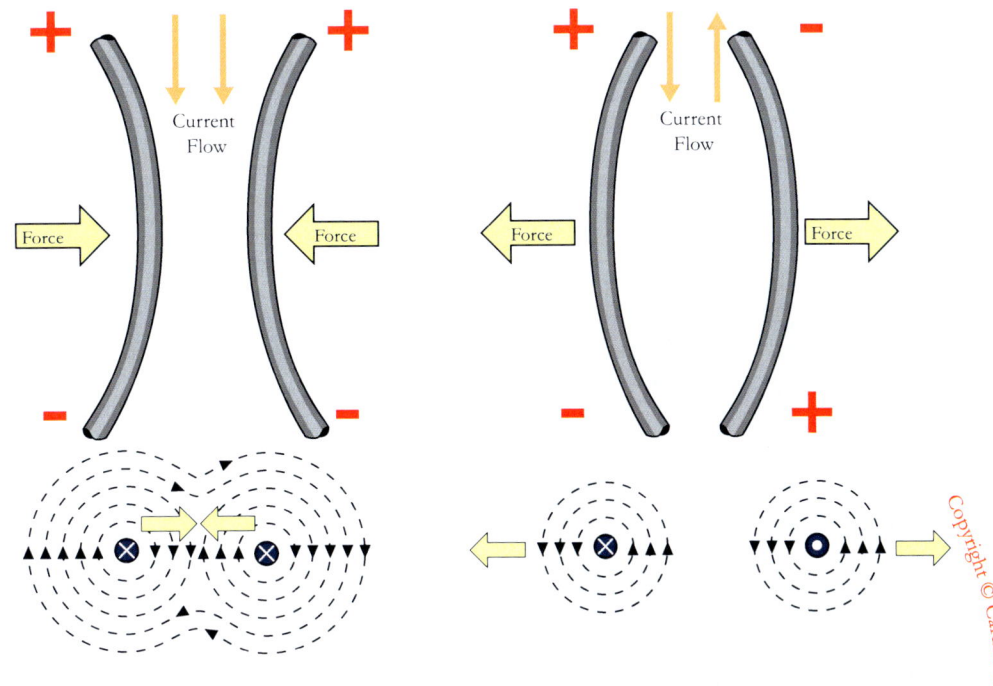

Figure 10.10 – Magnetic Force Between Electric Currents

As we can see, this replicates the example using bar magnets in figure 10.8.

Ampere also discovered that the force between two long straight parallel conductors with current flowing through them was *inversely proportional* to the distance between them and *proportional* to the strength of the current flowing in each.

The cross and dot in figure 10.10, are used to show the direction of conventional current flow. These are always likened to someone throwing a dart. The cross, rather like the flight on a dart, shows current flowing into the page away from you, whilst the dot, like the point on a dart, shows current flowing out of the page towards you.

To find the direction of the magnetic field made by an electrical current, we use a technique called the *right-hand gripping rule* or cork screw rule, figure 10.11.

CHAPTER TEN
MAGNETISM

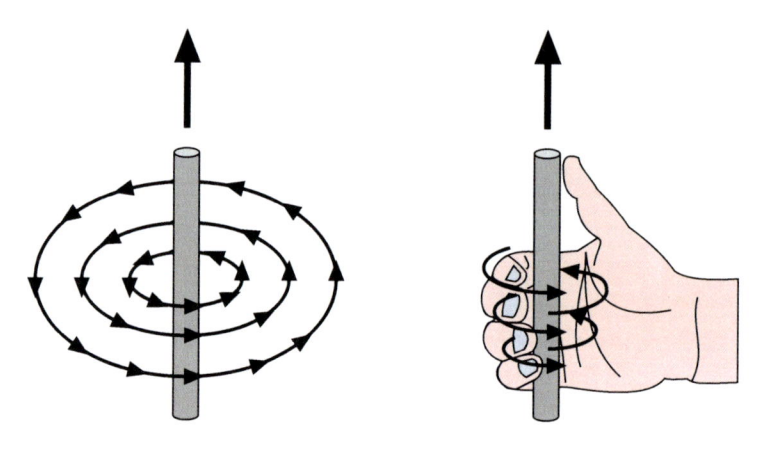

Figure 10.11 – Right Hand Gripping Rule

To find the direction of the magnetic field, place the right hand with the thumb parallel to the wire carrying the current, pointing in the direction of the electrical current using *conventional* current flow. When wrapped around the conductor, the fingers now indicate the direction of the magnetic field around the wire.

Figure 10.12 shows the magnetic field around current carrying conductors of different forms.

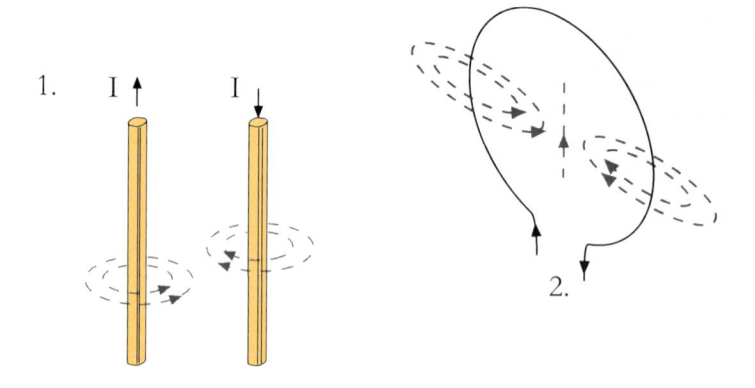

Figure 10.12 – Magnetic Fields Around Various Conductors

Illustration ① shows the magnetic field of straight conductors, as in figure 10.11, while ② shows the field around a single coil of wire. With ②, the resultant magnetic field is simply the result of addition of the field due to the turn in the wire conductor. Figure 10.13, represents a *solenoid*, which is a coil of wire with a length significantly larger than its diameter. In this case, the magnetic fields of each turn add up to give a resultant magnetic field with lines of flux entering one end of the coil and leaving at the other.

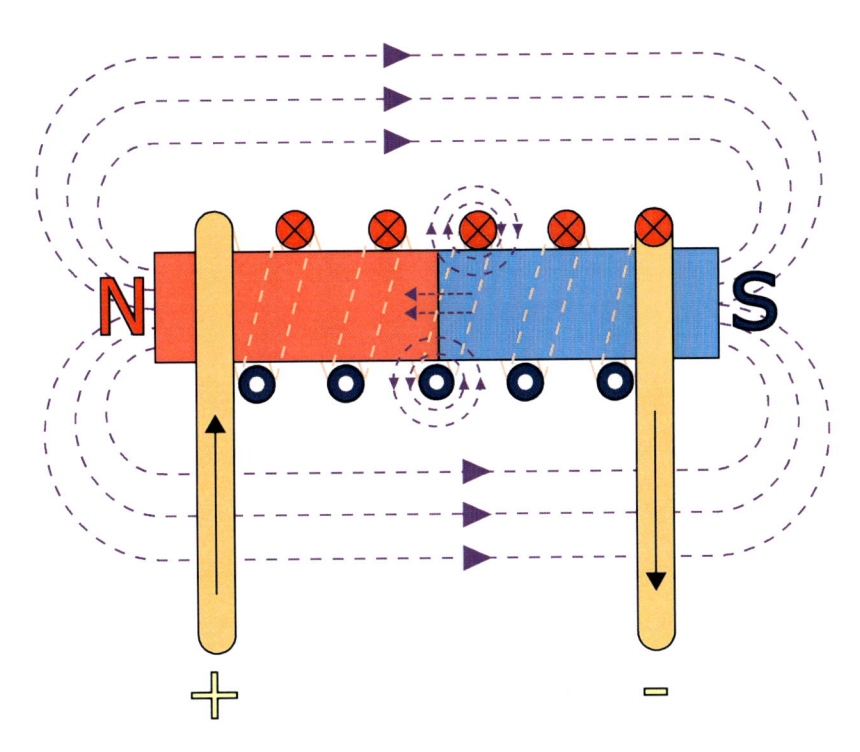

Figure 10.13 – A simple Electromagnet

For clarity, only the field for one loop is show, however it should be obvious that this resultant field is just like a permanent magnet and as a result, the solenoid can be thought of as having magnetic poles, ie it becomes polarised.

We can also use the right hand gripping rule to determine the poles of an electromagnet. Figure 10.14.

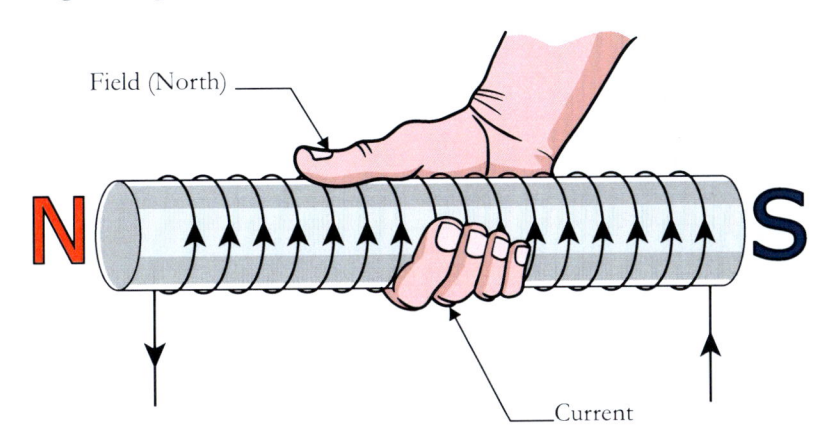

Figure 10.14 – Determining the Poles of an Electromagnet

Firstly you grip the coil with your right hand with the fingers pointing along the conductors in the direction of conventional current flow. If you now extend your thumb it indicates the end of the coil that has a magnetic north polarity.

Another method used to find out which end of the coil is acting as the North Pole and which the South Pole is by observing the direction of current flow at each end. This is called the End Rule or sometimes, the clock rule.

Looking at the end elevation of a solenoid and using conventional current flow, if the current flow is counter-clockwise, that end is a North Pole and if clockwise, a South Pole, figure 10.15.

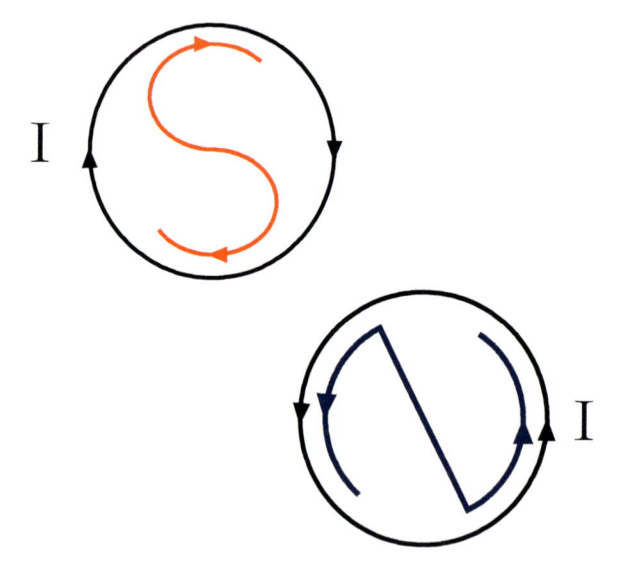

Figure 10.15 – Poles created by a Solenoid Coil

In many ways, there are many similarities between Electricity and Magnetism, for example:

- Magnetic poles give rise to magnetic forces - Electric charges give rise to electric forces

- Both Magnetic and Electric forces can be attractive and repulsive

However, there are also some distinct differences. As shown in figure 10.16, with permanent magnets, North and South magnetic poles always occur together, even when they break as already illustrated in figure 10.5 above. However, with electrical forces, positive and negative electric charges can exist independently.

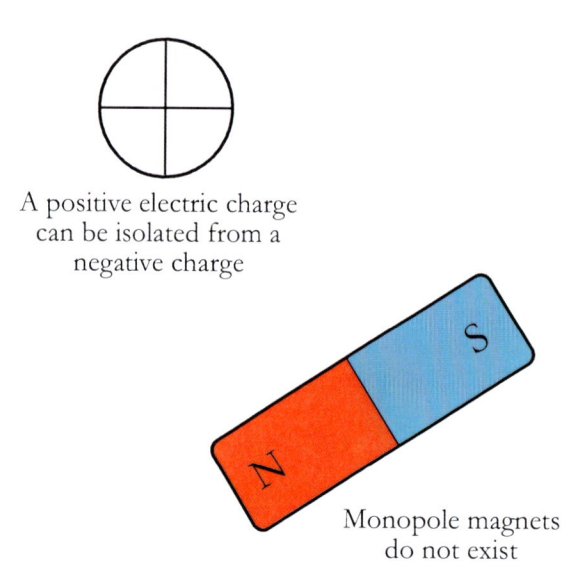

A positive electric charge can be isolated from a negative charge

Monopole magnets do not exist

Figure 10.16 – Differences Between Magnetic & Electric Force

Magneto Motive Force (MMF)

In the text above, we have established the characteristics of lines of force or flux. In an electric circuit, current is a result of *Electromotive Force (EMF)* and by using the same analogy, in a magnetic circuit, the magnetic flux is due to the existence of a *Magneto Motive Force (MMF)* due to current flowing through one or more turns and is given the symbol 'F'.

The value of MMF is directly proportional to the size of the current flowing through the conductor or coil and the number of turns in the coil. Therefore, if a current of I amperes is flowing in a coil of N turns, then the MMF is expressed as:

$$MMF(F) = NI \text{ Ampere Turns}$$

However, for the purposes of analysis, this is usually expressed, as simply *'Amperes (A)'* as the number of turns 'N' is dimensionless.

Magnetic Field Strength

If a current I is flowing in a coil of N turns then the mmf is the total current linked with the circuit, i.e. IN Amperes. If the magnetic circuit is homogeneous, i.e. uniform, with the same cross sectional area, the mmf per metre length is called the *Magnetic Field Strength*, represented by the symbol 'H'. Therefore, if the length of the circuit has a mean of 'l' metres, ie the flux path, the magnetic filed strength is calculated using the formula:

$$H = \frac{IN}{l} \text{ A/m}$$

This formula can be simplified further, as mmf = IN, to:

$$H = \frac{mmf}{l} \text{ A/m}$$

Permeability or Magnetic Constant

A material's permeability is a measure of its ability to accept or pass lines of magnetic flux. In essence, there are three (3) measures of permeability:

a. Permeability of free space (μ_0)

b. Absolute permeability of a material (μ)

c. Relative permeability (μ_r)

CHAPTER TEN
MAGNETISM

Permeability of Free Space (μ_0)

Permeability of free space is based on a vacuum and is defined as the magnetic flux density (B) divided by the magnetic field strength (H).

$$\mu_0 = \frac{B}{H} \text{ H/m}$$

The permeability of free space (μ_0) is a constant value of $4\pi \times 10^{-7}$ H/m. Which equals approximately 1.2µH/m

However, for those who would not rather just accept this. Let us take a real scenario as in figure 10.17.

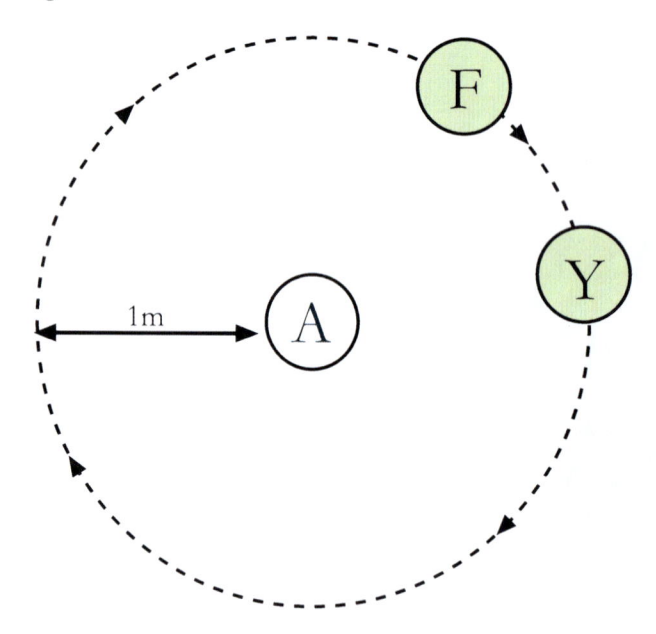

Figure 10.17 – Magnetic Field of a Conductor

Suppose conductor 'A' represents the cross sectional area of a long straight conductor, located in a vacuum, carrying a current of 1A towards the page; and in this scenario, assume that the current return path is far away from 'A' so as to have no affect of the magnetic field generated at 'A'. Then the lines of flux surrounding the conductor will be symmetrical concentric circles, one of which is represented by the dotted line. Now conductor 'A' and its return path form one turn, so the mmf acting on point 'F' is number of turns (N) multiplied by Current (1A) which equals 1A. We can calculate the length of the circuit as it is formed from a circle of radius 1m. Its length is $2\pi r = 2\pi$ metres. From the formula H=mmf/l, we can then see the magnetic field strength (H) at a radius of 1m is:

$$\frac{1}{2\pi} \text{ A/m}$$

If the flux density at point 'F' is B Teslas, then the force per metre on conductor 'Y' also carrying 1A is given by:

$$\text{Force per metre length} = BIL$$
$$= B \times 1m \times 1A$$
$$= B \text{ Newtons}$$

However, the definition of an ampere is given as a force of 2.7×10^{-7} Newtons (N) and therefore the flux density at 1m radius from a conductor carrying 1A is 2.7×10^{-7} T. Therefore the Permeability of free space is:

$$\frac{B}{H} = \frac{2 \times 10^{-7} T}{\frac{1}{2\pi} A/m} = 4\pi \times 10^{-7} \text{ H/m}$$

For all practical purposes, this value is the same for in a vacuum, in air or in any other non-magnetic material such as water, wood oil, etc.

The unit of permeability is the *Henry/metre*, abbreviated to *H/m*, which is discussed in more detail in the section on Inductors.

Absolute permeability of a material (μ)

A material's absolute permeability is defined as the magnetic flux density 'B' divided by the magnetic field strength 'H' in the material.

$$\mu = \frac{B}{H} \text{ H/m}$$

As with the permeability of free space, the unit of absolute permeability is the *Henry/metre*, abbreviated to *H/m*.

Relative permeability (μ_r)

A material's relative permeability is defined as the ratio of a material's absolute permeability to the permeability of free space, that is:

$$\mu_r = \frac{\mu}{\mu_0}$$

As this is a ratio, there are no units for relative permeability.

Rearranging the above formula for μ_r, gives the absolute permeability of a material as:

$$\mu = \mu_0 \mu_r$$

However, since:

$$\mu = \frac{B}{H} = \mu_0 \mu_r$$

then the flux density formula can be written as:

$$B = \mu_0 \mu_r H$$

The relative permeability of a vacuum is one (1), but for some forms of nickel-iron alloys it is as high as 100,000. However, the value of relative permeability for ferromagnetic materials is not a constant but varies considerably with magnetic field strength.

Reluctance

As its name implies, **Reluctance** is the opposition to the creation of magnetic lines of flux in a magnetic circuit. It is calculated by dividing the mmf in the circuit by the resulting magnetic flux and is given the SI symbol 'S'. To explain this further, let us look at a simple magnetic circuit, figure 10.18.

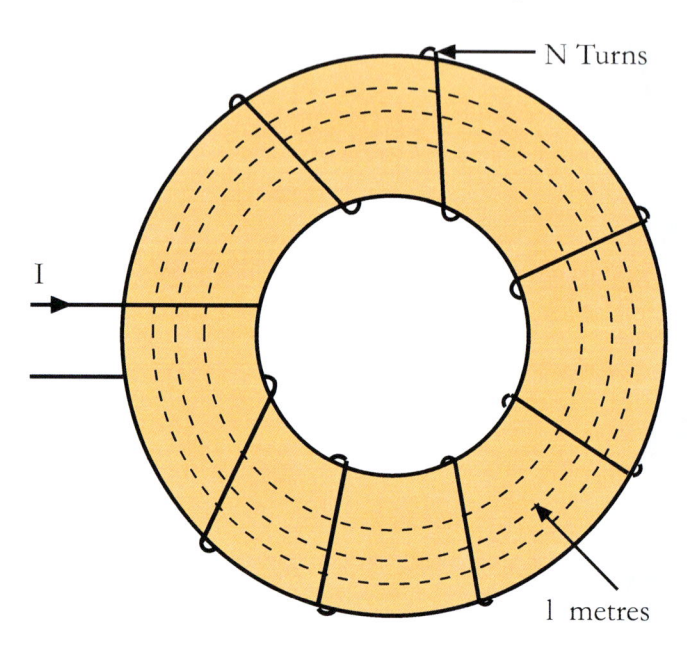

Figure 10.18 – Typical Magnetic Circuit, a Toroid

A Toroid is a simple ferromagnetic ring and this one has a cross-sectional area 'A' m^2, and a mean circumference of 'l' metres. It is wound with 'N' turns and has a current 'I' Amperes flowing through it. Therefore, the total flux (Φ) is equal to the flux density multiplied by the area, i.e.

$$\Phi = BA$$

In addition, the mmf is equal to the magnetic field strength multiplied by length, i.e.

$$F = Hl$$

So, dividing these two equations, we get:

$$\frac{\Phi}{F} = \frac{BA}{Hl}$$

$$\text{but } \mu_r\mu_0 = \frac{B}{H}$$

$$\text{so } \frac{\Phi}{F} = \mu_r\mu_0 \times \frac{A}{l}$$

Rearranging these formulas, we get:

$$\Phi = \frac{F\mu_r\mu_0 A}{l} \quad \text{and therefore: } \quad \frac{F}{\Phi} = \frac{l}{\mu_r\mu_0 A} = S$$

The unit of reluctance (S) is the **Ampere per Weber**, abbreviated to ***A/Wb***.

From the formulae above, we can see that:

1. Reluctance *increases* as the length of the magnetic circuit *increases*
2. Reluctance *decreases* as the permeability of material *increases*
3. Reluctance *decreases* as the cross sectional area *increases*

Where a magnetic circuit has different materials, eg a toroid with an air gap, the total reluctance (S_t) is calculated by adding the different reluctance values together, similar to resistors in series.

Air gaps are usually a necessity in magnetic circuits for construction purposes and so to avoid distortion, air gaps must be kept to a minimum. Air gaps also create another problem as not all the flux cross from one side to the other and *flux leakage* results.

CHAPTER TEN
MAGNETISM

Further Comparisons of Electric & Magnetic Circuits

One important difference between electric and magnetic circuits is that energy *must* be supplied to maintain the flow of electricity in an electric circuit; whereas in a magnetic circuit, once the magnetic flux is set up, it does not require any further energy input.

To simplify the comparisons, the major elements are listed in table 10.1 below.

Electric Circuit		Magnetic Circuit	
Quantity	Unit	Quantity	Unit
EMF	Volt	MMF	Ampere
		Magnetic Field strength	Ampere/metre
Current	Ampere	Magnetic Flux	Weber
Current density	Ampere/m^2	Magnetic flux density	Tesla
Resistance	Ohm	Reluctance	Ampere/Weber
Current(I) $=\dfrac{V}{R}$		Flux(F) $=\dfrac{F}{S}$	

Table 10.1 – Comparison of Electricity and Magnetism

B/H Characteristics

A magnet circuit's B/H characteristics are best represented as a graph. Figure 10.19 shows a magnetic circuit of a coil wrapped around a soft iron core.

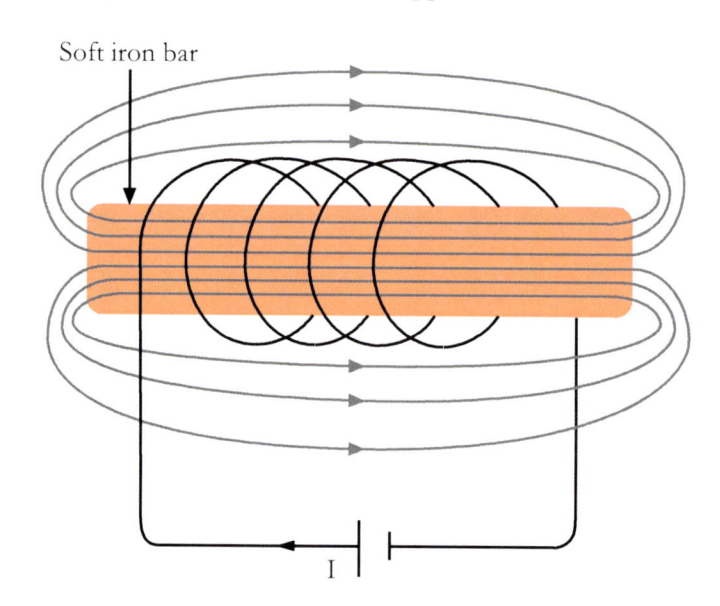

Figure 10.19 – Simple Magnetic Circuit

When current flows through the coil, it creates magnetic flux in the soft iron bar and it becomes magnetised. From the discussions above, we know that the magnetic flux density (B) in the iron core depends on the magnetic field strength (H). Figure 10.20 shows a graph of the initial, ie *first time*, magnetisation curve for a circuit containing unmagnetised material.

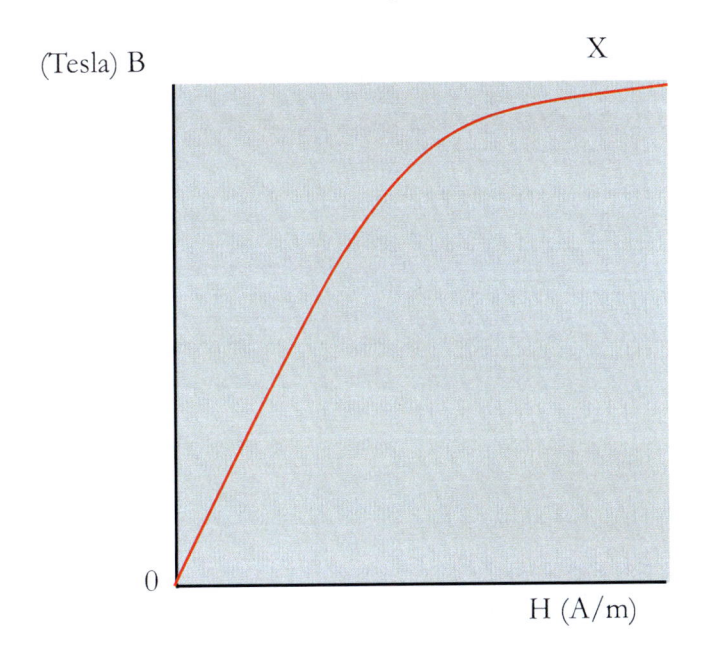

Figure 10.20 – Typical Initial B/H Curve

As shown in figure 10.20, as H increases, so does B up until point 'X' where the graph starts to level off. At this point, there is no further increase in flux density B, even if the field strength H continues to rise. Point 'X' is called the *flux saturation point*, and this indicates where a material cannot accept any more lines of flux. For obvious reasons, this graph is known as the *B-H curve*.

However, this graph only shows the initial magnetisation. If H is now reduced, we discover that the graph does not retrace the path X - 0 but follows the curve X - Y, shown in figure 10.21aterial is 0Y and is called the *remnant flux density*. **Remanance** is the magnetic induction that remains in a material immediately after removal of the magnetizing field. It is sometimes called **Retentivity**. Magnetism that remains in a material for a considerable time after the magnetizing force has been removed, is termed **Residual Magnetism**.

If H is increased in the reverse direction, the flux density decreases until at some value, 0W, the flux has been reduced to zero. The magnetic field strength 0W required to wipe out the residual magnetism is called the *coercive force*. Further increase of H in the reverse direction, causes the curve to grow in the reverse direction represent by the curve W to V.

If the magnetic field is now varied from 0M to 0N, the flux density follows a curve VUX, which is similar to curve XYWV, and the closed loop, XYWVUX is called the *Hysteresis Loop*.

CHAPTER TEN
MAGNETISM

Figure 10.21 – Hysteresis Loop

Hysteresis Loss

As with all work done, energy is required when magnetising and demagnetising materials and as things are rarely 100% efficient, there is energy lost. In a magnetised material, this appears as heating and is wasted energy, which is proportional to:

1. The area of the hysteresis loop

2. The frequency of magnetising and demagnetising

From this, we can deduce that the material used in a magnetic circuit needs to be chosen for its properties, depending on its particular use. These are characterised into **soft** and **hard** magnetic materials, collectively termed **Ferro-Magnetic**. Soft magnetic materials have smaller narrower hysteresis loops than hard materials and so they require less energy to magnetise and demagnetise.

Soft Magnetic Materials

The most common soft magnetic materials include soft iron, silicon, steel, nickel-iron alloys and soft ceramic ferrite compounds. Soft magnetic compounds have the following characteristics:

1. High permeability
2. Low coercive force
3. Low values of remnant flux
4. Saturate at low H values
5. Low hysteresis loss
6. Easy to magnetise/demagnetise

The most common hard magnetic materials include carbon steel, tungsten steel, cobalt steel and hard ferrites. Hard magnetic materials have the following characteristics:

1. Low permeability
2. High coercive force
3. High values of remnant flux and residual magnetism
4. Saturate at high H values
5. High hysteresis loss
6. Difficult to magnetise/demagnetise

Due to their high values of remnant flux and coercive force, hard magnetic materials are useful for making permanent magnets and are found in motors, generators, moving coil microphones and moving coil meters.

When a magnet is placed near an unmagnetised steel pin, the pin first becomes magnetised by induction. It is then attracted to the magnet. **Induction therefore precedes attraction.** This explains why unmagnetised iron and steel are attracted to magnets. Steel objects, such as pliers, railings, or girders in buildings, are often found to be magnetised. This is due to induction by the earth's magnetic field.

All materials can theoretically be magnetised to some extent, and as well as soft and hard iron materials we should consider the following:

Para-Magnetic Materials

Paramagnetic materials such as wood, glass, water and platinum are extremely difficult to magnetise. Any attempt to magnetise these materials will normally result in their destruction.

Dia-Magnetic Materials

Diamagnetic materials have a very weak and negative susceptibility to magnetic fields, by which they actually oppose the magnetising force. When placed in the magnetic field, they decrease the field strength. If freely suspended in a magnetic field they will take up a position at right angles to the field.

Examples are: Copper, Brass, Bronze, Mercury, Bismuth.

Eddy-Current Loss

There is an unfortunate loss associated with the hysteresis loop. The varying flux in the core of a magnetic circuit induces emfs, and therefore currents, in the core of the material. This generates I^2R losses called **eddy-current** losses and the sum of these and the hysteresis loss is called the **core losses**.

Owing to the low resistance of most cores, as they are usually made of metal, eddy currents can be considerable and can cause a large loss of power and excessive heating of the core if not controlled.

To overcome this, the core is often made of laminations that are insulated from each other. In this way, the eddy currents are limited to each sheet and the eddy current loss is reduced. Let us take a closer look at a theoretical example in order to understand things more easily. Figure 10.22 shows a solid core, ①, and one with five (5) laminations, ②.

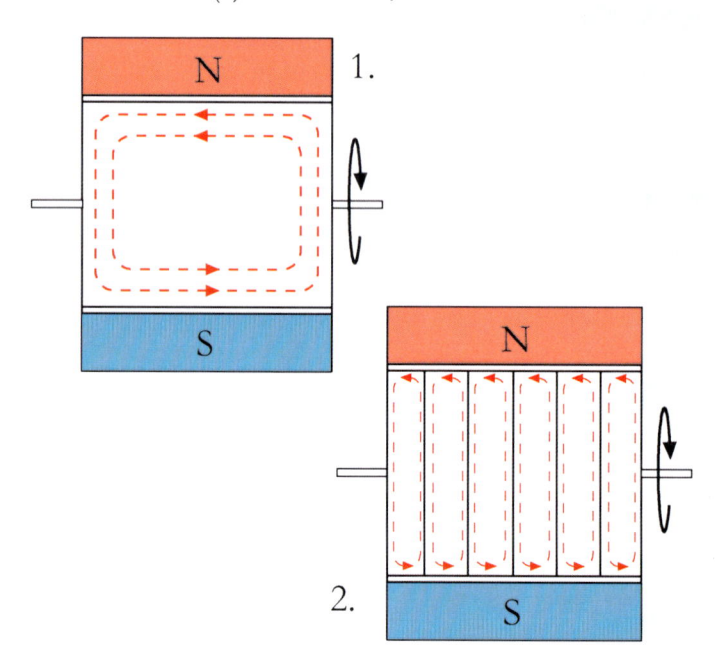

Figure 10.22 – Eddy Current in a Solid and Laminated Core

If the core is split into five laminations as in ② above, the emf generated per lamination will only be a-fifth (1/5) of that generated in the solid core. In addition, the cross sectional area per path is reduced by a fifth so that the resistance per path is approximately five times that of the solid core. Consequently, the current per path is about one twenty fifth (1/25) of that in the solid core.

Therefore:

$$\frac{I^2R\ Loss/Lamination}{I^2R\ Loss\ in\ Soild\ Core} = \frac{I^2R_i}{I^2R_C}$$

$$= \left(\frac{1}{25}\right)^2 \times 5$$

$$= \frac{1}{125}$$

However, since there are five laminations, this becomes:

$$\frac{5}{125} = \left(\frac{1}{5}\right)^2$$

From this, we can see that the eddy current loss is approximately proportional to the square of the thickness of laminations, and so can be reduced to any value. However, if the thickness of the lamination is reduced to below about 0.4 mm, then the reduction is loss does not justify the extra cost of production.

Eddy current losses can also be reduced considerably by using a silicon-iron alloy.

Retentivity

Retentivity is the capacity for a body to remain magnetised after the magnetising field has ceased to exert an effect. When applied to magnetic media, e.g. cassette tapes, video, etc, it measures how well a particular medium retains or remembers the field that it is subjected to. Although magnetic media are sometimes depended upon to last forever, think of the master tapes of phonograph records, the stored magnetic fields begin to degrade as soon as they have been recorded. A higher retentivity ensures a longer life for the signals recorded on the medium.

No practical magnetic material has perfect retentivity, however, as the random element of modern physical theories ensure that. Even the best hard disks slowly deteriorate with age, showing an increasing number of errors as time passes after data has been written.

In order to avoid such deterioration magnetically stored information should be periodically refreshed.

Magnetic Shielding

Magnetic shielding is often required to protect electronic equipment, dc and low frequency ac circuits from magnetic fields, sometimes called *magnetic interference*. A *magnetic shield* provides a low reluctance magnetic path for the interference field by attracting flux lines to it to divert the magnetic field away from the sensitive component or components Figure 10.23.

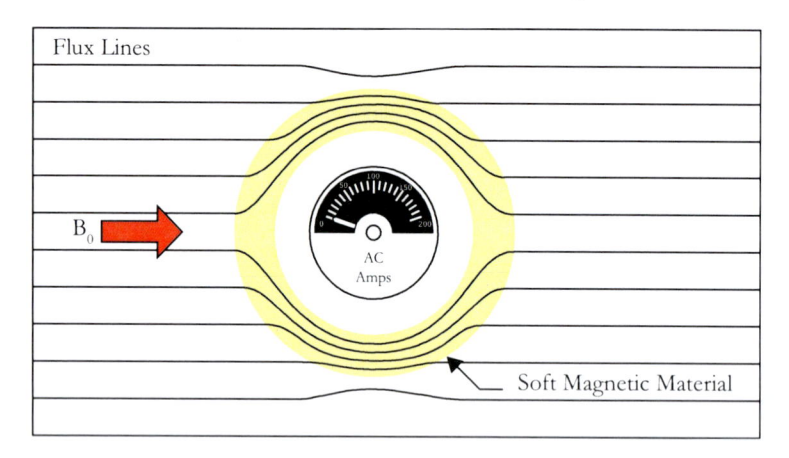

Figure 10.23 – Magnetic Sheilding

Shielding is usually provided by using very high-permeability materials, eg nickel-iron based alloys. All these materials must be treated at very high temperatures, typically 1100°C, after shaping.

Some amorphous metals, which have the same magnetic properties of the nickel-iron alloys and are easier to use and less sensitive to plastic deformations, but they are only available in the form of very thin sheets of limited width and these high-permeability materials have low saturation induction and therefore must only be used in the presence of weak stray fields, i.e. smaller than a few milli teslas (mT).

In the presence of stronger magnetic fields a supplementary shield, surrounding that made of the high-permeability material, must be used. Typical ferromagnetic materials that can be used for shielding are low carbon steel, silicon steel, 50%/50% Nickel Ferrous alloys, metallic glass, etc.

These materials are formed in the shape of a completely or almost completely closed box surrounding the equipment or of plates or cylinders placed near the equipment. While in the first case the magnetic field inside the box vanishes completely or reduces to negligible values, in the second case an effective shielding is limited to a small region near the shield and is proportional to its size.

The optimum shield is a closed, spherical configuration, but in practice, effective shields can be cans, open-ended cylinders, five-sided boxes, U or L shaped brackets, and even flat plates. A flat plate is effective if both the length and width of the shield exceed the distance separating the source and receiver of magnetic interference. Cables carrying currents in one direction only cannot be shielded; therefore the input and output cables should be placed together in order to avoid the generation of field lines around them.

Demagnetising or Degaussing

Magnetism is sometimes an unwanted phenomenon as over time, certain ferrous objects and equipment that contains ferrous materials can unintentionally become magnetic. If this is a tool such as a screwdriver, it can adversely affect aircraft components with which it comes into contact. Such unintentional magnetism can also affect the correct operation of magnetic tape recorders, like those found in older versions of *Cockpit Voice Recorder (CVR)* or *Flight Data Recorder (FDR)*. It does this by inducing a permanent magnetism on the recording heads, which can effectively wipe data from parts of the magnetic tape. In addition, aircraft components that contain ferrous or steel elements may also become magnetised, especially after a lightning strike or if the aircraft has been parked on one heading for some considerable time, i.e. 28 days or more.

When this happens, it can unduly influence some of the aircraft's systems, eg the Magnetic Compass System, CVR and FDR, causing them to provide erroneous information. In an aircraft, it is commonly seen in the cockpit as a distortion of one of the *Electronic Flight Instrument System (EFIS)* screens. It usually manifests itself as a '*rainbow*' effect on some or all of the screen, sometimes caused by engineers using magnetic screwdrivers in the vicinity of the instrument panel.

If this occurs, the component or system must be *demagnetised* or *degaussed*, a process named after *Carl Friedrich Gauss (1777 to 1855)*, a mathematician and physicist who studied and worked with *Wilhelm Weber (1804 to 1891)*, who has the measurement of magnetic flux named after him, on electromagnetic fields.

Technically, *Degaussing* is the process of removing any permanent magnetism, also called *magnetic hysteresis*, from an object. It is achieved by passing the object through a magnetic field that oscillates with a reducing amplitude as it moves along.

The British Royal Navy was the first to use degaussing in WWII to protect ships against magnetic mines. An electromagnetic cable was inserted on the inside and around each warship and each time it came into harbour, it was degaussed by passing an electrical current through the cable to neutralise any magnetic field.

To remove unintentional magnetism from an aircraft or its components, an engineer will need a demagnetising tool, more commonly called a *Degausser*. Indeed, some modern aircraft components have these devices built in. In aircraft with *Cathode Ray Tube (CRT)* screens, most monitors automatically degauss the CRT whenever it is turned on. In some case, CRTs have a manual degauss button that performs a more thorough degaussing of the monitor, if required.

For other components, or CRTs that do not have an automatic degaussing coil, in order to remove unwanted magnetism, they must go through a powerful magnetic field that is strong enough to rearrange the magnetic particles.

CHAPTER TEN
MAGNETISM

A degausser consists of a coil of wire, usually copper, of predetermined length and wire gauge that is wound on a circular former of the required diameter. The supply voltage and frequency and desired magnetic field strength set the number of wire turns of the assembly, figure 10.24.

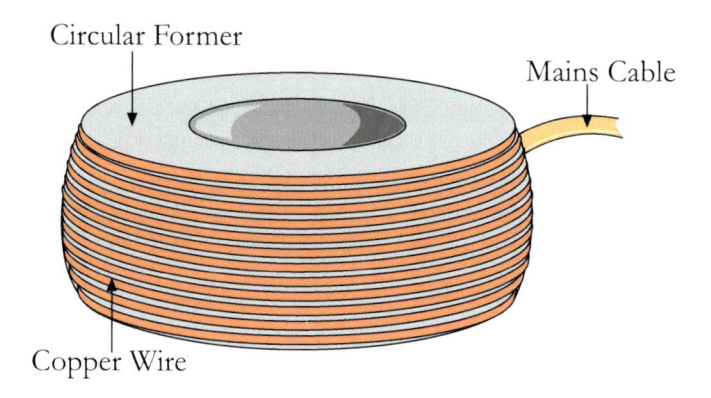

Figure 10.24 – Standard Coil for Small Tools

The degaussing tool must be supplied with an AC voltage. The AC supply frequency influences the design of the tool, as the coil's reactance must be taken into consideration. When fed with AC, the coil produces an alternating magnetic field that fluctuates at the same rate as the supply frequency. The constantly reversing magnetic field is designed to disrupt the field/polarity in the affected component so rearranging its molecules so that it magnetism disperses.

Time is an important element in the degaussing process; too little and there may be residual magnetism left in the affected component; too much and the degaussing coil will overheat. For simple small components and tools, ten (10) seconds should be enough for a successful demagnetisation.

Degaussing Operation for Small Objects

With power applied, the object is placed in the coils centre for a few seconds and is then quickly withdrawn well away from the degausser. As any object that is inserted into a degaussing tool is going to be subjected to a strong alternating magnetic field, it should not be attached to anything or touch any other metal during this process.

If is part of a larger component or system, it should be removed from it before degaussing as this process will induce large voltages that can damage other electronic components if they are left in circuit.

Degaussing CRTs

As mentioned earlier, some CRTs have built in degaussing coils. For those that do not, a less powerful degaussing tool may be constructed with less wire turns and smaller current draw. This may be used like a 'wand' over the CRT face to remove any magnetised areas, figure 10.25.

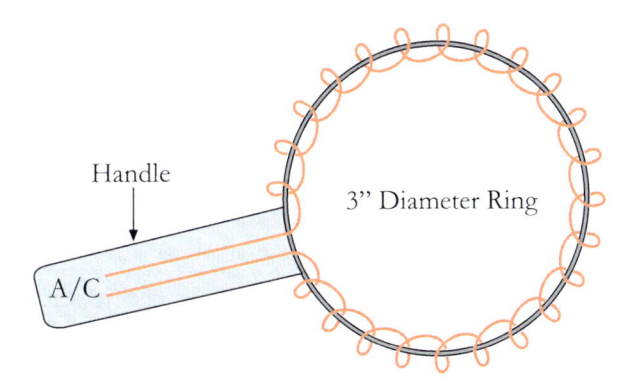

Figure 10.25 – Coil for Degaussing CRTs

However, if using such a device, ensure that the coil's power is not removed before moving well away from the CRT.

Normal degaussing tools should not be used on CRTs as their powerful magnetic fields can disrupt the internal shadow mask, making the unit unserviceable.

Locating Areas of Unwanted Magnetism

If unwanted magnetism is suspected on an aircraft, its position can be identified using a ***Magnetometer***. Once located, a portable degaussing tool, which is usually constructed with an open-ended head that has a long lead with a standard current rating of ten amperes (10A), may be used to demagnetise affected components, figure 10.26.

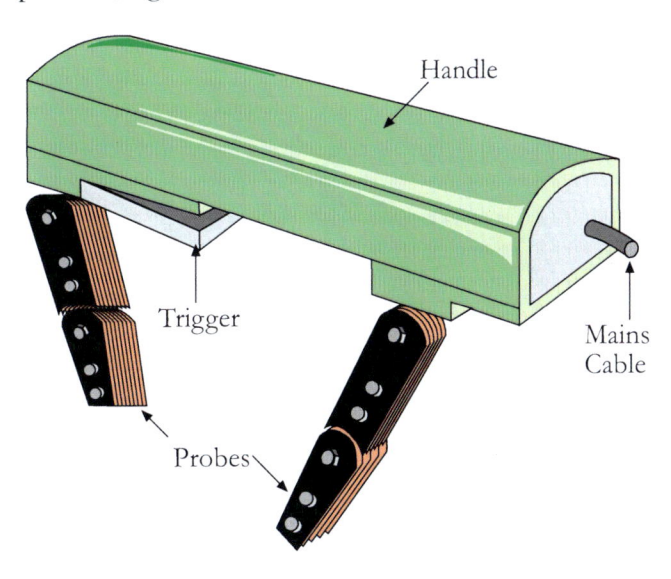

Figure 10.26 – Degaussing Aircraft

Precautions for Care & Storage of Magnets

Magnets, like any other brittle or sensitive object should be treated carefully. However, as long as you take a common sense approach, they should come to no harm; the following bullet points give a few simple rules:

- *Never* allow the two magnets to come together abruptly as the attraction can be so strong that the magnets may chip or break

- Take care that your finger or a sensitive fold of skin does not get caught between two magnets; it can be painful and a blood blister or minor abrasion could result

- *Never* allow magnets near computer disks, EFIS screens, recording tapes of any kind, credit cards, bank cards, or any other device that uses magnetic tape to record information

Sometimes it is very difficult to separate the two magnets by hand; the technique is to slide them apart rather than by a direct pull. Storage is not usually a problem, but magnets should never be stored near sources of high temperature or high humidity as this can reduce their magnetism and they should not be stored near magnetically sensitive materials.

Transporting Magnets

In order for magnets to be shipped by air, Flux measurements of all packages containing magnets must measure less than 0.00525 gauss 15 feet from the package. As such powerful magnetic components and assemblies need to be shielded so that magnetic fields will satisfy air transport requirements. Magnets are often shipped in a steel-lined box in order to remain below this limit.

Revision

Magnetism

Questions

1. If two magnets are placed in close proximity to each other with opposite poles head to tail, their magnetic fields will:

   a) Repel each other

   b) Cancel each other out

   c) Attract each other

2. The figure below shows a conductor with current flowing in it as shown. With current flowing, the magnetic field will be rotating:

   a) Anti-clockwise

   b) Clockwise

   c) Pulsating back and forth clockwise/anti-clockwise

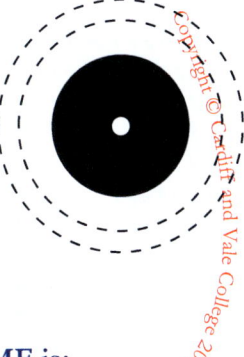

3. If a current of 5 A is flowing in a coil of 15 turns, its MMF is:

   a) 75 AT

   b) 3 AT

   c) 10 AT

4. If the length of the conductor with a current of 5 A flowing in a coil of 15 turns is 10 metres, the circuit's magnetic field strength is:

   a) 7.5 A/m

   b) 30 A/m

   c) 1 A/m

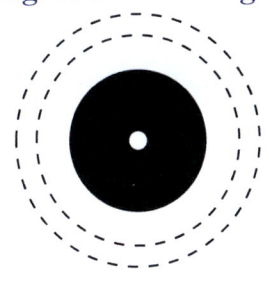

5. As the length of a magnetic circuit increases, its reluctance

   a) Increases

   b) Remains unchanged

   c) Decreases

6. As the permeability of a material decreases, its reluctance

   a) Increases

   b) Remains unchanged

   c) Decreases

7. As the cross-sectional area of a material decreases, its reluctance

   a) Increases

   b) Remains unchanged

   c) Decreases

8. The magnetic field used to wipe out residual magnetism in a material is termed:

   a) The negative flux

   b) The coercive force

   c) The saturation flux

9. Soft magnetic compounds have:

   a) Low permeability

   b) High permeability

   c) Variable permeability

10. **Hard magnetic materials are:**

    a) Difficult to magnetise/demagnetise

    b) Easy to magnetise/demagnetise

    c) Cannot be magnetised

11. **In a electromagnetic core, eddy current losses are improved by:**

    a) Increasing its cross-sectional area

    b) Reducing the number of laminations

    c) Increasing the number of laminations

12. **Magnetic recording materials should have:**

    a) Low permeability

    b) Low retentivity

    c) High retentivity

13. **When a conductor is in a magnetic field, in order to generate a current there must be:**

    a) An applied voltage

    b) Relative motion

    c) A resistive load

14. **Electrons orbit an atom's nucleus and have:**

    a) A neutral charge

    b) A positive charge

    c) A negative charge

CHAPTER TEN
MAGNETISM

Revision

Magnetism

Answers

1. **C**
2. **A**
3. **A**
4. **A**
5. **A**
6. **A**
7. **A**
8. **B**
9. **B**
10. **A**
11. **C**
12. **C**
13. **B**
14. **C**

Inductors & Inductance

Introduction

An *Inductor*, according to the dictionary, is a component in an electric or electronic circuit that possesses inductance; pretty obvious but does not mean much. Inductors are electronic devices that consist of a conducting wire, usually wound into the form of a coil that opposes rapid changes in circuit current. When a current passes through the coil, a magnetic field is set up around it that *tends to oppose* rapid changes in current intensity.

Inductance is a constant relating the current flowing in a circuit to the magnetic flux, i.e. the amount of magnetism, produced by the current flow. Or put more simply, Inductance is the ability of a coil of wire to store energy in the magnetic field that surrounds it when current flows in the wire, Figure 11.1.

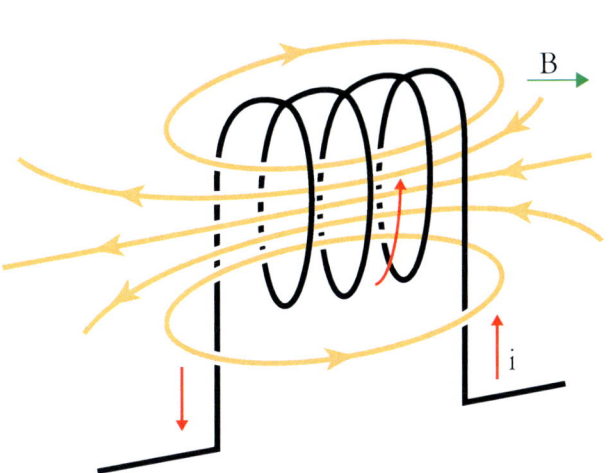

Figure 11.1 – Basic Inductor

The unit of magnetic inductance is the *Henry (H)*, named in honour of the 19th-century American physicist *Joseph Henry*, who first recognized the phenomenon of *Self-Induction*. One (1) Henry is equivalent to one volt divided by one ampere per second; i.e. if a current changing at a rate of one (1) ampere/second induces an emf of one (1) volt, the circuit has an inductance of one (1) Henry, a relatively large inductance; more on this later. The generation of an emf and current by a changing magnetic field is called *Electromagnetic Induction*.

CHAPTER ELEVEN
INDUCTORS & INDUCTANCE

Inductor History

As mentioned earlier, Joseph Henry, who was one of the first great American scientists after Benjamin Franklin, first investigated inductors and inductance. He helped ***Samuel Morse***, who created the Morse code, in the development of the telegraph and discovered several important principles of electricity, including ***Self-Induction***.

While working with electromagnets in New York in 1829, he experimented and made important design improvements. He found that by insulating the wire instead of the iron core, he was able to wrap a large number of turns of wire around it, greatly increasing the magnet's power. He went on to make an electromagnet for Yale College that supported 2,086 lb, a world record at the time. During these studies at Yale, he first noticed the principle of self-induction in 1832, and three years later, he devised and constructed the first electric motor. In addition, when continuing his research, he discovered the laws upon which the transformer is based.

However, it is ***Michael Faraday***, a British physicist and chemist, who is accredited with making the major advances in the study of magnetism, electricity, and the chemical effect of a current. Working as an analytical chemist, he discovered Benzene in 1825 and prepared the first known compounds of Carbon and Chlorine. He also investigated the composition of alloy steels and optical glasses, but his greatest achievements were in electromagnetism.

In 1821, he constructed a simple form of electric motor, applying Hans Christian Oersted's discovery that electric currents produce a magnetic effect. After much research, he demonstrated that the converse was also true, i.e. that a magnetic field can induce an electric current. In 1831, he published his ***Laws of Electromagnetic Induction*** and put them to practical use in the dynamo and transformer, two inventions that are still fundamental to the large-scale generation and supply of electricity. His laws of electrolysis, published in 1834 and named after him, described the changes caused by electric current passing through liquids and other discoveries include diamagnetism and the rotation of light waves by strong magnetic fields.

Certainly one of the most outstanding experimental scientists of the millennium, he always refused a knighthood and even the Presidency of the Royal Society, as he feared that such honours would undermine his integrity and his intellectual freedom.

Faraday's Law of Electromagnetic Induction

Faraday knew that a coil of wire with an electric current flowing through it generated a magnetic field, and so he hypothesized that a ***changing*** magnetic field could do the reverse and induce a current in a coil.

Faraday demonstrated that his hypothesis was correct by moving a simple bar magnet back and forth inside a coil, figure 11.2, observing that a current was induced in the coil *only* when the magnet was in motion. He also observed that a current was induced in the coil when the coil itself was moved near a stationary permanent magnet.

In other words, he discovered that it is the *relative motion* between a conductor and a magnetic field that produces current. To generate current, either the conductor can move through the field, or the field can move past a conductor, ie it is necessary for there to be a *change* of magnetic flux in order for electromagnetic induction to occur.

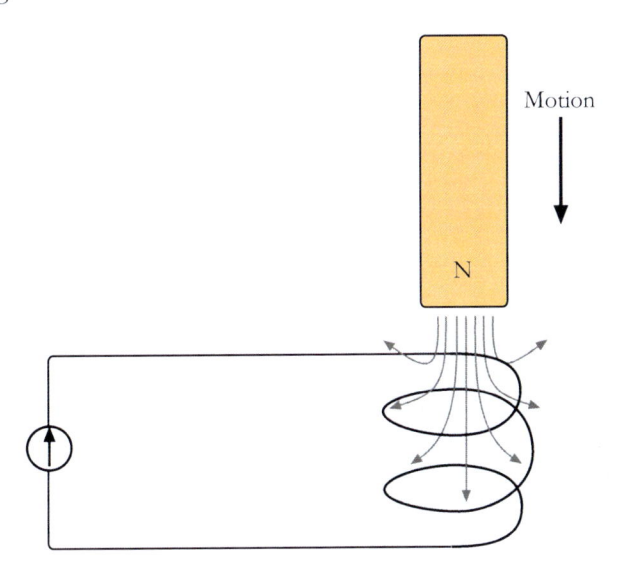

Figure 11.2 – Faraday's Demonstration of Electromagnetic Induction

Faraday's second experiment was more sophisticated and enhanced his understanding of electromagnetic induction by using two coils of wire wound around opposite sides of a ring of soft iron similar that illustrated in figure 11.3 below.

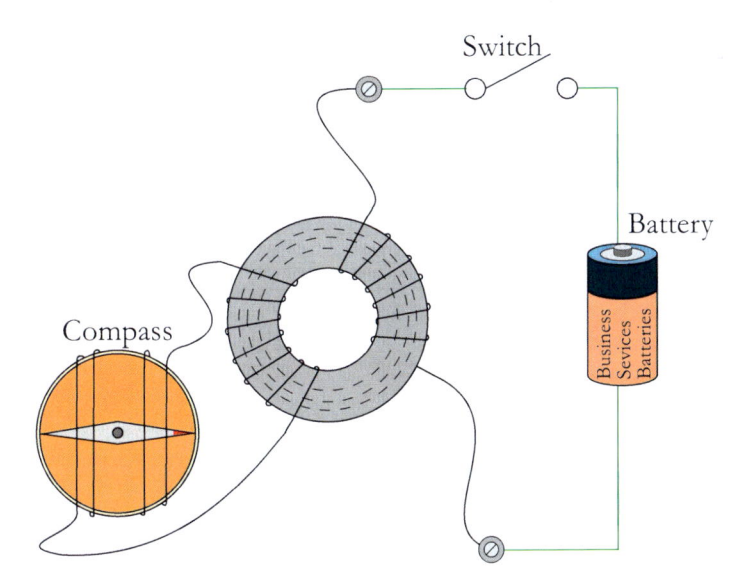

Figure 11.3 – Representation of Faraday's Second Experiment

CHAPTER ELEVEN
INDUCTORS & INDUCTANCE

In figure 11.3, the coil on the right is wrapped around an iron core and connected across a battery. The second coil is wound around a compass, which acts as a *galvanometer* to detect current flow. When the switch closes, current flows through the first coil and the iron ring becomes magnetised. When the switch is first closed, the compass in the second coil deflects momentarily and then returns to its original position. The deflection of the compass is an indication that an electromotive force (emf) was induced causing current to flow briefly in the second coil. Faraday also observed that when the switch is opened, the compass again deflects momentarily, but in the *opposite* direction.

As already mentioned above, Faraday knew that a coil of wire with an electric current flowing through it generates a magnetic field, and this experiment proved his hypothesis that a *changing* magnetic field induces a current in the second coil; i.e. the closing and opening of the switch causes a magnetic field to expand and collapse respectively.

From these experiments came Faraday's law, which can be summarised as:

> *Whenever the magnetic flux of an electrical circuit varies, an emf is induced into it and the size of the induced emf is proportional to the rate of change of flux.*

Put into more sensible English, this is:

> *The greater the number of magnetic field lines cutting across the conductor, the greater the induced voltage. The faster the field lines cut across a conductor or the conductor cuts across the field lines, the greater the induced voltage.*

Mathematically, Faraday's law is written as:

$$E = \frac{\Delta \Phi}{\Delta t}$$

Where:

'E' is the induced electromotive force (emf) in volts

$\Delta \Phi$ is the change in magnetic force in Webers

Δt is the amount of time in seconds in which the change in magnetic force takes place

This can then be put into a more logical statement of:

> *A circuit has an inductance of one (1) Henry (H) if an emf of one (1) volt is induced in the circuit when the current varies uniformly at the rate of one (1) Ampere/second.*

Therefore, if a circuit has an inductance of 'L' Henrys and if the current increase from i_1 to i_2 Amperes in t seconds, the average rate of current change is:

$$\frac{i_1 - i_2}{t} \text{ A/s}$$

and the average induced emf is:

$$L \times \frac{i_1 - i_2}{t} \text{ Volts}$$

If we now consider instantaneous values, if ΔI equals the change in current in a time of Δt seconds, then the induced emf (e) is given by the equation:

$$e = L \frac{\Delta i}{\Delta t} \text{ Volts}$$

However, although this equation gives the emf's *magnitude*, it does not give its polarity. This is usually overcome by looking at the direction of the current change. For example, if a current through a coil with an inductance of 2 H is reduces from 8A to 3A in 0.5 seconds then the average rate of change of current is:

$$\frac{3 - 8}{0.05} = -100 \text{ A/s}$$

From the equation given above, we can now see that the induced emf is:

$$e = 2 \times (-100) = -200V$$

From this, we can see that an inductor *opposes* the increase of current in a circuit by acting against the applied voltage using an *opposing* induced emf.

Put another way, when an emf opposes the *increase* in current in the inductor, energy is *taken into* the inductor and when it opposes the *decrease* in current energy is *supplied from* the inductor back into the circuit.

This relationship between changing magnetic flux and induced electromotive force is now known as *Faraday's Law of Electromagnetic Induction*.

The most important feature of Faraday's experiments, and his subsequent law, is that to induce an emf the lines of flux *must* cut the coil.

Lenz's law

Experiments in electromagnetism continued throughout the 19th century all over Europe and the USA and in 1834, a Russian physicist, *Heinrich Friedrich Emil Lenz*, discovered another feature. He found that an induced electric current flows in a direction such that the current *opposes* the change that induced it; or more formally:

> *An induced electromotive force generates a current that induces a counter magnetic field that opposes the magnetic field generating the current.*

For example, thrusting a pole of a permanent bar magnet through a coil of wire, as in figure 11.2, induces an electric current in the coil and the current in turn sets up a magnetic field around the coil, making it a magnet. Because like magnetic poles repel each other, Lenz's law states that when the *north* pole of the bar magnet is approaching the coil, the induced current flows in such a way as to make the side of the coil nearest the pole of the bar magnet itself a *north* pole to *oppose* the approaching bar magnet. When withdrawing the magnet from the coil, the induced current reverses itself, and the near side of the coil becomes a *south* pole to produce an *attracting* force on the receding bar magnet.

This can be seen in Figure 11.4, When the bar magnet is moved towards the coil as shown in 2(a), the strength of the magnetic field induced in the coil will increase producing a current flow in the loop. However the current flowing in the coil will produce its own magnetic field in the opposite direction to that of the bar magnet, which will then oppose the increase in current flow. When the bar magnet is moved away from the coil, 2(b), the current flow and induced magnetic field change direction. The direction of induced emf and current flow can also be controlled by reversing the orientation of the bar magnet, 2(c).

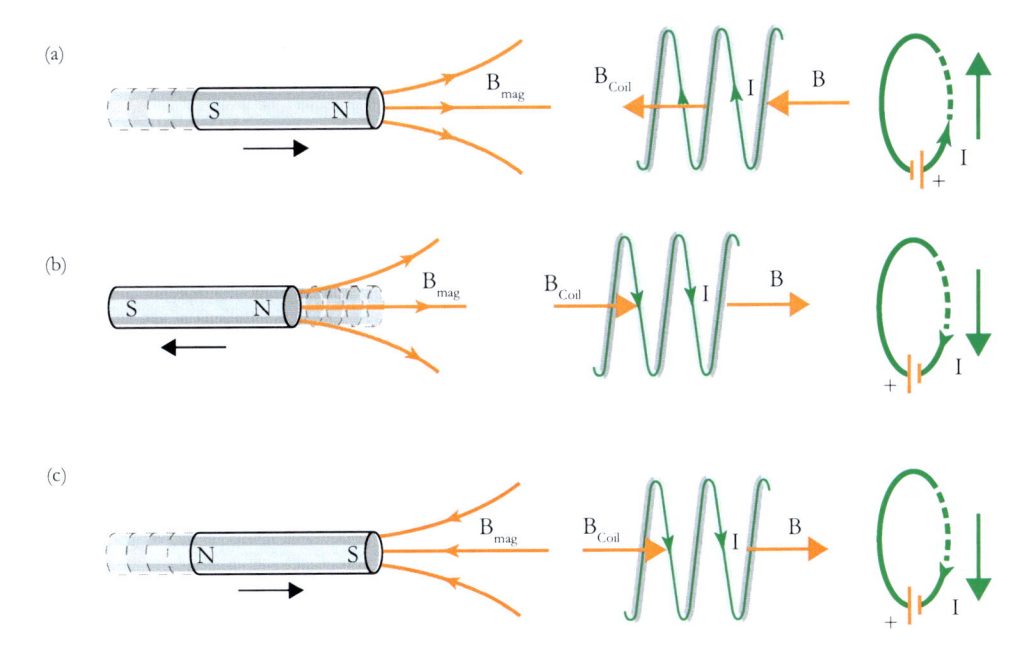

Figure 11.4 – Lenz's Law

It follows then that a small amount of work is done in pushing the magnet into the coil and in pulling it out against the magnetic effect of the induced current. This small amount of energy manifests itself as a slight heating effect as the result of the induced current encountering resistance in the material of the coil. This small amount of work done helps in the proof of Lenz's law as it upholds the general principle of the ***conservation of energy***. The law of conservation of energy states that energy may neither be created nor destroyed. Therefore the sum of all the energies in the system is a constant. If the current were induced in the opposite direction, its action would spontaneously draw the bar magnet into the coil in addition to the heating effect, which would violate the conservation of energy rule.

Induction Principles

We have already seen from Faraday and Lenz's laws that a moving conductor in a magnetic field induces an emf in the conductor and their laws can be represented as:

> Induced emf (e) α minus (-) rate of change of flux (Φ) with time (t)

The minus (-) sign indicates that the induced emf is in a direction as to oppose the change producing it. If we take the constant of proportionality as one (1) this equation can be simplified to:

$$e = -\frac{\Delta \Phi}{\Delta T}$$

CHAPTER ELEVEN
INDUCTORS & INDUCTANCE

We already know from chapter 10 of this module that the unit of flux is the Weber (Wb) and if the flux changes by 1Wb/s then the induced emf is 1V.

We also know that the amount of magnetic flux density passing at right angles through a given unit area is called the magnetic flux density B, i.e.:

$$B = \frac{\Phi}{A}$$

If we look at a simple conductor of length 'L' moving through a magnetic field with **velocity (v)**, figure 11.5, in a **time 't'**, the conductor will move a distance of '*vt*' and the area through which the flux passes will change by '*Lvt*'.

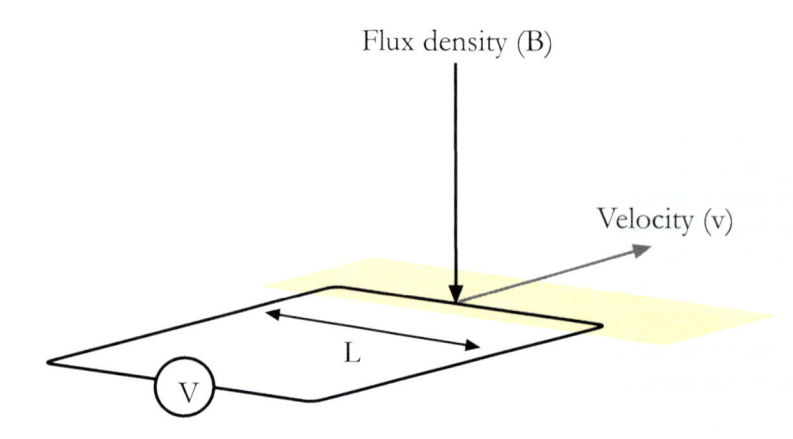

Figure 11.5 – Conductor Moving in Magnetic Flux

Therefore, the flux change will be '*BLvt*' in a time '*t*' and so there is an induced emf of:

$$E = BLv$$

Example

A conductor of 0.1 metres moves at right angles to a magnetic field of flux density 0.1 tesla at a velocity of 5 m/s. What is the emf induced between the ends of the conductor?

$$e = 0.1 \times 0.1 \times 5$$
$$= 0.05 \text{ or } 50 \text{ mV}$$

From the formula, we can see that there are three (3) factors that affect the magnitude of the induced emf; these are:

- The flux density of the magnetic field (B)
- The length 'L', in metres, of the conductor in the magnetic field
- The velocity (v) at which the inductor cuts the magnetic field

Note: As already mentioned above, it is important to remember that the conductor **must** cut the magnetic field; if it is parallel to it, there will be no induced emf.

Another British scientist, John Fleming, came up with a simple method of remembering the relationship between the three (3) factors of current, motion and magnetic field direction, figure 11.6. With Fleming's right hand rule, the thumb gives the direction of motion; the index finger gives the direction of the magnetic field; and the second finger gives the direction of conventional current flow.

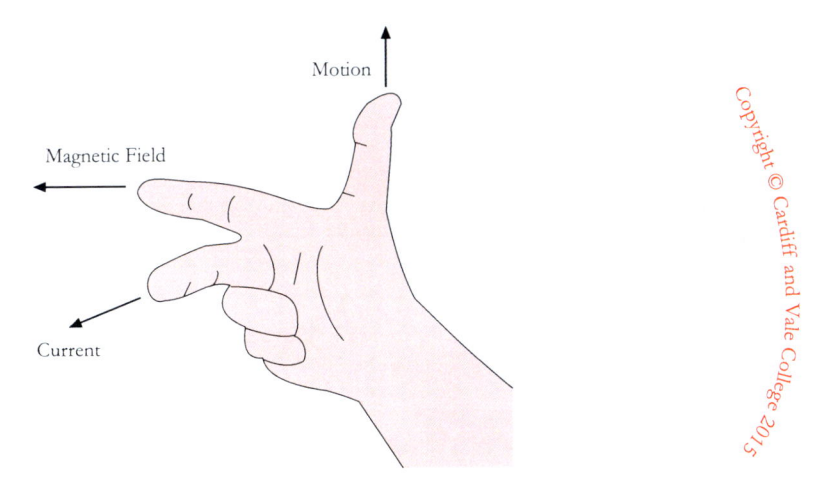

Figure 11.6 – Fleming's Right Hand Rule

Self Inductance

If an inductor is shaped in a coil, a current passing through it produces a magnetic flux that links each turn of the coil. Therefore, as the current flowing through the coil changes so the flux linked by the coil also changes and an emf is induced. This phenomenon is called *Self-Inductance* or more usually just *Inductance* because the induced emf is in the same coil that carries the changing current, i.e. itself.

CHAPTER ELEVEN
INDUCTORS & INDUCTANCE

As a textbook definition, this is usually stated as:

Self-inductance is the property of a circuit that induces an emf in itself due to a changing current in, and a changing flux around, the same circuit.

The symbol and measurement unit of self-inductance are the same as those for an inductor, i.e. 'L' and the Henry (H).

By Faraday's law, the induced emf is proportional to the rate of change of linked flux. However, the flux produced by the current is proportional to its size and so its rate of change is proportional to the rate of change of current, i.e.:

$$e \alpha \frac{\Delta \Phi}{\Delta t}$$

Therefore, we can produce the formula:

$$e = L \frac{\Delta I}{\Delta t}$$

Where 'L' is the inductance of the circuit in Henrys. This can then be put into the more logical statement already mentioned of:

A circuit has an inductance of one (1) Henry (H) if an emf of one (1) volt is induced in the circuit when the current varies uniformly at the rate of one (1) Ampere/second.

The effect of inductance on the current in a circuit is that when an applied voltage is switched on or off, it does not immediately rise to its maximum value or fall to zero, as appropriate.

This is because when the voltage is first switched on and the current starts to rise from zero, the changing current induces an emf. This emf is in a direction to oppose the growing current and slow its growth as dictated by Lenz's law, and so is often called a **back emf**. When the voltage is switched off, the current starts to fall and so induces an emf that opposes the current fall, and so the current takes longer to fall to zero.

In practical terms, it is impossible to produce a totally non-inductive circuit, i.e. a circuit that generates no flux when a current flows. However, for most purposes, a circuit that is not shaped in a coil can be looked at as being practically non-inductive. In circumstances where a circuit's inductance must be as low as possible, the wire is bent back on itself, as illustrated in figure 11.7, so that the magnetising effect of one conductor is neutralised by that of the other conductor. The conductor can then be coiled around an insulator without increasing the inductance.

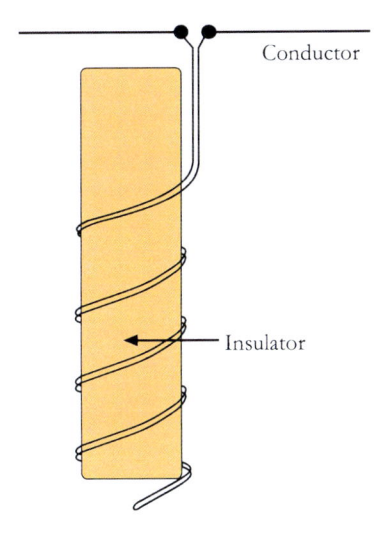

Figure 11.7 – Non-Inductive Resistor

Inductance in Terms of Flux-Linkages

Inductance can also be explained and calculated in terms of flux-linkages/ampere.

Suppose a current of 'I' amperes flows through a coil of 'N' turns to produce a flux of 'Φ' Webers, assuming the reluctance of the circuit stays constant and that the inductance of the coil is 'L' Henrys. If the current increases from zero to 'I' amperes in 't' seconds then the average rate of change of current is shown as:

$$\frac{I}{t} \text{ Amperes / second}$$

$$\therefore e = \frac{LI}{t} \text{ Volts}$$

As already discussed, the value of the emf induced in a coil is equal to the rate of change of flux linkages/second. Therefore, when the flux increases from zero to Φ Webers in t seconds, the average rate of change of flux is Φ/t and the average induced emf is shown in equation 1:

$$e = \frac{N\Phi}{t} \text{ Volts} \qquad \text{Equation (1)}$$

If we now consider instantaneous values and Δϕ is the increase in flux in Webers due to an increase of ΔI Amperes in Δt seconds, then the rate of change of flux is:

$$N \frac{\Delta i}{\Delta t} \text{ and}$$

$$e = N \frac{\Delta \Phi}{\Delta t} \qquad \text{Equation (2)}$$

If we now equate formulas (1) and (2) for instantaneous values we get equation 3:

$$L \frac{\Delta i}{\Delta t} = N \frac{\Delta \Phi}{\Delta t}$$

$$\therefore L = N \frac{\Delta \Phi}{\Delta i} \qquad \text{Equation (3)}$$

Effects on Circuit Inductance

Consider a uniform coil wrapped around a non-magnetic ring of uniform section, shown here in figure 11.8.

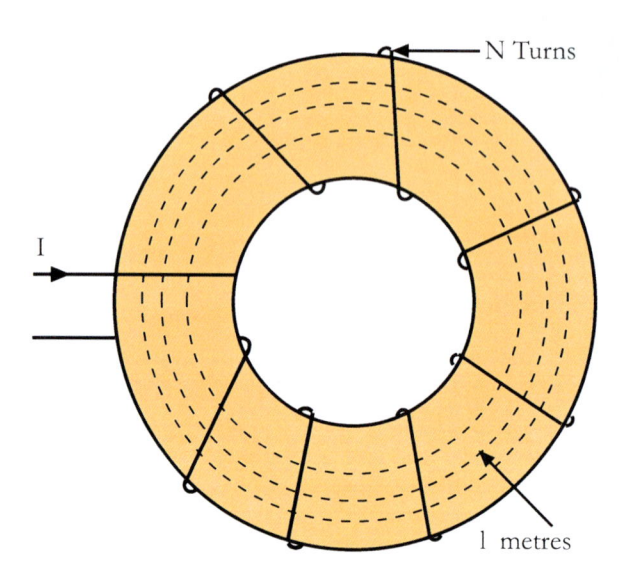

Figure 11.8 – A Toroid Ring

Without going into the explanation, as you do not need it, for a toroid:

$$L = 4\pi \times 10^{-7} \frac{N^2 A}{l}$$

From this equation, we can see that the factors that affect inductance are:

- The cross-sectional area of the core, sometimes referred to as cross-sectional area of the coil.
- The number of turns in the coil squared, i.e. N^2
- The length of the coils magnetic field

However, this assumes a non-magnetic core. If a coil is wound on a closed ferromagnetic core the problem of defining the inductance becomes involved due to the hysteresis of the core. Without going into the explanation, the equation for an inductance with a ferromagnetic core is:

$$L = \frac{\mu_0 \mu_r A N^2}{l}$$

From this second formula for inductance, we can see that in addition to the factors mentioned above, the following features also affect its value:

- Permeability of the core material
- Space between the turns of a coil
- The shape of the coil

Effects on circuit induced voltage

We can now see that there are four (4) factors affecting the magnitude of the induced emf, and these are:

- The flux density of the magnetic field (B)
- The length 'L', in metres, of the conductor in the magnetic field
- The velocity (v) at which the inductor cuts the magnetic field
- The current flowing in the circuit

Mutual Inductance

If two (2) coils, A & B, are mounted adjacent to each other but not connected as shown in figure 11.9, then when switch 'S' is closed some of the flux induced in 'A' by the flowing current becomes linked with the 'B' and the induced emf in 'B' circulates a momentary current through the galvanometer 'G'. Similarly, if the switch 'S' is then opened, the collapse of the flux in coil 'A' induces an emf in 'B' in the reverse direction.

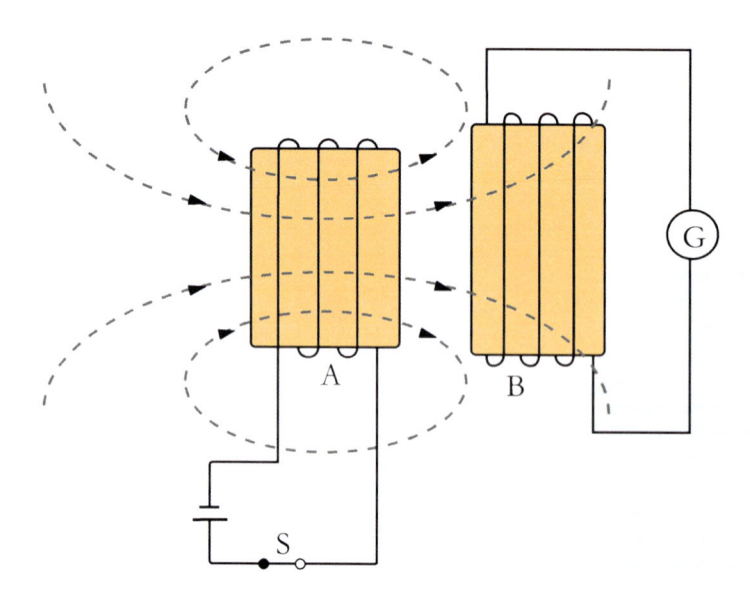

Figure 11.9 - Mutual Inductance

As a current change in one coil is accompanied by a change of flux linked with the other coil and therefore an emf is induced in it, the two coils are said to have *Mutual Inductance*. The unit of mutual inductance is the same as for self-inductance, ie the Henry, and the textbook definition is:

Two (2) coils have a mutual inductance of one (1) Henry if an emf of one (1) Volt is induced in one coil when the current in the other coil varies uniformly at a rate of one (1) Ampere/second.

If two circuits possess a mutual inductance of 'M' Henrys and the current in the primary circuit alters by Δi Amperes in Δt seconds, the induced emf in the secondary circuit is given by:

$$e = M \frac{\Delta i}{\Delta t} \text{ Volts}$$

The induced emf tends to circulate a current in the secondary circuit that opposes the increase of flux due to the increase of current in the primary circuit.

If $\Delta\phi$ Webers is the increase of flux linked with the secondary circuit due to Δi in the primary circuit, then the emf induced in the secondary is given by:

$$e = N_2 \frac{\Delta\Phi}{\Delta t} \text{ Volts}$$

Where N_2 is the number of turns in the secondary.

By combining these two equations we get:

$$M \frac{\Delta i}{\Delta t} = N_2 \frac{\Delta\Phi}{\Delta t}$$

$$\therefore M = N_2 \frac{\Delta\Phi}{\Delta i} \text{ Henrys}$$

This equates to:

$$\frac{\text{Change of flux linkages with the secondary}}{\text{Change of current in the primary}}$$

If the relative permeability of the magnetic circuit remains constant, then the ration of $\Delta\phi/\Delta i$ must also remain constant and is equal to the flux/Ampere and so:

$$M = \frac{N_2 \Phi_2}{I_1} \text{ Henrys}$$

Another formula for mutual inductance, which we do not have to prove here, thank goodness, is:

$$M = \frac{\mu_0 \mu_r A N_1 N_2}{l}$$

From these formulae for determining mutual inductance, we can see that the factors that affect it are:

- The permeability of the core material
- Cross sectional area of the coils
- Number of turns in each coil
- The length of the coils

Other factors that also influence mutual inductance are:

- Spacing between the turns of the coils
- Shape of the coils
- Position of the coils with respect to each other

Coefficient of Coupling

As we have seen, the mutual inductance of two adjacent coils is dependent upon the physical dimensions of the two coils, the number of turns in each coil, the distance between the two coils, the relative positions of the axes of the two coils and the permeability of the cores.

However, there is another factor that must be taken into account, which is the *Coefficient of Coupling* between the two coils. The coefficient of Coupling is given the SI symbol '*K*' and is equal to the ratio of the flux cutting one coil to the flux originating in the other coil.

If two coils are positioned so that all of the flux of one coil cuts all of the turns of the second, then the coils have a K of unity or one (1). In reality it is not possible to get a K of one (1), but it does approach this value in certain efficient coupling devices.

The mutual inductance (M) between two coils, L_1 and L_2, is expressed in terms of the inductance of each coil and their coefficient of Coupling (K), i.e.:

$$M = K\sqrt{L_1 L_2}$$

Where:

M is the Mutual Inductance in Henrys

K is the Coefficient of Coupling

L_1 and L_2 are the inductance values of each coil in Henrys

Example

Two (2) coils of 3H and 4H are connected in series and are physically close enough to each other so that their Coefficient of Coupling is 0.5. What is the Mutual Inductance between the coils?

$$M = K\sqrt{L_1 L_2}$$
$$M = 0.5\sqrt{3 \times 12}$$
$$M = 0.5\sqrt{36}$$
$$M = 0.5 \times 6$$
$$M = 3H$$

Inductors in Series

If the inductors in a given circuit are shielded or are far enough apart to prevent mutual inductance, as in figure 11.10, then the total inductance of the circuit is cumulative and can be calculated in the same way as resistors in series, i.e.

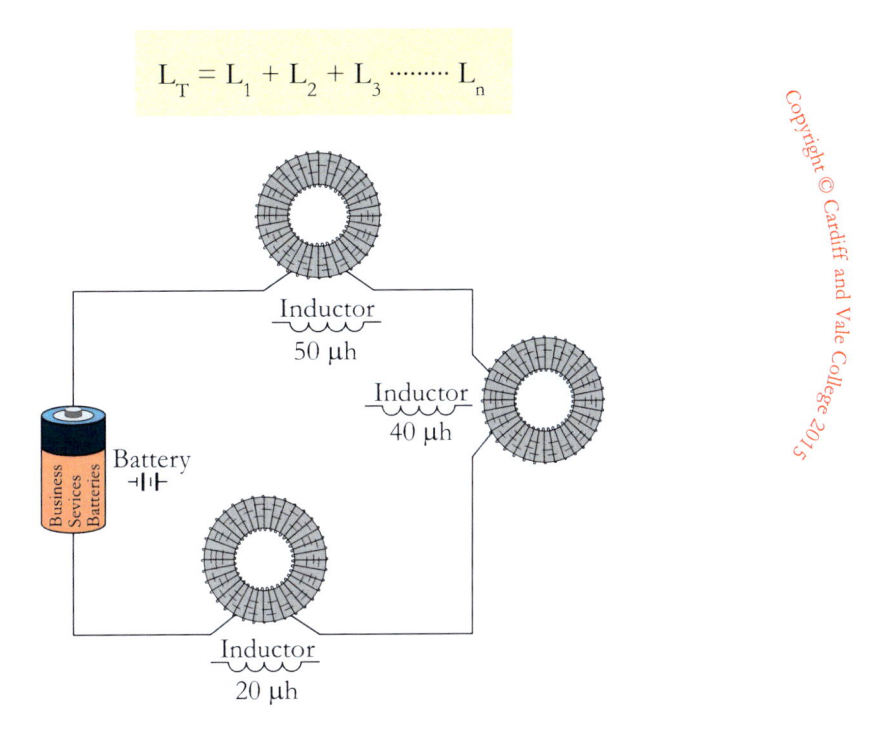

$$L_T = L_1 + L_2 + L_3 \cdots\cdots L_n$$

Figure 11.10 – Inductors in Series

Using this formula, the total inductance of the series circuit shown in figure 11.10 is:

$$L_T = 50 + 40 + 20$$
$$\therefore L_T = 110\mu h$$

Series Inductors with Magnetic Coupling

When two inductors in series are arranged so that they have mutual inductance, their combined inductance is given by:

$$L_T = L_1 + L_2 \pm 2M$$

Where:

L_T is the total inductance

L_1 and L_2 are the individual inductance values

M is the Mutual Inductance between the coils

The plus (+) sign is used with M when the magnetic fields of the two inductors are **aiding** each other, the minus (-) sign is used with M when the magnetic fields of the two inductors **oppose** each other, while the factor 2M accounts for the influence of L_1 on L_2 and L_2 on L_1.

Inductors in a Parallel Circuit

Placing inductors in parallel, as in figure 11.11, always decreases the total inductance of the circuit.

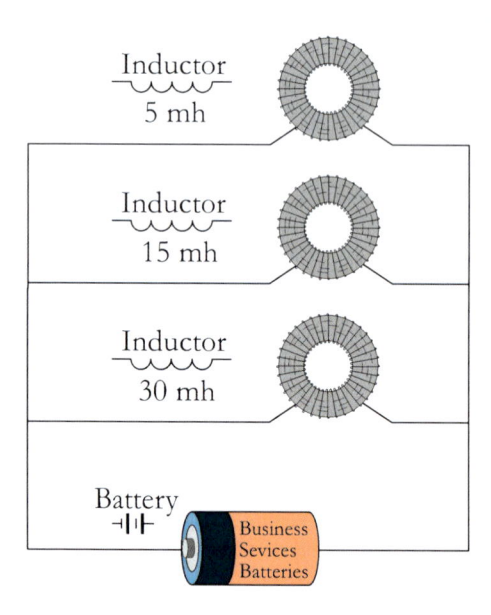

Figure 11.11 – Inductors in Parallel

The circuit shown in figure 11.11 has three inductors in parallel and if they are shielded, or far enough apart to prevent mutual inductance, the total inductance of the circuit can be calculated using the same formula for resistors in parallel, i.e.

$$\frac{1}{L_T} = \frac{1}{L} + \frac{1}{L_2} + \frac{1}{L_3} + \cdots \cdots \frac{1}{L_n}$$

Using this formula, the total inductance of the series circuit shown in figure 11.11 is:

$$\frac{1}{L_T} = \frac{1}{5} + \frac{1}{15} + \frac{1}{30}$$

$$\frac{1}{L_T} = \frac{6+2+1}{30} = \frac{9}{30}$$

$$L_T = \frac{30}{9} = 3.3\dot{3} \text{ mh}$$

Calculation of mutually coupled inductors in parallel is beyond the scope of these notes.

Rise & Fall of 'I' in an Inductive Circuit

If we look at an inductive circuit like that shown in figure 11.12 below that has an inductor 'L' wrapped around a ferromagnetic core 'C' in parallel with a resistor 'R' across a voltage source 'E'.

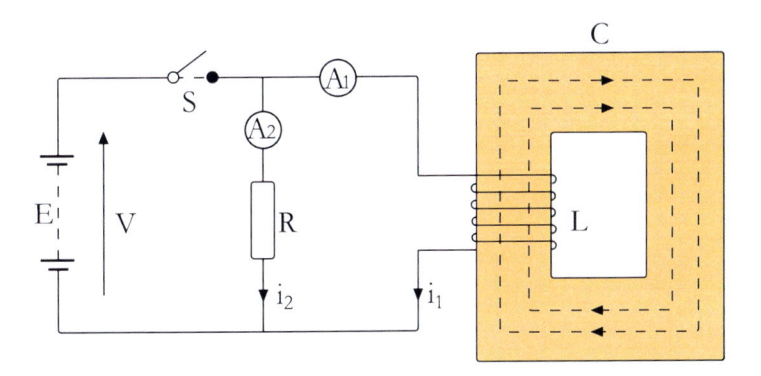

Figure 11.12 – Typical Resistor/Inductor Circuit

CHAPTER ELEVEN
INDUCTORS & INDUCTANCE

In the analysis that follows and in the graph in figure 11.13 below, capital letters represent fixed values while those in lower case represent changing values. Also, please note that although I_1 and I_2 look different maximum values in figure 11.13, this is only to keep them apart in the graph, in reality they are both approximately the same.

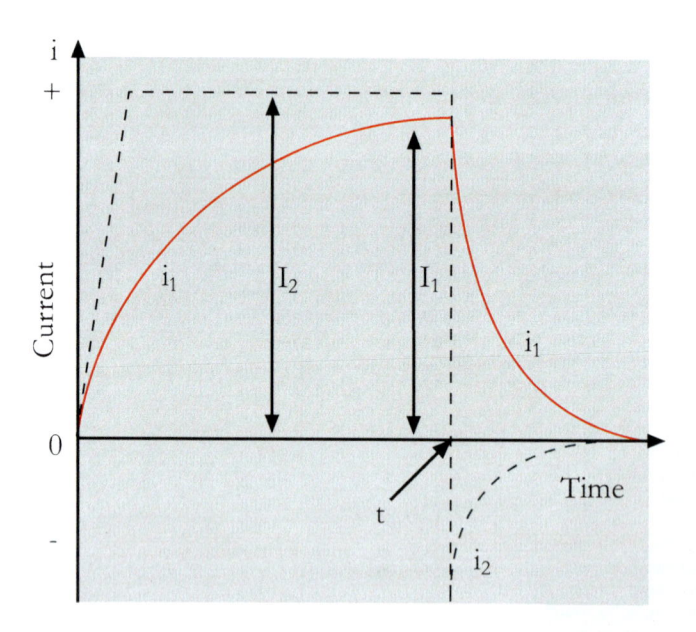

Figure 11.13 – Changing Currents at Switch On and Off

When switch 'S' in figure 11.12 is closed, the current through the resistor 'R' rises almost immediately to its maximum value I_2. However, the current i_1 flowing in the coil 'L' takes an appreciable time to grow to its maximum value.

When the switch 'S' is opened at time 't', the current i_1 through 'L' decreases slowly while the current in the resistor i_2 instantly reverses and becomes the same current as i_1, ie 'L's current is flowing through 'R'. So, what is going on?

The growth of current in 'L' when the switch is closed is accompanied by an increase of flux in the core 'C', which by Lenz's Law must induce an emf that opposes the change of current in 'L'. In the resistor part of the circuit, the induced flux is negligible and so the current grows instantaneously. When 'S' is opened, the currents in both 'L' and 'R' tend to decrease, but any decrease of i_1 is accompanied by a decrease of flux in the core 'C' and therefore by an induced emf in 'L' that opposes the decrease in current. This means the induced emf is acting in the same direction as the current and after 'S' has opened, the only path for the current to flow is through 'R' and this is why i_1 and i_2 are the same current.

If this experiment were performed again without the resistor, then when 'S' was closed the result would be the same, but when 'S' was opened there would be considerable '*arcing*' at the switch due to the maintenance of the current due to the induced emf in 'L'. This is why it is dangerous to quickly break the full excitation of an unprotected magnetic circuit such as a solenoid or dc machine.

Resistive-Inductive Series Circuit Time Constant

Whether it is a parallel resistor-inductor circuit as shown in figure 11.13 above, or a series circuit as in figure 11.14 below, when the switch 'S' is closed to position ①, the current flowing through the inductor, I_1 does not increase to its maximum value immediately but instead, grows slowly and non-linearly.

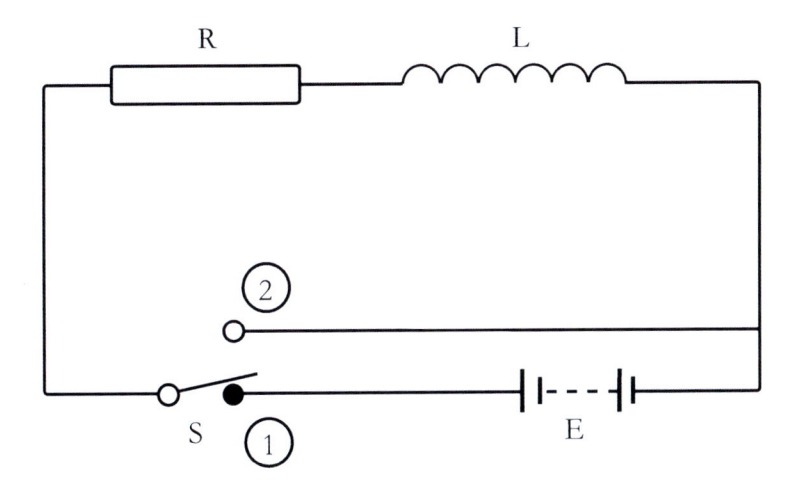

Figure 11.14 – Simple Resistor-Inductor Circuit

As already explained, this is due to the induced emf in the inductor that by Lenz's Law opposes the increasing current. As with the capacitor, the time taken for the current to reach maximum value is dependent on the values of the inductor and resistor.

The current reaches 63.2% of its limit in approximately one fifth of the time it takes to get to maximum and this is called the circuit's **RL Time Constant**, as illustrated by the first half of I_1 in figure 11.13 above. The RL time constant of a circuit is calculated by using the formula:

$$t = \frac{L}{R}$$

Where:

t is time in seconds

L is inductance in Henrys

R is resistance in ohms

In an RL circuit, it takes *five (5) times the RL time constant* for the current to reach its maximum value.

When switch is moved to position ②, the current does not fall immediately to zero as already discussed above, but falls exponentially to approximately 37% of its maximum value in one RL time constant and to zero in approximately five RL as illustrated in figure 11.15.

CHAPTER ELEVEN
INDUCTORS & INDUCTANCE

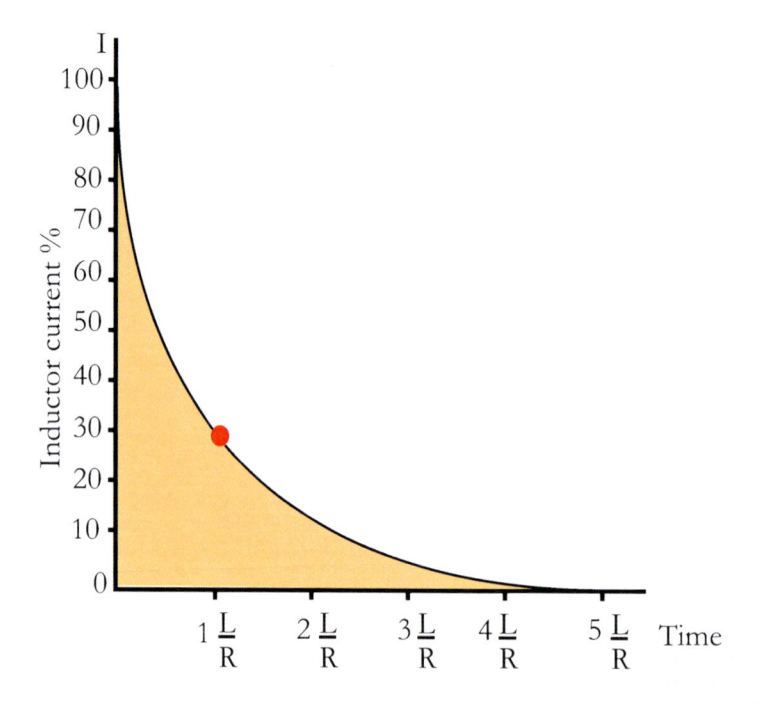

Figure 11.15 – Exponential Decay of Inductor Current

As with the growth of the current, the decay is modified due to the self-induced emf opposing the current change in accordance with Lenz's Law.

Energy Stored in an Inductive Circuit

The ideal inductor does not dissipate energy but stores it as a magnetic field and the stored energy (W) is given by the formula:

$$\text{Energy (W)} = \frac{1}{2} LI^2 \text{ joules}$$

This can be explained as follows:

If the current in a coil of constant inductance 'L' grows at a uniform rate from zero to 'I' Amperes in 't' seconds, the average value of the current is ½I and the emf induced in the coil is L × I/t. The power absorbed by the magnetic field is the product of the current and that component of the applied emf, which neutralises the induced emf. Therefore the average power absorbed by the magnetic field is:

$$\frac{1}{2} I \times \frac{LI}{t}$$

Given then that the total energy absorbed by the magnetic field is average power × time:

$$\frac{1}{2} I \times \frac{LI}{t} \times t$$

$$\therefore W = \frac{1}{2} LI^2$$

Inductor Types

There are several types of inductors and their construction will depend on the application they are used in, frequency of operation and maximum current carrying capacity. They are generally made with a fixed value but some are variable.

However, unlike capacitors and resistors, inductors cannot be considered as pure elements. Inductors obviously introduce inductance into a circuit but they also always bring in resistance due to their construction.

Inductors are made of coils of wire that cannot be of too large a cross-sectional area, and so the resistance is at least a few ohms and can be several thousand.

Inductors fall into two (2) main categories:

- Air Core
- Ferromagnetic Core

Air Core

Air core inductors have the advantage of having a linear B/H curve characteristic, which means that the inductance is the same no matter what the current. However, as the relative permeability of air is '1' it means they have low values of inductance, typically in the range µH to mH. They are used in **high frequency** circuits up to 250Mhz; e.g. radio transmitters and receivers, televisions, etc.

Ferromagnetic Core

Ferromagnetic core inductors produce very high values of inductance but their B/H curve characteristics are not linear and so the circuit inductance varies with the current flow. However, many of the modern inductors with sintered ferromagnetic cores, i.e. those constructed of powered alloy, have almost linear characteristics and so are ideal for modern electronic circuits. They are typically used at mains and audio frequencies, i.e. up to 20kHz, where high values of inductance are required; i.e. up to 10H.

Variable inductors have a core that is mounted on a screw assembly that can be moved in and out of the coil to vary the inductance. Figure 11.16 shows the circuit symbols used for these various inductors.

Air Core Inductor

Ferromagnetic Inductor

Variable Inductor

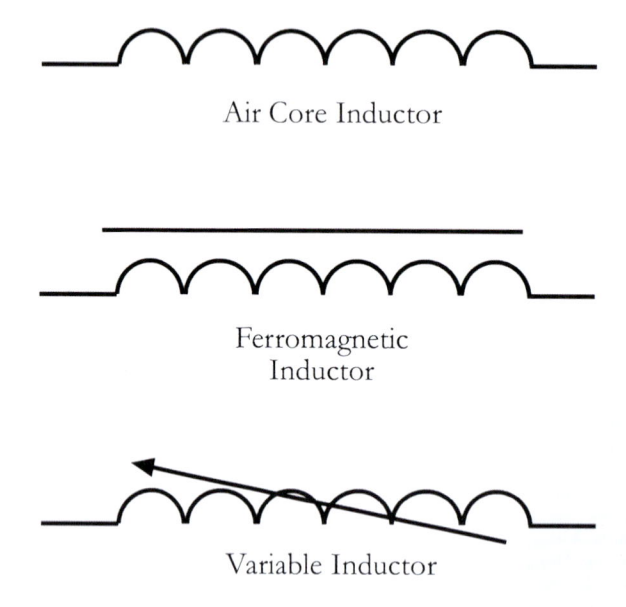

Figure 11.16 – Typical Inductor Circuit Symbols

Inductor Applications

Inductor properties make them very useful for various applications. For example, as they oppose any changes in current, inductors can be used to protect circuits from surges of current. Inductors are also used to stabilise direct current (dc) and to control or eliminate alternating current (ac). Inductors used to eliminate alternating current above a certain frequency are often called *chokes* as they *choke* the increase of current in a coil and if a coil is wound on a soft iron core, it more effectively *chokes* the increase of a current than the same coil with an air core. Some typical examples of inductor use are detailed below:

Generators

One of the most common uses of electromagnetic inductance is in the generation of electric current, which is covered in a later chapter of this module.

Radio Receivers

Inductors can be used in circuits with capacitors to generate and isolate high frequency currents. For example, inductor coils are used with capacitors in tuning circuits of radios. In figure 11.17 below, a variable capacitor is connected to an antenna transformer circuit.

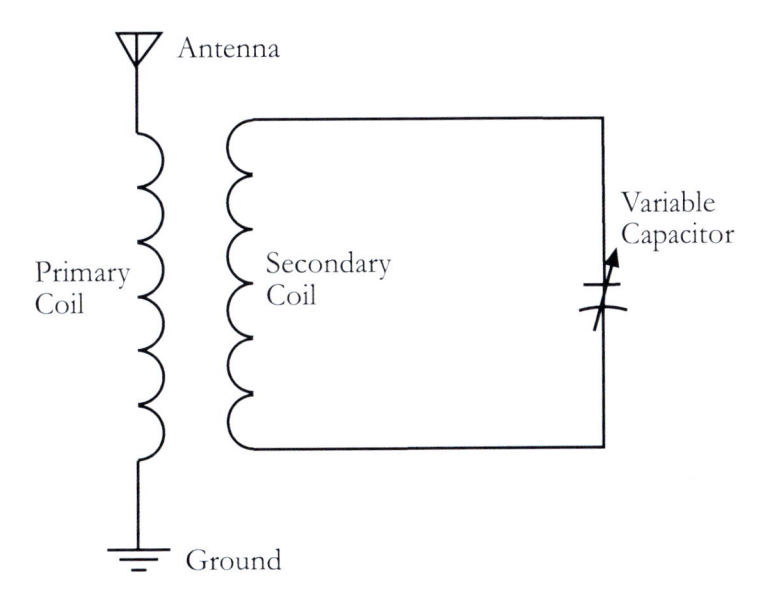

Figure 11.17 – Typical Radio Circuit

Transmitted radio waves cause an induced current to flow in the antenna through the primary inductor coil to ground. A secondary current in the opposite direction is induced in the secondary inductor coil and this current flows to the capacitor.

The surge of current to the capacitor induces a counter emf called ***capacitive reactance*** and the induced flow of current through the coil induces a counter emf called ***inductive reactance***. So we end up with both capacitive and inductive reactance in the circuit.

At higher frequencies, inductive reactance is greater, capacitive reactance is smaller, and at lower frequencies, the opposite is true. A variable capacitor is utilised to equalise the inductive and capacitive reactance and the condition at which the reactance are equal is called ***resonance***. The particular frequency isolated by the equalised reactance is called the circuit's ***resonant frequency***.

A radio circuit is tuned by adjusting the capacitance of the variable capacitor to equalise the inductive and capacitive reactance of the circuit for the desired resonant frequency, i.e. to tune in the desired radio station.

CHAPTER ELEVEN
INDUCTORS & INDUCTANCE

Metal Detectors

Metal detector operation is based on the principle of electromagnetic induction. Metal detectors contain one or more inductor coils and when metal passes through the magnetic field generated by the coil or coils, the field induces electric currents in the metal. These currents are called eddy currents, which in turn induce their own magnetic field that generates current in the detector to power a signal indicating the presence of the metal.

Moving Coil Meter

A moving coil meter is used to measure small voltages and currents and uses the motor principle, i.e. when a current carrying inductor is placed in a magnetic field the conductor moves. The moving coil is wound on a soft iron core and they are then mounted between poles of a strong magnet, figure 11.18.

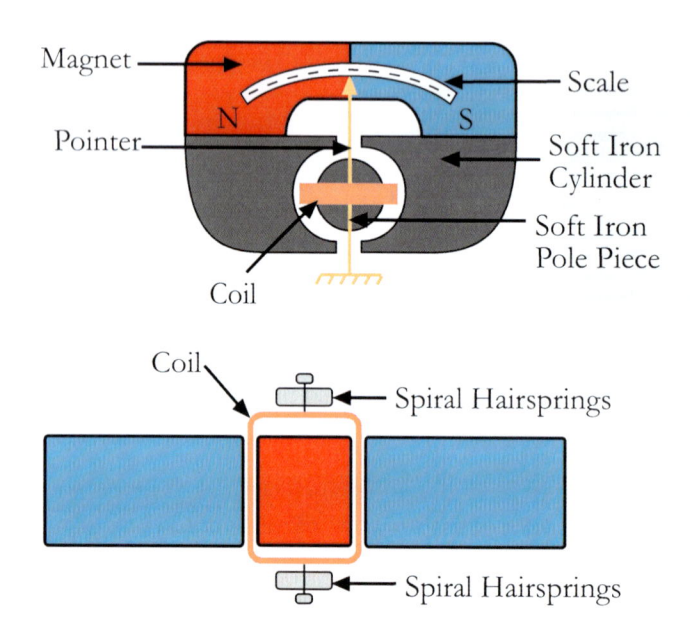

Figure 11.18 – Simple Moving Coil Meter

The front elevation and sectional plan in figure 11.18 shows a permanent magnet with attached soft iron pieces. As most permanent magnetic materials are hard and difficult to machine accurately, this arrangement allows the soft iron material to be machined to give exact air-gap dimensions. This can be achieved by a resin-bonding technique where the soft iron pieces are attached in one piece to the grinded flat surface of the magnet.

The moving coil is insulated copper wire wound onto a light frame, e.g. aluminium, fitted with polished steel pivots resting on jewelled bearings. Current feeds into and out of the coil via the spiral hairsprings, which are also used to control the torque and bring the pointer to rest on the scale and return the pointer to zero when the measured value is removed. The coil, wound on its own soft iron cylinder, is free to move in the air gap between the soft iron pieces. The cylinder helps intensify the magnetic field by reducing the length of the air gap and gives a radial magnetic flux of uniform density so that the scale is linear.

The meter works because when current flows in it, a magnetic field is developed around the coil. This reacts with the permanent magnetic field, a turning force, i.e. torque, is placed on the coil, and it and its attached pointer rotate.

The pointer then moves across the scale and indicates the measured value. Its displacement is directly proportional to the current flowing in the coil.

Moving Coil Microphone & Speaker

Sound waves travel through the air and microphones convert this mechanical movement into electrical signals or vice versa for the speaker. Figure 11.19 shows a typical construction that can be used for a speaker or microphone; indeed, in some applications, e.g. police radios, the microphone and speaker are the same thing.

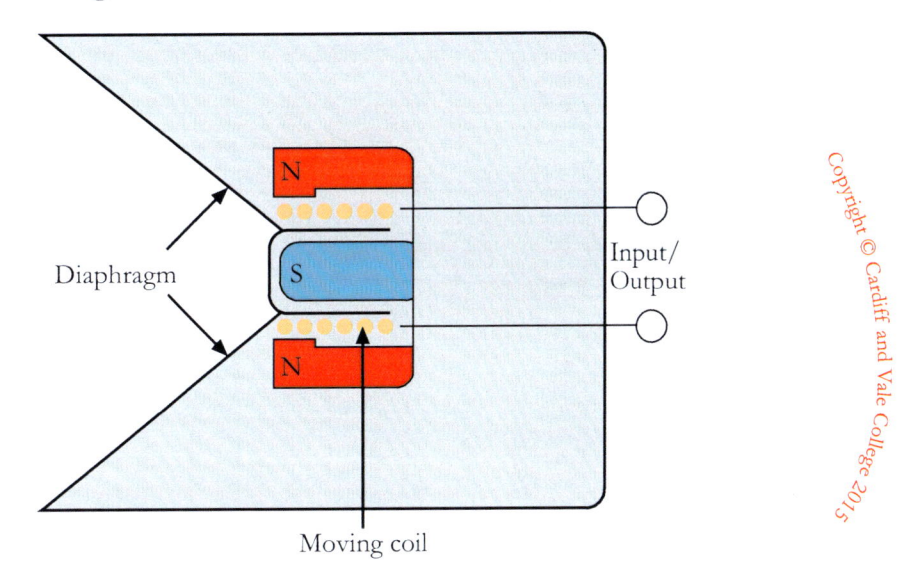

Figure 11.19 – Simple Microphone and/or Speaker

When operating as a microphone, the vibrating sound waves cause the diaphragm to vibrate in unison with them. The moving coil, wrapped around a tube that encases the centre pole of the permanent magnet, also moves in unison with the sound waves as it is attached to the diaphragm. The now moving coil in the magnetic field induces an emf, which has the same amplitude and frequency, within limits, as that of the sound wave. The induced emf is then available as an output to other circuits.

When operating as a speaker, the electrical signals enter the coil and a magnetic field forms around it. This interacts with the permanent magnetic field, causing the coil to move in and out between the magnetic poles so also causing the diaphragm to vibrate. This disturbs the air around the diaphragm so setting up a sound wave that varies in frequency and amplitude, within limits, to that of the electrical signal input.

CHAPTER ELEVEN
INDUCTORS & INDUCTANCE

Summary

Inductance is the property of a conductor, often in the shape of a coil, which is measured by the *size of the emf* induced in it, *compared with the rate of change of the electric current* that produces the voltage. A *steady current* produces a *stationary* magnetic field; a *steadily changing* current, *alternating current*, or *fluctuating direct current* produces a *varying* magnetic field, which in turn, induces an emf in a conductor that is present in the field. The size of the induced emf is proportional to the rate of change of the electric current. The proportionality factor is called the inductance and is defined as the value of the emf induced in a conductor divided by the magnitude of the rate of change of the current causing the induction.

If the emf is induced in a conductor that is different from the one in which the current is changing, the phenomenon is called *mutual induction*. A changing magnetic field caused by a varying current in a conductor, however, also induces an emf in the very conductor that carries the changing current and such a phenomenon is called *self-induction*.

A self-induced emf opposes the change that brings it about. Consequently, when a current begins to flow through a coil of wire, it undergoes an opposition to its flow in addition to the resistance of the metal wire. On the other hand, when an electric circuit carrying a steady current and containing a coil is suddenly opened, the collapsing, and hence diminishing, magnetic field causes an induced electromotive force that tends to maintain the current and the magnetic field and may cause a spark between the contacts of the switch. The self-inductance of a coil, or simply its inductance, may therefore be thought of as *electromagnetic inertia*, a property that opposes changes both in current and in magnetic field.

Revision

Inductors & Inductance

Questions

1. If a current through a coil with an inductance of 4 H is increased from 0A to 5A in 0.5 seconds, the induced voltage is:

    a) 25 V

    b) 20 V

    c) 40 V

2. If the velocity of a conductor moving in a magnetic field is increased, the induced emf:

    a) Increases

    b) Decreases

    c) Remains the same

3. If a circuit has three inductors in series that are separated from each other and have inductances of 45 mH, 30 mH and 15 mH respectively, the circuit's overall inductance is:

    a) 90 mH

    b) 90 µH

    c) 8.2 mH

CHAPTER ELEVEN
INDUCTORS & INDUCTANCE

4. If a circuit has an inductance of 75 mH and a resistance of 25 kΩ, its circuit time constant will be:

    a) 3 µs

    b) 15 µs

    c) 6 µs

5. The picture shown in the figure below is a:

    a) Air core Inductor

    b) Ferro magnetic inductor

    c) Variable inductor

6. An inductor opposes the change of:

    a) Current

    b) Voltage

    c) Magnetism

7. A capacitor opposes the change of:

    a) Current

    b) Voltage

    c) Magnetism

8. The induced emf in an inductor that opposes the applied voltage is called:

    a) A self emf

    b) Back emf

    c) Reverse emf

Revision

Inductors & Inductance

Answers

1. **C**   EMF = L x (I ÷ t)

2. **A**

3. **A**

4. **A** $t = l \div r = 0.075 \div 25000 = 3 \times 10^{-6} = 3\mu s$

5. **B**

6. **A**

7. **B**

8. **B**

DC Generator & Motor Theory

Introduction

DC machines were the first electrical machines invented in the 1800's when in 1839 an elementary motor was used to drive a locomotive in Edinburgh. Most electric machines convert energy by using a magnetic field that allows force to be transmitted from a stationary to a moving part without physical connection. There are two basic principles utilised in generator and motor operation. The first, originally discovered by the French physicist André-Marie Ampere, states that:

> *An electrical conductor carrying a current at right angles to a magnetic field will experience a force at right angles to both the field and the current.*

The second principle, formulated on the observations made by Michael Faraday, states that:

> *A potential difference, or voltage, will be established between the ends of an electrical conductor that moves across or perpendicular to a magnetic field.*

These principles apply for a moving conductor in a stationary magnetic field or equally for a *stationary* conductor with a *moving* magnetic field.

Basic Generator Principles

A generator is a machine that converts mechanical energy into electrical energy by using the principle of magnetic induction. This can be summarised as:

Whenever a conductor is moved within a magnetic field in such a way that the conductor cuts across magnetic lines of flux, voltage is generated in the conductor.

CHAPTER TWELVE
DC GENERATOR & MOTOR THEORY

The amount of voltage generated depends on:

- The strength of the magnetic field

- The angle at which the conductor cuts the magnetic field

- The speed at which the conductor is moved

- The length of the conductor within the magnetic field

- The polarity of the voltage depends on the direction of the magnetic lines of flux and the direction of movement of the conductor

To determine the direction of the conventional current in a given situation, we can use **Fleming's right-hand rule for generators**, illustrated in figure 12.1.

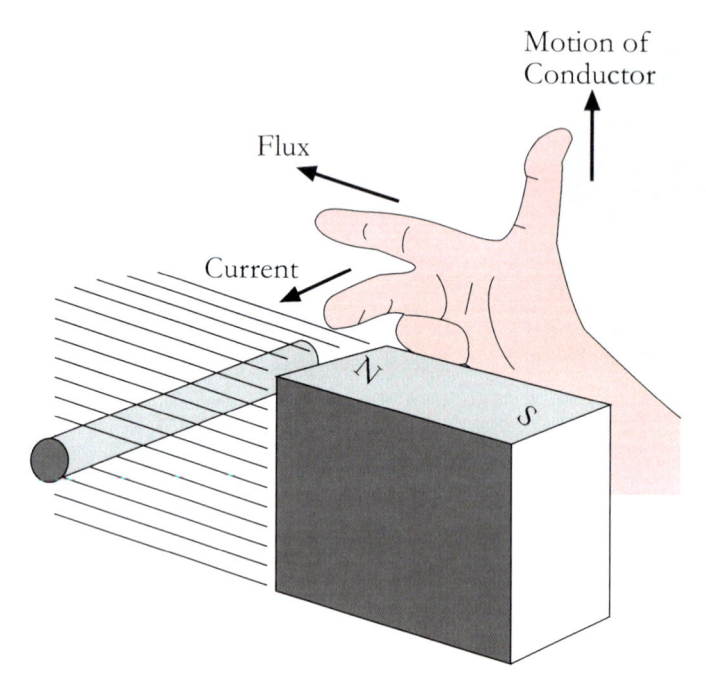

Figure 12.1 - Fleming's Right-Hand Rule for Generators

The rule is applied by using the thumb and first two fingers of the right hand, arranged perpendicular to each other as in figure 12.1. If the *thumb* is pointed in the direction of conductor movement; the *forefinger* in the direction of magnetic flux from north to south; then the *middle* finger points in the direction of current flow in an external circuit to which the voltage is applied.

Note: *Conventional* current flow is used in figure 12.1; if *electron* current flow is used, it becomes the *left-hand* rule, which is widely used in the USA.

A simple way of remembering which way round these rules are is that in the UK we have the MG car, i.e. motor is left hand and generator is right hand, while in the USA they have GM cars.

256

The Simple Generator

The simplest generator is an ac generator and basic generating principles are more easily explained by looking at this type first; the DC generator will then be discussed later. A simple generator, figure 12.2, consists of a wire loop positioned so that it can rotate in a stationary magnetic field and as it does will produce an induced emf in the loop. Sliding contacts, usually called *brushes*, connect the loop to an external circuit load in order to *pick up* or use the induced emf.

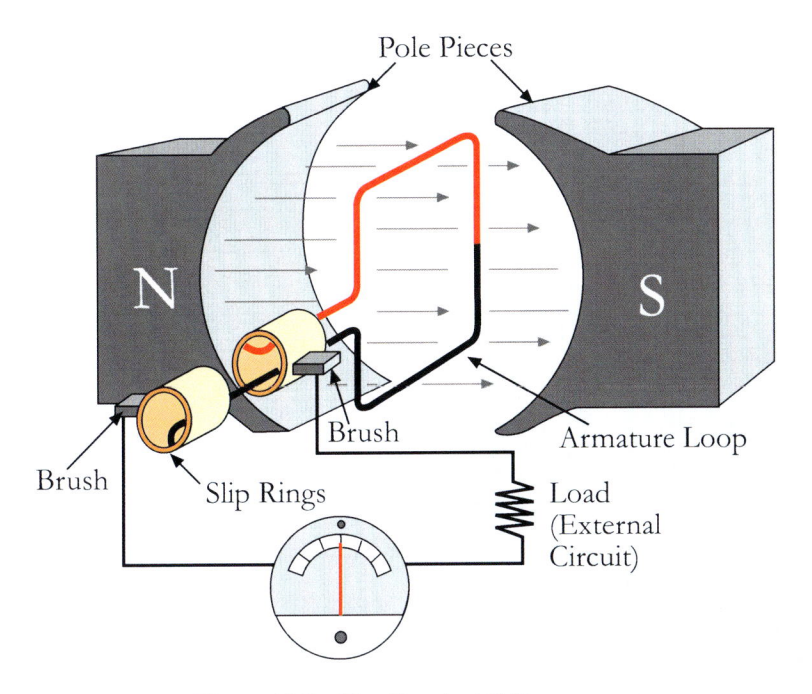

Figure 12.2 - The Simple AC Generator

In figure 12.2 the pole pieces provide the magnetic field and are shaped and positioned as illustrated to concentrate the magnetic field as close as possible to the wire loop. The rotating wire is called the *Armature* and its ends are connected to rings, more correctly called *Slip Rings*, which rotate with the armature. Brushes, usually made of carbon, ride against the slip rings and have wires attached to them that connect to the external load. Any generated voltage appears across these brushes; this is demonstrated graphically in figure 12.3.

CHAPTER TWELVE
DC GENERATOR & MOTOR THEORY

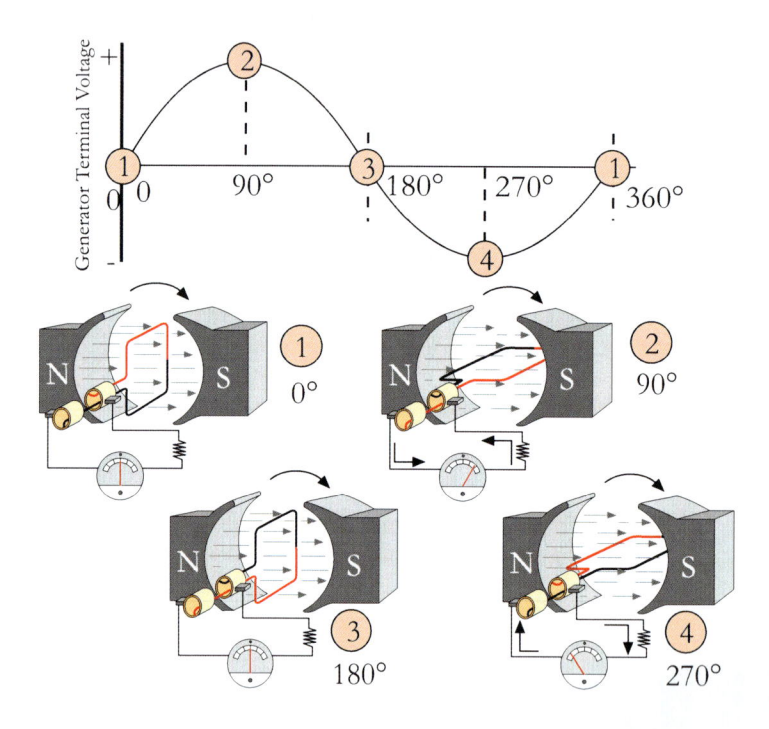

Figure 12.3 – Voltage Generation

The simple generator illustrated in figure 12.3, produces a voltage as the armature loop rotates in a clockwise direction. For the purposes of this explanation, the initial or starting point is shown with the loop vertical, i.e. at position ①, called the 0° position. As illustrated in figure 12.3, at 0° the armature loop is perpendicular to the magnetic field and the loop's red and black conductors are moving parallel to the magnetic field.

At this point, indeed at any time the conductors are moving parallel to the magnetic field, they do not cut any lines of flux, so no emf is induced and the meter at position ① indicates zero. This position is called the **Neutral Plane**. As the armature loop now rotates from position ① to ②, i.e. 0° to 90°, the conductors cut through more and more lines of flux, at a continually increasing angle. At 90°, they are cutting through maximum flux lines and at a maximum angle. This results in the induced emf going from zero to maximum between 0° and 90°.

Note that from 0° to 90°, the **red** conductor cuts **down** through the field and at the same time the **black** conductor cuts **up** through the field. This means that the induced emfs in the conductors are **series-adding** and the resultant voltage across the brushes, i.e. the terminal voltage, is the sum of the two induced voltages and the meter at position ② reads maximum value.

As the armature loop continues rotating from 90° to 180°, position ③, the conductors that were cutting through a maximum of flux lines at position ② now cut through fewer lines and they are again moving parallel to the magnetic field at position ③, no longer cutting through any flux lines. As the armature continues the rotation from 90° to 180°, the induced voltage decreases to zero in the same way that it increased during the rotation from 0° to 90° and the meter reads zero again.

As the armature has rotated from 0° to 180°, the conductors have been moving through the magnetic field in the same direction and so the polarity of the induced voltage has remained the same, illustrated by points A to C on the graph in figure 12.3. As the loop rotates beyond 180° through 270°, position ④, and back to the initial starting point, position ①, the direction of the conductors' cutting reverses through the magnetic field. Now the *red* conductor cuts *up* through the field while the *black* conductor cuts *down* through the field and as a result, the polarity of the induced voltage reverses, as shown in figure 12.3. The terminal voltage will be the same as it was from ① to ③, except that the polarity is reversed as shown by the meter deflection at position ④. The voltage output waveform for a complete revolution of the loop is shown on the graph in figure 12.3.

The Simple DC Generator

With the DC generator, we need to look at a single-loop generator again, but this time with each terminal connected to a section of a two-segment metal ring. The two segments are insulated from each other, forming a *Commutator* that replaces the slip rings of the ac generator, Figure 12.4.

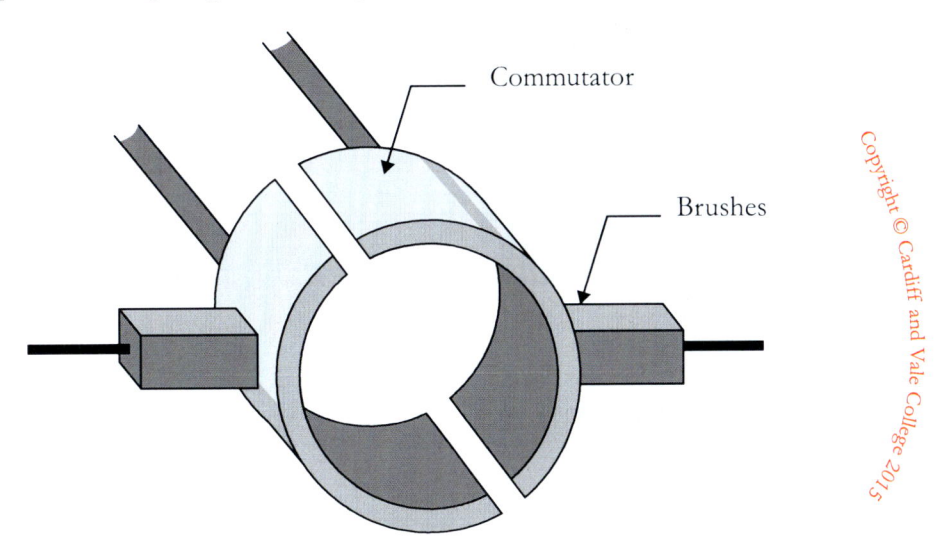

Figure 12.4 – Commutator Segments

The commutator *mechanically* reverses the armature loop connections to the external circuit and this occurs at the same time that the armature loop voltage reverses. Through this process, known as *commutation*, the commutator changes the generated ac voltage to a pulsating DC voltage as shown in the graph of figure 12.5.

CHAPTER TWELVE
DC GENERATOR & MOTOR THEORY

Figure 12.5 – Results of Using a Commutator

When the armature loop rotates clockwise from ① to ②, a voltage is induced in the armature loop that causes a current in a direction that deflects the meter to the right. Current flows through loop, out of the negative brush, through the meter and the load, and back through the positive brush to the loop. Voltage reaches its maximum value at ② on the graph and the generated voltage and the current fall to zero at ③. At this instant, each brush makes contact with both segments of the commutator and as the armature loop continues to rotate to position ④, a voltage is again induced in the loop, but is the opposite polarity. The voltages induced in the two sides of the coil at ④ are in the reverse direction to that shown at ②.

In this case, the current is flowing from the **black** side to the **red** side in position ② and from the **red** side to the **black** side at ④. However, because the segments of the commutator have rotated with the loop and are contacted by opposite brushes, the direction of current flow through the brushes and the meter remains the same as at ②.

The voltage developed across the brushes is now pulsating and unidirectional, i.e. in one direction only and varies twice during each revolution between zero and maximum; this variation is called **Ripple**.

Obviously, for most applications, a pulsating voltage is inappropriate. Therefore, in practical generators, more armature loops, i.e. coils, and more commutator segments are used to produce an output voltage waveform with less ripple. The effects of additional coils may be illustrated by adding a second coil to the armature, figure 12.6.

CHAPTER TWELVE
DC GENERATOR & MOTOR THEORY

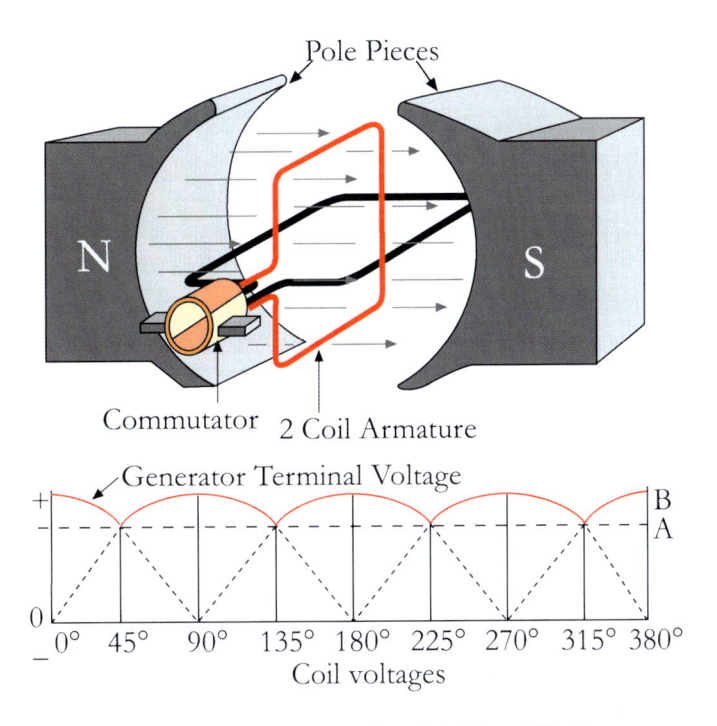

Figure 12.6 – Effect of an Additional Coil

The commutator in figure 12.6 must now be divided into four (4) parts since there are now four (4) coil ends. As the assembly rotates in a clockwise direction, the voltage induced in the **black** coil, **decreases** for the next 90° of rotation.

In other words, from maximum to zero and the voltage induced in the **red** coil *increases* from zero to maximum at the same time. Since there are four (4) segments in the commutator, a new segment passes each brush every 90° instead of every 180°. This allows the brush to switch from the black coil to the red coil at the instant the voltages in the two coils are equal. The brush remains in contact with the **red** coil as its induced voltage increases to maximum, level 'B' in the graph and then decreases to level 'A', 90° later. At this point, the process repeats itself.

The graph in figure 12.6 shows the ripple effect of the voltage when two armature coils are used. Since there are now four (4) segments in the commutator and only two brushes, the voltage cannot fall any lower than at point 'A' and therefore, the ripple is limited to the rise and fall between levels 'A' and 'B' on the graph. Adding more armature coils can further reduce the ripple effect, and decreasing ripple in this way increases the **effective** voltage of the output.

The **Effective voltage** is the equivalent level of DC voltage that will cause the same average current flow through a given resistance. By using additional armature coils, the voltage across the brushes is not allowed to fall to as low a level between peaks. Practical generators use many armature coils and usually also use more than one pair of magnetic poles. The additional magnetic poles have the same effect on ripple, as does the additional armature coils. In addition, the increased number of poles provides a stronger magnetic field, ie greater flux lines, and this in turn, allows an increase in output voltage because the coils cut more lines of flux per revolution.

261

Electromagnetic Poles

Nearly all-practical generators use electromagnetic poles instead of the permanent magnets used in the simple generator illustrated earlier. The electromagnetic field poles consist of coils of insulated copper wire wound on soft iron cores, as shown in figure 12.7.

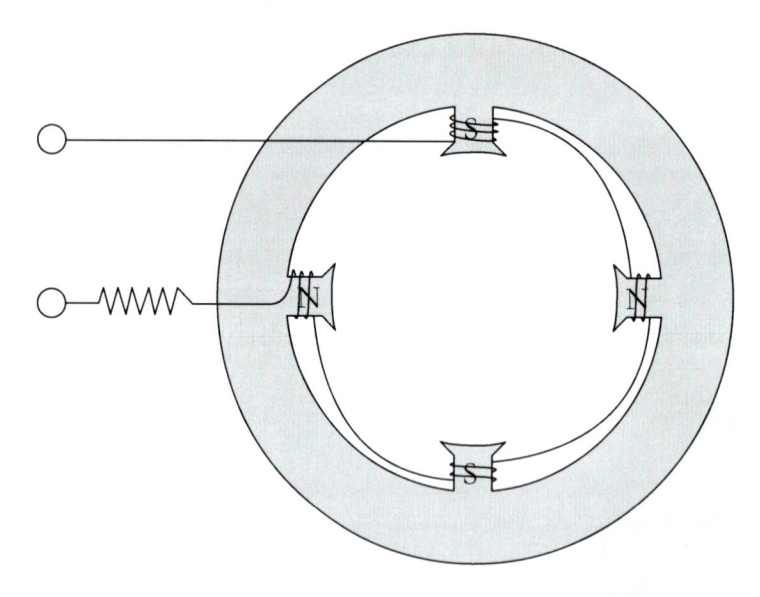

Figure 12.7 – Four-Pole Electromagnetic Generator

The main advantages of using electromagnetic poles are that there is increased field strength and now it can be controlled; i.e. by varying the input voltage, the field strength can vary; and by varying the field strength, the output voltage of the generator can be controlled.

Commutation

Commutation, as discussed earlier, is the process by which a DC voltage output is taken from an armature that has an ac voltage induced in it. As shown with the example of the simple generator in figure 12.5, the commutator *mechanically* reverses the armature loop connections to the external circuit.

This occurs at the same instant that the voltage polarity in the armature loop reverses. A DC voltage is applied to the load because the output connections are reversed as each commutator segment passes under a brush, the segments being insulated from each other.

In figure 12.8, commutation occurs simultaneously in the two coils that are briefly short-circuited by the brushes. Coil 'B' is short-circuited by the negative brush while coil 'Y', the opposite coil, is short-circuited by the positive brush. However, the brushes are positioned on the commutator so that each coil is short-circuited as it moves through its own electrical *neutral* plane, i.e. when there is no voltage generated in the coil, so preventing sparking between the commutator and brush. Sparking between the brushes and commutator is an indication of improper commutation, usually caused by improper brush placement.

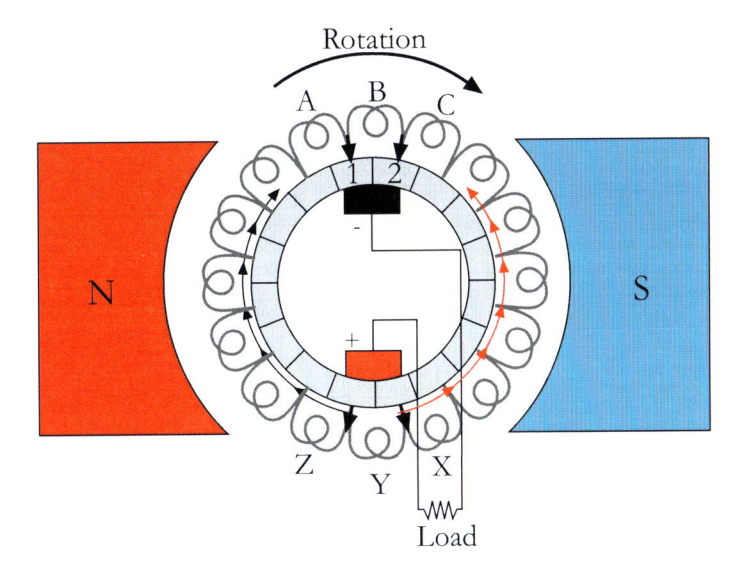

Figure 12.8 - Commutation of a DC Generator

Armature Reaction

From previous chapters, you know that all current-carrying conductors produce magnetic fields. The magnetic field produced by current in the armature of a DC generator affects the flux pattern and distorts the main field, causing a shift in the neutral plane, which affects commutation. This change in the neutral plane and the reaction of the magnetic field is called *Armature Reaction*.

As previously discussed for proper commutation, the coil short-circuited by the brushes must be in the neutral plane. If we look at the operation of a simple two-pole DC generator, figure 12.9, view 'A' shows the field poles and the main magnetic field and the armature is shown in a simplified view in views 'B' and 'C' with the cross section of its coil represented as small circles.

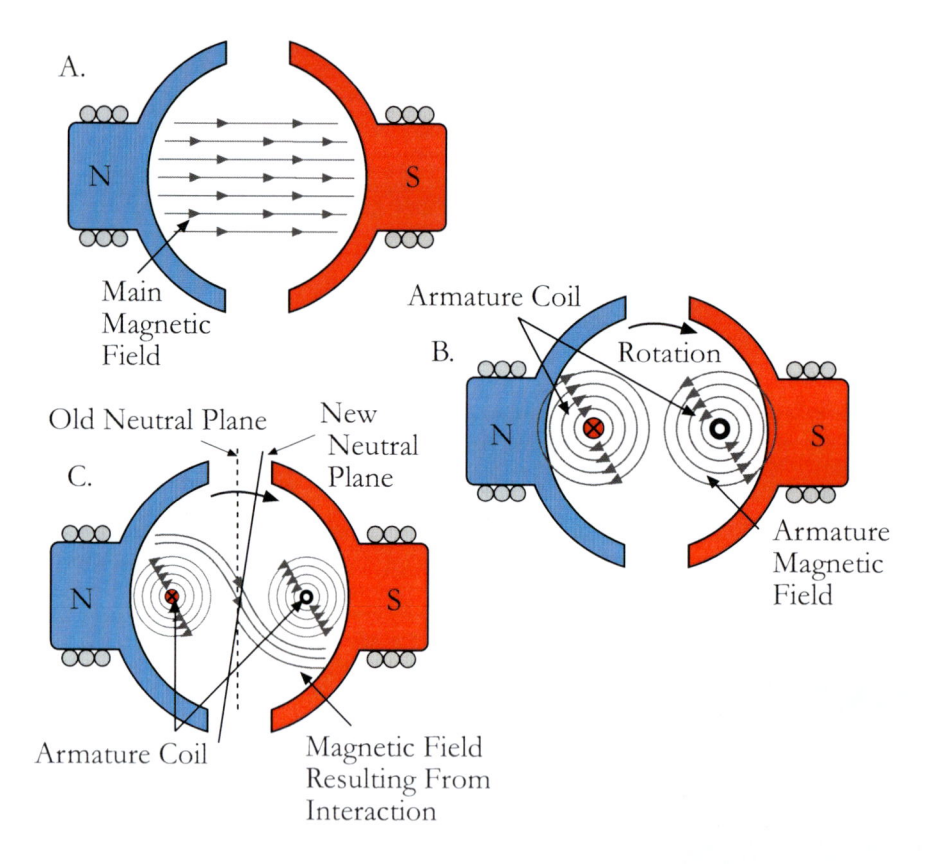

Figure 12.9 - Armature Reaction

The symbols within the circles represent arrows with the dot representing the point of an arrow coming towards the front, and the cross represents the tail, or feathered end, going away to the back. When the armature rotates clockwise, the sides of the coil to the left will have current flowing towards the front, as indicated by the dot and the other side of the coil will have current flowing away. The field generated around each side of the coil is shown in view 'B' of figure 12.9 and this increases in strength for each wire in the armature coil, setting up a magnetic field almost perpendicular to the main field.

Now there are two (2) fields, i.e. the main field, view 'A', and the field around the armature coil, view 'B'. View 'C' shows how the armature field distorts the main field and how the neutral plane is shifted in the direction of rotation. Now if the brushes remain in the old neutral plane, they will be short-circuiting coils that have voltage induced in them and so there will be arcing between the brushes and commutator. To prevent this arcing in a practical generator, the brushes must therefore be shifted to the new neutral plane.

However, shifting the brushes to the advanced position, i.e. the new neutral plane, does not completely solve the problems of armature reaction as its effect varies with the load current. Therefore, each time the load current varies, the neutral plane shifts and this means the brush position must be changed each time the load current varies.

In small generators, the effects of armature reaction are reduced by physically shifting the position of the brushes. However, this is not practical for large generators and other means are used eliminate armature reaction.

Compensating Windings and Interpoles

In larger generators, *Compensating Windings* or *Interpoles* are used to overcome armature reaction, figure 12.10.

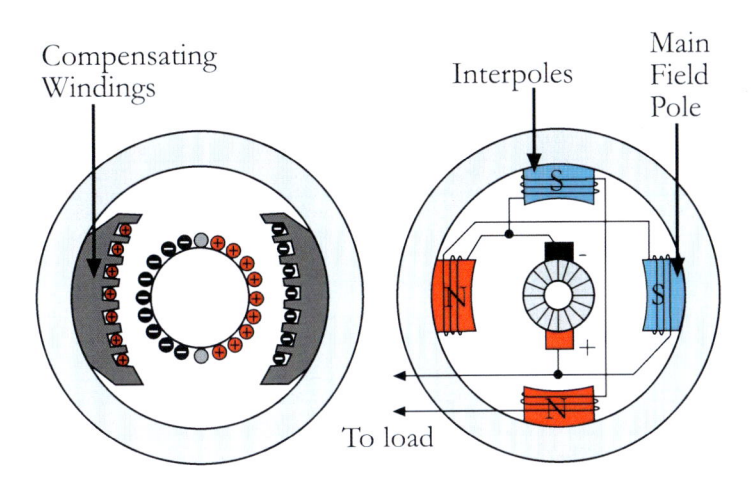

Figure 12.10 – Compensating Windings & Interpoles

Compensating windings consist of a series of coils embedded in slots in the pole faces and these are connected in series with the armature. The series-connected compensating windings produce a magnetic field that varies *directly* with armature current. However, because the compensating windings are wound to produce a field that opposes the magnetic field of the armature, they tend to cancel the effects of the armature magnetic field. The neutral plane will therefore remain stationary and in its original position for all values of armature current. As a result, once the brushes have been set correctly, they do not have to be moved again.

Another way to reduce the effects of armature reaction is to place small auxiliary poles, called *Interpoles*, between the main field poles. These have a few turns of wire and are connected in series with the armature. They are wound and placed so that each one has the same magnetic polarity as the next main pole ahead of it, in the direction of rotation, and the generated field then produces the same effect as the compensating winding. This field, in effect, cancels the armature reaction for all values of load current by causing a shift in the neutral plane opposite to the shift caused by armature reaction. The amount of shift created by the interpoles will equal the shift caused by armature reaction since both are a result of armature current.

Generator Motor Reaction

When a generator delivers current to a load, the armature current creates a magnetic force that opposes the rotation of the armature called the *Motor Reaction*. To simplify matters again, if we take a single armature conductor, figure 12.11 view 'A', when it is stationary, no voltage is generated and no current flows and so no force acts on the conductor. When the conductor is moved downwards, view 'B' figure 12.11, and the circuit is completed through an external load, conventional current flows through the conductor in the direction indicated setting up lines of flux around the conductor in a clockwise direction.

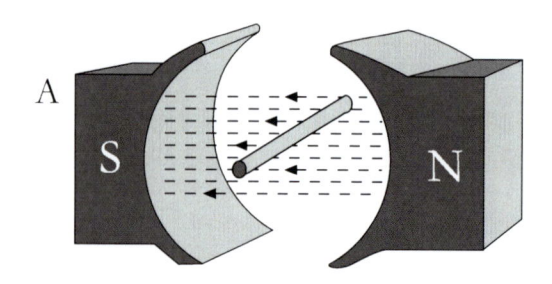

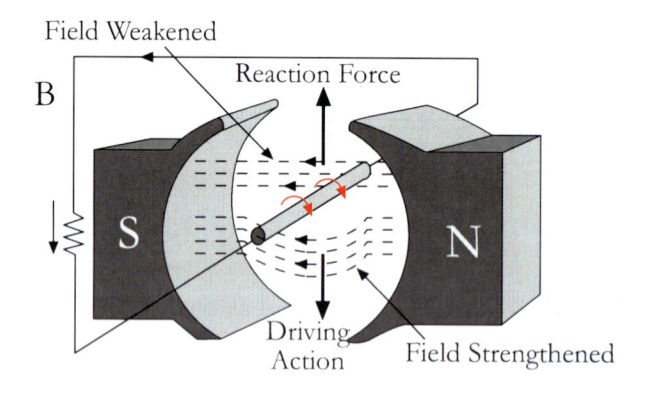

Figure 12.11 - Motor Reaction in a Simple Generator

The interaction between the conductor field and the main field of the generator weakens the field above the conductor and strengthens the field below the conductor. The main field consists of lines that now act like stretched rubber bands and so an upward reaction force is produced that acts in opposition to the downward driving force applied to the armature conductor. If the current in the conductor increases, the reaction force increases and therefore, more force must be applied to the conductor to keep it moving.

With no armature current, there is no magnetic, i.e. motor, reaction and so the force required to turn the armature is low. As the armature current increases, the reaction of each armature conductor against the rotation also increases. The generator's *actual* force is multiplied by the number of conductors in the armature and the driving force required to maintain the generator armature speed must be increased to overcome the motor reaction. The force applied to turn the armature must overcome the motor reaction force in all DC generators and the device that provides the turning force applied to the armature is called the *Prime Mover*. The prime mover may be an electric motor, a petrol engine, a steam turbine, or any other mechanical device that provides turning force.

Armature Losses

In DC generators, as in most electrical devices, certain unwanted forces act to decrease its efficiency. These forces, as they affect the armature, are considered as *losses* and may be defined as one of three (3) types:

1. Copper or I^2R loss in the winding
2. Eddy current loss in the core
3. Hysteresis loss

Copper Losses

There is power lost in the armature winding of a generator due to heat and this is known as the *Copper Loss*. Heat is generated any time current flows in a conductor and copper loss is an I^2R loss, i.e. $P=I^2R$, which increases as current increases. The amount of heat is also proportional to the resistance of the conductor, which varies directly with its length and inversely with its cross-sectional area (csa). Copper loss is minimised by using large diameter wire that can easily carry the current.

Eddy Current Losses

The generator armature core is made from soft iron, which is a conducting material with desirable magnetic characteristics. As we already know, any conductor will have currents induced in it when it is rotated in a magnetic field and these currents are called *Eddy Currents*, already discussed in previous chapters. As with any current, these produce heat, which is considered a loss.

The resistance of the material in which the currents flow affects eddy currents, just like any other electrical currents. As mentioned above, the resistance of any material is inversely proportional to its cross-sectional area.

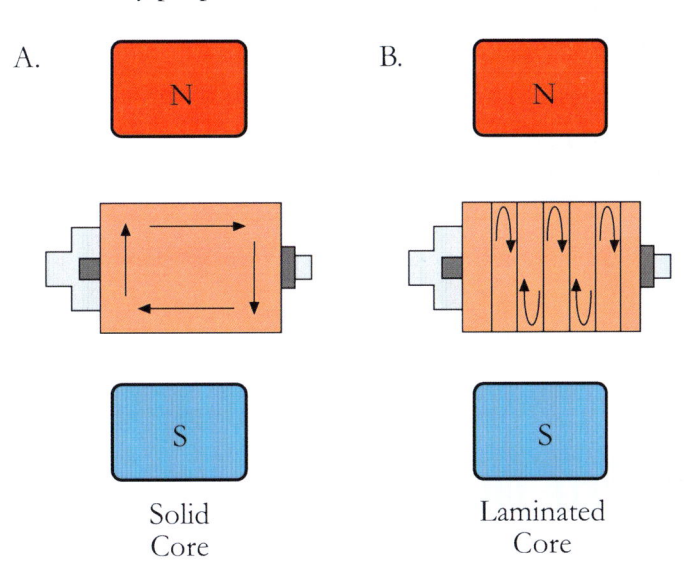

Figure 12.12 – Eddy Currents in Different Cores

Figure 12.12 view 'A', shows the eddy currents induced in an armature core that is a solid piece of soft iron while view 'B', shows a soft iron core of the same size, but made up of several small pieces insulated from each other. This process is called *lamination*.

As illustrated in view 'B', the currents in each piece of the laminated core are considerably less than in the solid core because the resistance of the pieces is much higher, i.e. resistance is inversely proportional to cross-sectional area. The currents in the individual pieces of the laminated core are so small that the *sum* of the individual currents is much less than the total of eddy currents in the solid iron core and so eddy current losses are kept low when the core material is made up of many thin sheets of metal.

Laminations in a small generator armature may be as thin as 1/64 inch and are insulated from each other by a thin coat of *lacquer* or, in some cases, simply by the oxidation of the surfaces.

Oxidation is caused by contact with the air while the laminations are being annealed and this provides enough insulation, as its value need not be high because the induced voltages induced are very small. Most generators use armatures with laminated cores to reduce eddy current losses.

Hysteresis Losses

We have already discussed Hysteresis and Hysteresis loss in a previous chapter and I am sure you will remember it is a heat loss caused by the magnetic properties of the armature.

When an armature core is in a magnetic field, the magnetic particles of the core tend to line up with the magnetic field, but when the armature core is rotating, its magnetic field keeps changing direction. The continuous movement of the magnetic particles as they try to align themselves with the magnetic field, produces molecular friction, which in turn, produces heat. This heat is transmitted to the armature windings causing the armature resistances to increase.

To compensate for hysteresis losses, most DC generator armatures use heat-treated silicon steel laminations. After the steel has been formed to the proper shape, the laminations are heated and allowed to cool. This annealing process reduces the hysteresis loss to a low value.

The Practical DC Generator

What we have discussed so far is the theoretical DC generator. The actual construction and operation of a practical DC generator differs somewhat from our simple generators and these differences include the construction of the armature, the manner in which the armature is wound, and the method of developing the main magnetic field.

As already shown earlier, a generator that has only one or two armature loops has high ripple voltage, which results in too little current to be of any practical use. To increase the amount of output current, several loops of wire are used and these eliminate most of the ripple.

The loops of wire, usually called *windings*, are evenly spaced around the armature so that the distance between each winding is the same.

The commutator in a practical generator is also different as it has several segments instead of two or four, as seen in our simple examples, with the number of segments equalling the number of armature coils. The following paragraphs describe several different methods of armature construction.

Gramme-Ring Armature

Figure 12.13 below, shows an end and composite view of a *Gramme-Ring* armature where each coil is connected to two commutator segments.

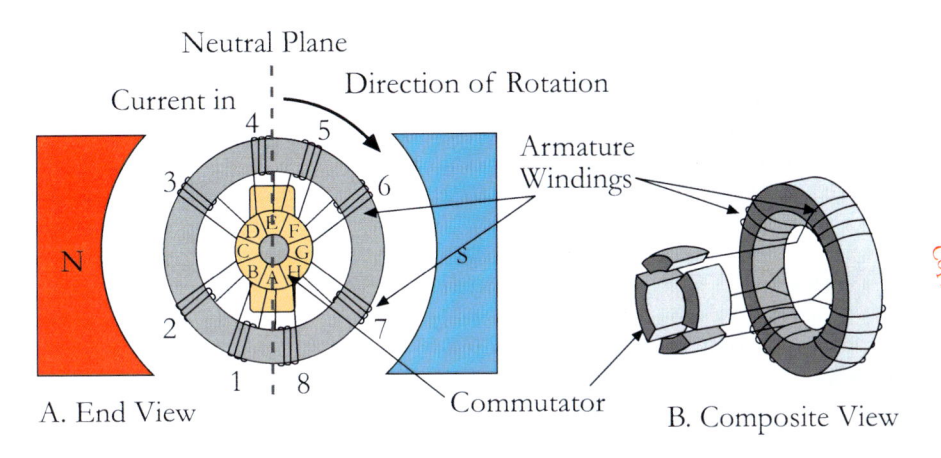

Figure 12.13 – Gramme Ring Armature

With this construction, one end of coil 1 goes to segment A, and its other end goes to segment B; one end of coil 2 goes to segment C, and the other end of coil 2 goes to segment B, etc, i.e. the rest of the coils are connected in a similar way in series, around the armature. To complete the series arrangement, coil 8 connects to segment A and therefore, each coil is in series with every other coil. Figure 12.13 view B shows a composite view and illustrates graphically the physical relationship of the coils and commutator locations.

The windings of a Gramme-ring armature are placed on an iron ring. A disadvantage of this arrangement is that the windings located on the inner side of the iron ring cut few lines of flux and so they have little, if any, voltage induced in them. For this reason, the Gramme-ring armature is not in wide use.

Drum-Type Armature

A drum-type armature, figure 12.14, has the armature windings placed in slots cut in a drum-shaped iron core. Each winding completely surrounds the core so that the entire length of the conductor cuts the main magnetic field. Therefore, the total voltage induced in the armature is greater than in the Gramme-ring and so is much more efficient. This type of armature now accounts for the majority of modern DC generators.

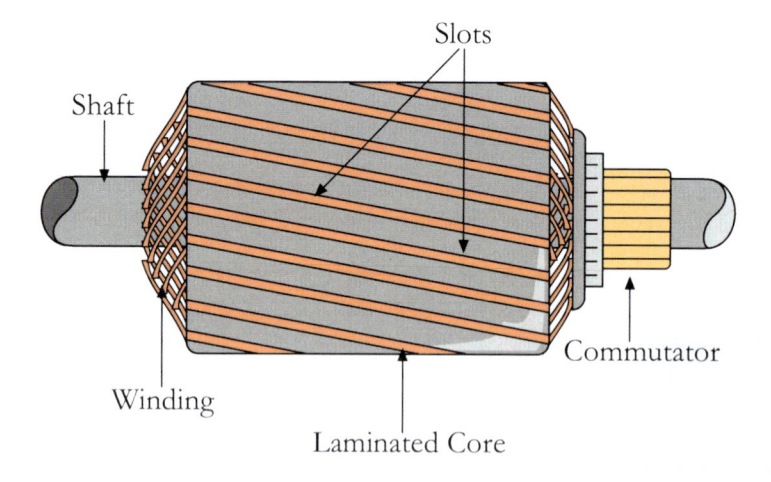

Figure 12.14 - Drum-Type Armature

Drum-type armatures are wound with either of two types of windings, **Lap** or **Wave** winding, depending on how their wires are connected to the armature.

With lap windings the two ends of any one coil are taken to adjacent segments, whereas in wave windings, the two ends of each coil are bent in opposite directions and are taken to segments some distance apart.

The lap winding illustrated in figure 12.15 view A, is used in DC generators designed for high-current applications. The windings are connected to provide several parallel paths for current in the armature and for this reason, lap-wound armatures require several pairs of poles and brushes.

Figure 12.15 view B shows a wave winding that is used in DC generators designed for high-voltage applications. Notice that the two ends of each coil are connected to commutator segments separated by the distance between poles. This configuration allows the series addition of the voltages in all the windings between brushes. This type of winding only requires one pair of brushes, but in practice, a practical generator may have several pairs to improve commutation.

CHAPTER TWELVE
DC GENERATOR & MOTOR THEORY

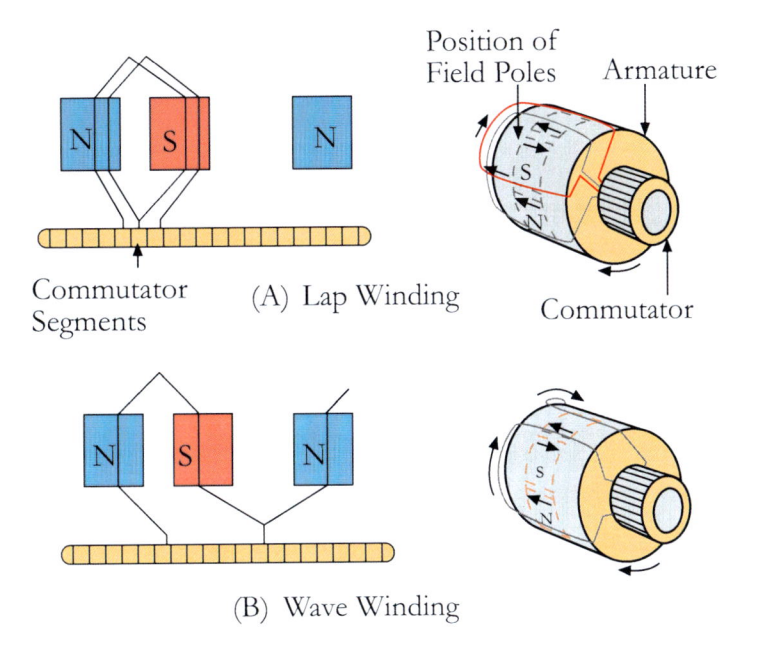

Figure 12.15 – Lap & Wave windings on Drum-Type Armatures

Lap windings have as many paths in parallel between the negative and positive brushes as there are poles. For instance, with an eight (8)-pole lap winding, the armature conductors form eight (8) parallel paths between the negative and positive brushes. Note that an eight pole lap winding has four pairs of poles.

A wave winding on the other hand has only two (2) paths in parallel, irrespective of the number of poles. Therefore, if a machine has 'P' pairs of poles:

Number of paths with a lap winding = 2P

Number of paths with a wave winding = 2

Field Excitation

When a DC voltage is applied to the field windings of a DC generator, current flows through the windings and sets up a steady magnetic field. This is called *Field Excitation*. This excitation voltage can be produced by the generator itself or it can be supplied by an outside source, such as a battery.

A generator that supplies its own field excitation is called a *Self-Excited Generator*, but is only possible if the field pole pieces have retained a slight amount of permanent magnetism, called *Residual Magnetism*. When the generator starts rotating, the weak residual magnetism generates a small voltage in the armature. This small voltage applied to the field coils causes a small field current. Although small, this field current strengthens the magnetic field and allows the armature to generate a higher voltage, which then increases the field strength, etc. This process continues until the output voltage reaches the generator's rated output.

Generator Classification

Self-excited generators are classed according to the type of field connection they use and there are three general types:

- Series wound
- Shunt wound
- Compound wound

Compound wound generators are further classified as cumulative compound and differential compound but are not discussed in this chapter.

Series Wound Generator

In the series wound generator, figure 12.16, the field windings are connected in series with the armature. Current that flows in the armature flows through the external load circuit *and* through the field windings.

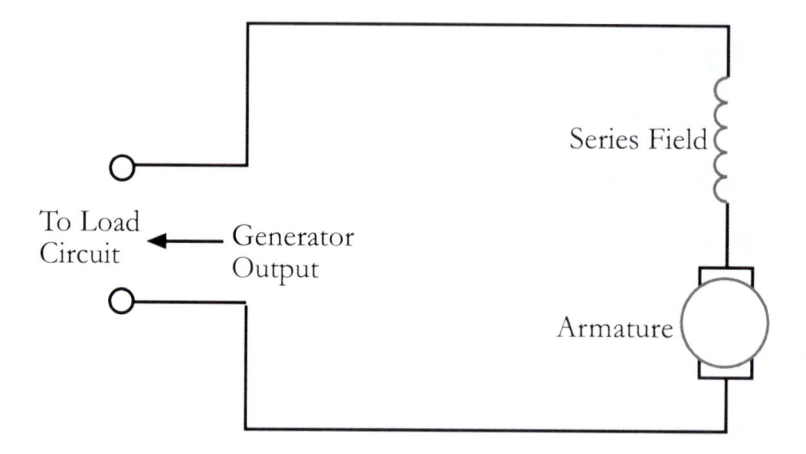

Figure 12.16 - Series-Wound Generator

The series-wound generator uses very low resistance field coils, which consist of a few turns of large diameter wire. The voltage output increases as the load circuit starts drawing more current and under low-load current conditions, the current flowing in the load and through the generator is small. Since small current means that the field poles set up a small magnetic field, only a small voltage is induced in the armature. If the resistance of the load decreases, by Ohm's Law, the load current increases and more current flows through the field. This increases the magnetic field and increases the output voltage. Therefore, a series wound DC generator's characteristic is that the output voltage varies with load current. This is undesirable in most applications, and so this type of generator is rarely used in practice.

Shunt Wound Generator

In a shunt-wound generator, figure 12.17, the field coils consist of many turns of small gauge wire that are connected in parallel with the load. In other words, they are connected across the output voltage of the armature.

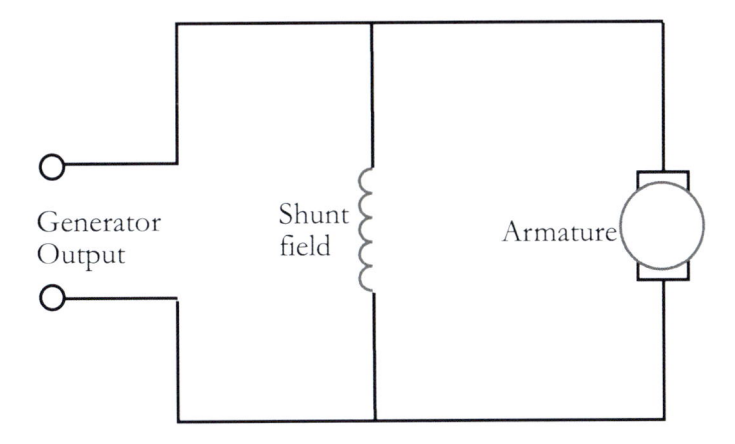

Figure 12.17 - Shunt-Wound Generator

It is vitally important that a shunt wound generator be allowed to build up to its correct output voltage before connecting a load. This is because the armature current of a shunt wound generator is shared by the field winding and the external load. During initial excitation, because of the low value of residual magnetism, only a small voltage is induced in the armature winding. Therefore, the armature current is small and, if a load is connected during initial excitation, the current in the field winding will be insufficient to assist the initial excitation. If this happens the voltage induced in the armature winding remains static, the output voltage fails to build up, and the generator fails to excite.

Once the correct output voltage has been reached, a load can be connected to the generator and load current will flow. The voltage drop across the armature winding resistance now reduces the output voltage which, in turn, reduces the field excitation current, thus weakening the main field and producing a further reduction in output voltage

In summary; the output voltage in a DC shunt wound generator varies *inversely* as load current varies, ie the output voltage decreases as load current increases because the voltage drop across the armature resistance increases.

Compound Wound Generators

To overcome the disadvantages of series and shunt wound generators, compound wound generators have a series field winding in addition to a shunt field winding, as shown in figure 12.18.

CHAPTER TWELVE
DC GENERATOR & MOTOR THEORY

Figure 12.18 – Compound Wound Generator

With the compound wound generator, the shunt and series windings are wound on the same pole pieces and when load current increases, the armature voltage decreases just as in the shunt wound generator. This causes the voltage applied to the shunt field winding to decrease, which results in a decrease in the magnetic field. This same increase in load current, since it flows through the series winding, causes an increase in the magnetic field produced by that winding.

Now by proportioning the two fields so that the decrease in the shunt field is just compensated by the increase in the series field, the output voltage remains constant.

This is illustrated in figure 12.19, which shows the voltage characteristics of the series, shunt and compound wound generators.

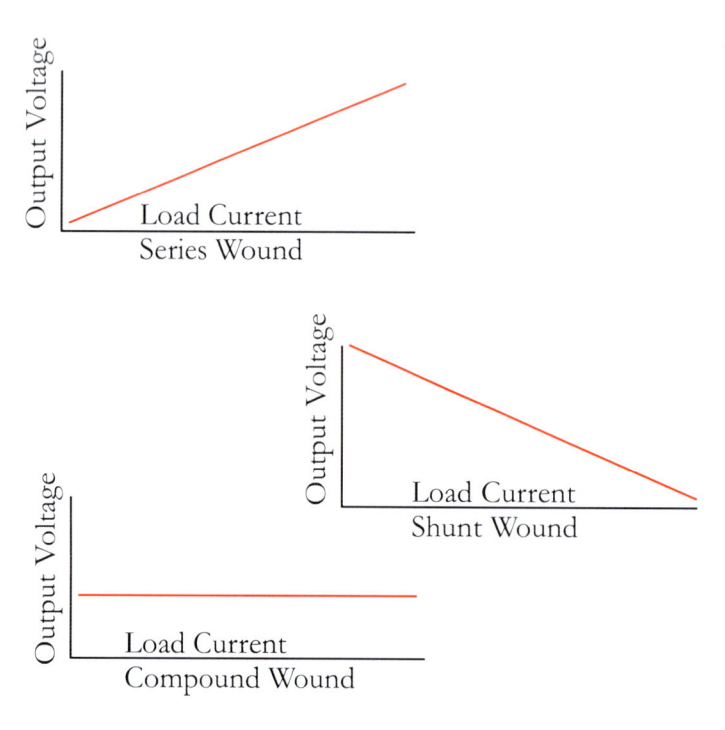

Figure 12.19 – Voltage Characteristics of Generators

As you can see, by proportioning the effects of the two fields, series and shunt, a compound-wound generator provides a constant output voltage under varying load conditions. However, the actual output voltage/load currents are seldom, if ever, as perfect as shown in figure 12.19.

Generator Construction

As with all machines, the DC generator can be broken down into several component parts, figure 12.20.

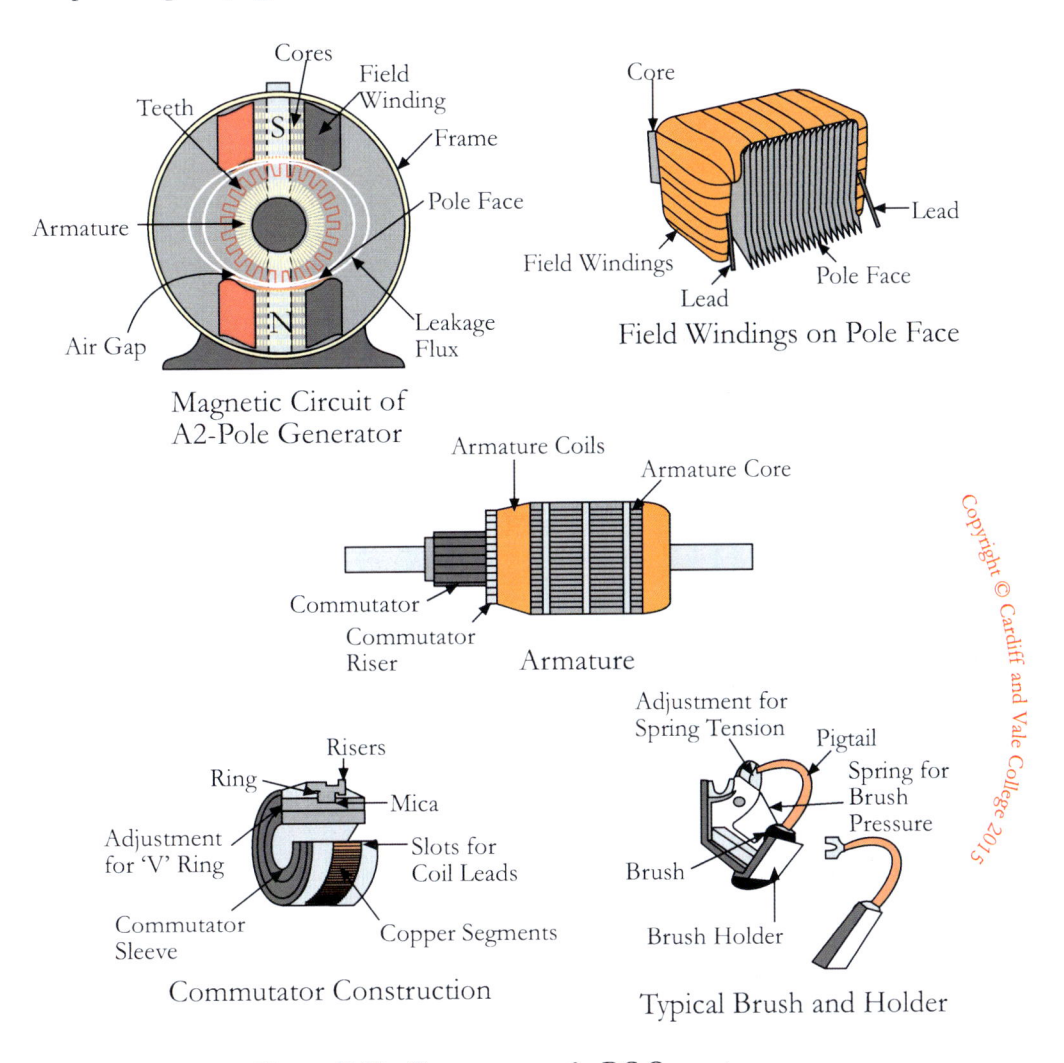

Figure 12.20 - Components of a DC Generator

Figure 12.21 shows a cutaway drawing of the generator with the component parts installed, illustrating the physical relationship of the components to each other.

CHAPTER TWELVE
DC GENERATOR & MOTOR THEORY

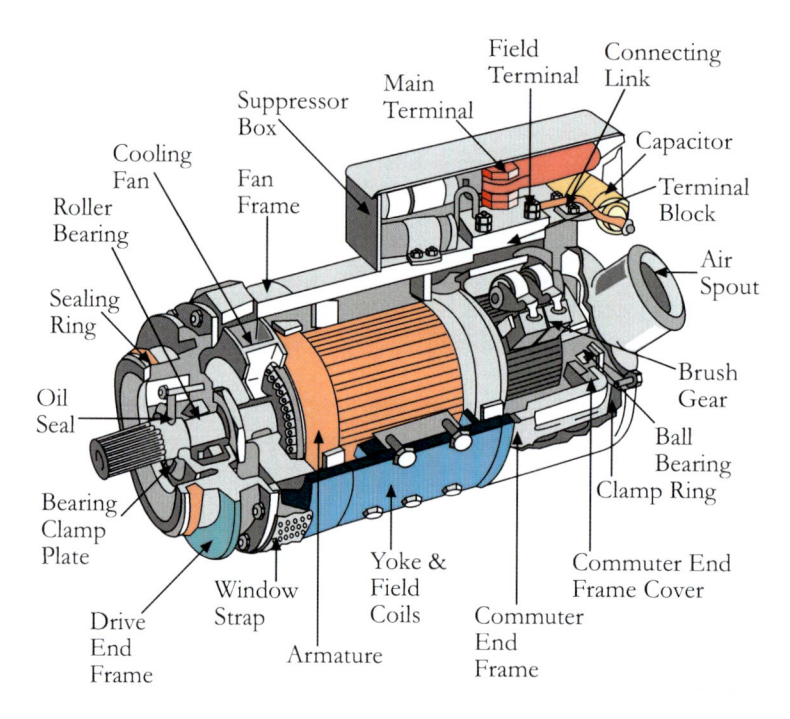

Figure 12.21 – Cutaway of a DC Generator

Voltage Regulation

Regulation of a generator refers to the **voltage change** that takes place as the load changes. It is usually expressed as the change in voltage from a no-load condition to a full-load condition, and is expressed as a percentage of full-load using the following formula:

$$\text{Percentage of Regulation} = \frac{E_{nl} - E_{fl}}{E_{fl}} \times 100$$

Where E_{nL} is the generator's no-load terminal voltage and E_{fl} its full-load terminal voltage. For example, to calculate the percent of regulation of a generator with a no-load voltage of 462 volts and a full-load voltage of 440 volts:

$$\text{Percentage of Regulation} = \frac{462 - 440}{440} \times 100$$

Percentage of Regulation = 5%

In practical terms, the lower the percent of regulation, the better the generator. In the above example, the 5% regulation represented a 22-volt change from no load to full load, whereas a 1% change would represent a change of 4.4 volts, which, of course, would be better still.

Voltage Control

Voltage control is either manual or automatic and in most cases, the process involves changing the resistance of the field circuit to control the field current. Controlling the field current in this way allows control of the output voltage. The major difference between the various voltage control systems is merely the method by which the field circuit resistance and the current are controlled.

Note: *Voltage Regulation* should not be confused with *Voltage Control* although in many technical manuals they are talked about as one and the same. As discussed earlier, *voltage regulation* is an internal action occurring within the generator whenever the load changes, whereas *voltage control* is an imposed action, usually through an external adjustment, for the purpose of increasing or decreasing the terminal voltage.

Manual Voltage Control

The hand-operated field rheostat, shown in figure 12.22, is a typical example of manual voltage control. The field rheostat is connected in series with the shunt field circuit, which provides the simplest method of controlling the terminal voltage of a DC generator.

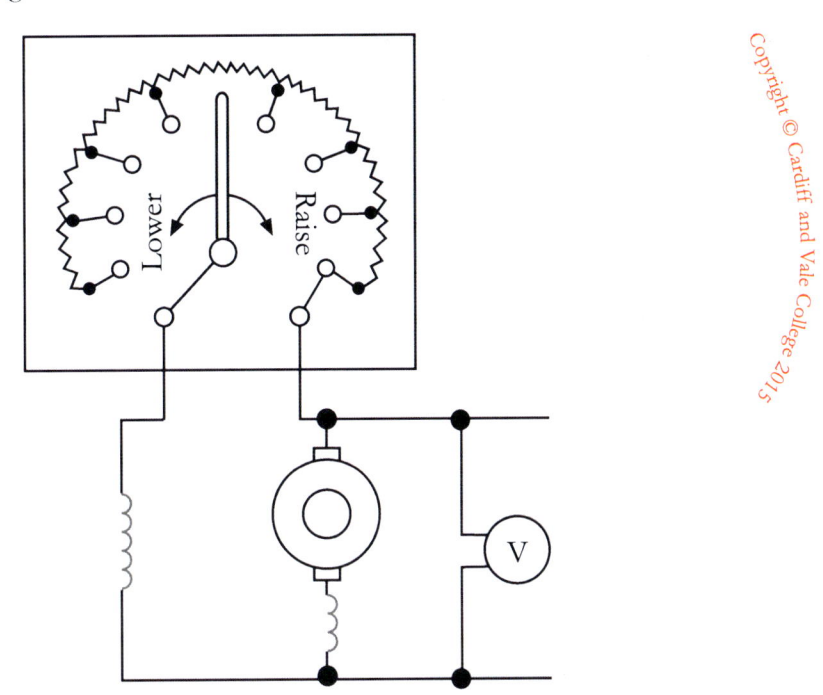

Figure 12.22 – Simple Hand-Operated Field Rheostat

This type of field rheostat contains tapped resistors with leads to a multi-terminal switch. The switch arm may be rotated to make contact with the various resistor taps and this varies the amount of resistance in the field circuit. Rotating the arm in the direction of the *Lower* arrow, i.e. counter-clockwise, increases the resistance and lowers the output voltage; while rotating the arm in the *Raise* direction decreases the resistance and increases the output voltage.

Most generator field rheostats use resistors of alloy wire as they have a high specific resistance and a low temperature coefficient. These alloys include copper, nickel, manganese, and chromium and they are often marked under trade names such as Nichrome, Advance, Manganin, etc.

Automatic Voltage Control

Automatic voltage control is used where load current variations exceed the built-in ability of the generator to regulate itself. It achieves this by *sensing* changes in output voltage causing a change in field resistance to keep the output voltage constant.

Basic DC Motor Principles

The DC motor is a *mechanical* workhorse that can be used in many different ways. Many large pieces of equipment depend on a DC motor for their power to move as the speed and direction of rotation of a DC motor is easily controlled. This makes it especially useful for operating equipment, such as winches, cranes, actuators, etc that must move in different directions and at varying speeds.

A DC motor's operation is based on the same principle as the generator, ie that a current-carrying conductor placed in a magnetic field, perpendicular to the lines of flux, tends to move in a direction perpendicular to the magnetic lines of flux. There is a definite relationship between the direction of the magnetic field, the direction of current in the conductor, and the direction in which the conductor tends to move, which is best explained by using *Fleming's left-hand rule for motors*, as illustrated in figure 12.23, using *conventional* current flow.

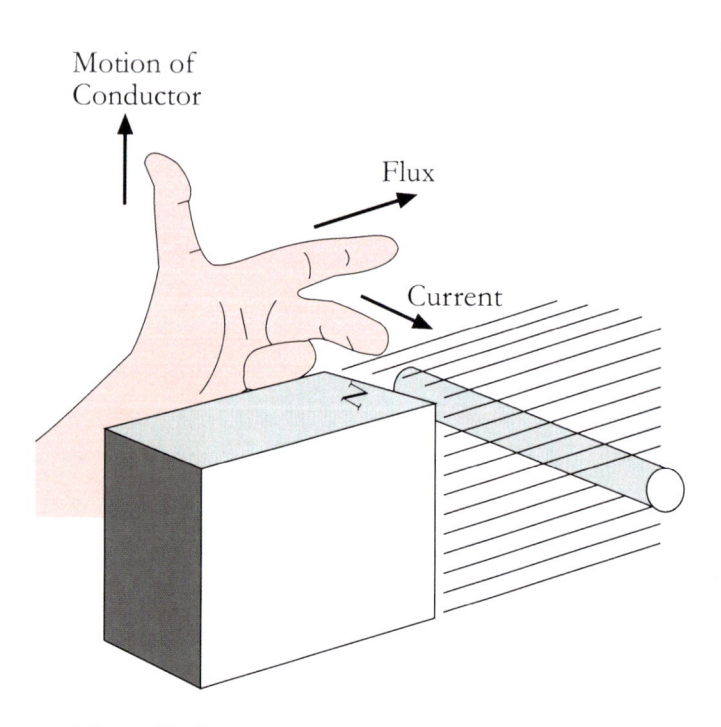

Figure 12.23 – Fleming's Left-Hand Rule for Motors

To find the direction of motion of a conductor, using this method, extend the thumb, forefinger, and middle finger of the right hand so they are at right angles to each other, as illustrated in figure 12.23. If the forefinger is pointed in the direction of magnetic flux, i.e. north to south, and the middle finger is pointed in the direction of current flow in the conductor, the thumb will point in the direction the conductor will move.

Put simply, a DC motor rotates to produce *mechanical* energy because of two magnetic fields interacting with each other. The armature of a DC motor acts like an electromagnet when current flows through its coils and since the armature is located within the magnetic field of the field poles, these two magnetic fields interact.

Now we know from earlier chapters and the discussions concerning generators above that like magnetic poles repel and unlike magnetic poles attract. The DC motor has field poles that are stationary and an armature that turns on bearings in the space between the field poles.

The armature of a DC motor has windings on it just like the armature of a DC generator and these are also connected to commutator segments. A DC motor consists of the same components as a DC generator and in fact, most DC generators can be made to act as motors, and vice versa.

To explain the DC motor principle, let us start-off again by looking at a simple DC motor, figure 12.24.

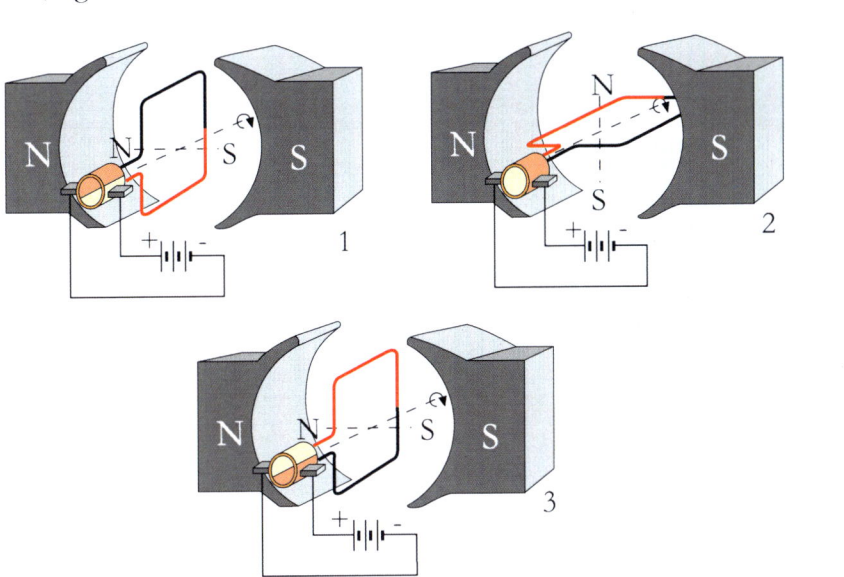

Figure 12.24 – Simple DC Motor

This simple motor has two field poles, one a north and one a south pole, and the magnetic lines of force extend across the opening between the poles north to south.

The armature in this simple DC motor is a single loop of wire, just as in the simple armature we discussed at the beginning of this chapter on DC generators. However, in this case, the loop of wire has current flowing through it from an external source, which produces a magnetic field, indicated by the dotted line through the loops.

The loop, i.e. armature, field is both attracted and repelled by the field from the permanent magnetic poles. Since the current through the loop goes around in the direction of the arrows, the north pole of the armature is at the upper left, and the south pole of the armature is at the lower right, as shown in figure 12.24 view (1) and of course, as the loop turns, these magnetic poles turn with it.

Now, as shown in the illustrations, as the loop rotates, the **north** armature pole is **repelled** from the **north field pole** and **attracted** to the **right** by the **south field pole.**

Likewise, the south armature pole is repelled from the south field pole and is attracted to the left by the north field pole. This action causes the armature to turn in a clockwise direction, as illustrated in figure 12.24 view (2).

After the loop has turned far enough so that its north pole is exactly opposite the south field pole, the brushes advance to the next segments, changing the direction of current flow through the armature loop and changing the polarity of the armature field, as shown in figure 12.24 view (3). The process now repeats itself as long as the current flows and the magnetic fields again repel and attract each other, causing the armature to turn continuously.

In this illustration of a simple motor, the momentum of the rotating armature carries it past the position where the unlike poles are exactly lined up. However, if these fields are exactly lined up when the armature current is initially turned on, there is no momentum to start the armature moving and in this case, the motor would not rotate. In order to start the motor rotating in this situation, it would be necessary to give it a mechanical spin, but in reality, this problem does not exist when there are more turns on the armature as there is more than one armature field.

Back Emf

While a DC motor is running, it acts somewhat like a DC generator as there is a magnetic field from the field poles, and a loop of wire is turning and cutting this magnetic field. For the moment, disregard the fact that there is current flowing through the loop of wire from the external source. As the loop sides cut the magnetic field, a voltage is induced in them, the same as it was in the loop sides of the DC generator and this induced voltage causes current to flow in the loop.

Now, consider the relative direction between **this** current and the current that causes the motor to run. Applying the right-hand rule for generators, discussed above, we can see that the direction of current flow caused by the generator action is in the direction opposite to that of the battery current. Since this *generator-action* voltage is opposite that of the power source, it is called a '*counter emf*' or back emf.

However, this is a bit oversimplified, as you may already suspect and in reality, only one current flows. Because the counter emf can never become as large as the applied voltage, and because they are of opposite polarity, the counter emf effectively cancels part of the armature voltage. The single current that flows *is* armature current, but it is greatly reduced because of the counter emf.

A counter emf is always developed in a DC motor, but it cannot be equal to or greater than the applied battery voltage; as if it were, the motor would not run. However, the counter emf opposes the applied voltage enough to keep the armature current from the battery to a fairly low value. If there were no such thing as counter emf, more current would flow through the armature, and the motor would run much faster. Unfortunately, there is no way of avoiding counter emf and so it is catered for in the motor design.

Motor Loads

Motors are used to turn mechanical devices, such as water pumps, grinding wheels, fan blades, and circular saws. When a motor is turning a water pump, the water pump is the load, i.e. the water pump is the mechanical device that the motor must move. Sorry to labour the point but this really is the definition of a motor load. As with electrical loads, the mechanical load connected to a DC motor affects many electrical quantities; e.g. power consumption, amount of current, speed, efficiency, etc, are all these are partially controlled by the size of the load. The physical and electrical characteristics of the motor must be matched to the requirements of the load if the work is to be done without of damage to either the load or the motor.

Practical DC Motors

As discussed previously, DC motors are electrically identical to DC generators and in fact, the same DC machine may be driven mechanically to generate a voltage, or it may be driven electrically to move a mechanical load. While this is not normally done, it does point out the similarities between the two machines.

In practice, motors and generators are usually used as dedicated devices, but their similarities will become evident below, as we discuss the series, shunt, and compound types of motor.

Series DC Motor

In a series DC motor, the field is connected in series with the armature and is wound with a few turns of large wire, because it must carry full armature current, figure 12.25.

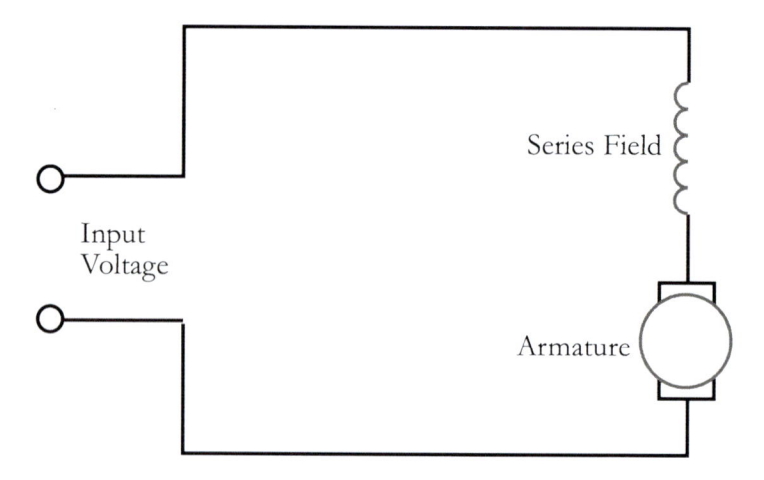

Figure 12.25 - Series-Wound DC Motor

This type of motor develops a very large amount of turning force, called **Torque**, from a standstill and so can be used to operate small electric appliances, portable electric tools, cranes, winches, hoists, and the like. Its speed varies widely between no-load and full-load conditions and so cannot be used where a relatively constant speed is required under varying load conditions.

The wide speed characteristic is a major disadvantage of the series motor as it can continue increasing with no load connected to it, to the point of destruction.

When this happens, usually the bearings are damaged or the windings fly out of the slots in the armature; a danger to both equipment and personnel, especially in the aircraft environment. Therefore, with a series motor, there must *always* be some load connected before it is turned on.

Shunt Motor

A shunt motor is connected in the same way as a shunt generator, figure 12.26, i.e. the field windings are connected in parallel, i.e. shunt, with the armature windings.

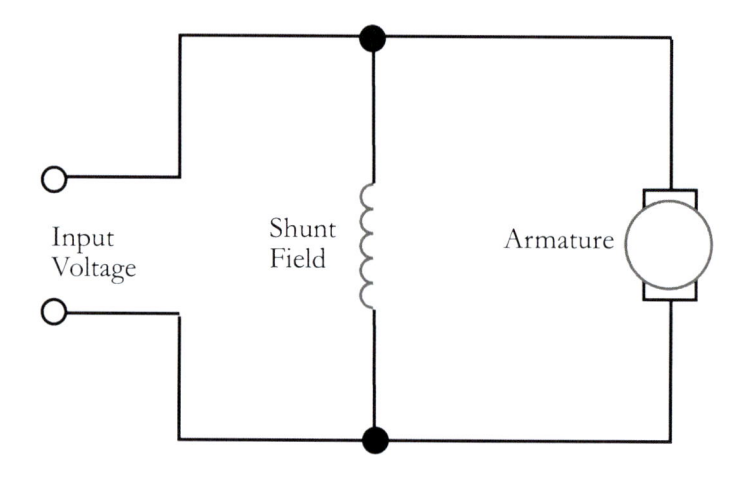

Figure 12.26 – Shunt Wound DC Motor

With a DC shunt motor, its speed remains relatively constant even under changing load conditions as the field flux also remains constant. A constant voltage across the field makes the field independent of variations in the armature circuit. For example if the motor load increases, it tends to slow down, which causes the counter emf generated in the armature to decrease. This then causes a corresponding decrease in the opposition to power source current flow through the armature and the armature current increases, causing the motor to speed up. The starting conditions are re-established, and the original speed is maintained. Conversely, if the motor load is decreased, it tends to increase speed; counter emf increases, armature current decreases, and the speed decreases.

In each case, all of this happens so rapidly that any actual change in speed is slight as there is instantaneous tendency to change rather than a large fluctuation in speed.

Because of the ability of the shunt motor to maintain an almost constant speed under a variety of loads, it is often called a *constant-speed motor*.

Compound Motor

A compound motor has two field windings, figure 12.27; one is a shunt field connected in parallel with the armature; the other a series field connected in series.

With the compound wound motor, the shunt field gives the constant speed advantage of a regular shunt motor, while the series field gives it the advantage of being able to develop a large torque when first started under a heavy load.

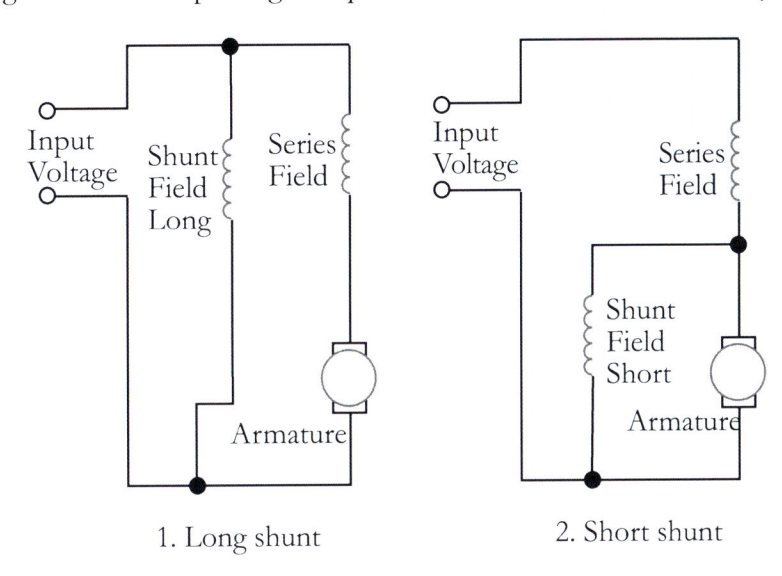

Figure 12.27 - Compound-Wound DC Motor

When the shunt field is connected in parallel with the series field and armature, it is called a *long shunt*, figure 12.27 view (1), and when just in parallel with the armature, a *short shunt*, figure 12.27 view (2).

CHAPTER TWELVE
DC GENERATOR & MOTOR THEORY

Types of Armatures

As with DC generators, DC motors can be constructed using one of two types of armatures. I have included a brief review of the Gramme-ring and drum-wound armatures here to emphasise the similarities between DC generators and DC motors.

Gramme-Ring Armature

The Gramme-ring armature is constructed by winding an insulated wire around a soft-iron ring, figure 12.28. Eight (8) equally spaced connections are made to the winding and each of these is connected to a commutator segment, but the brushes touch only the top and bottom segments.

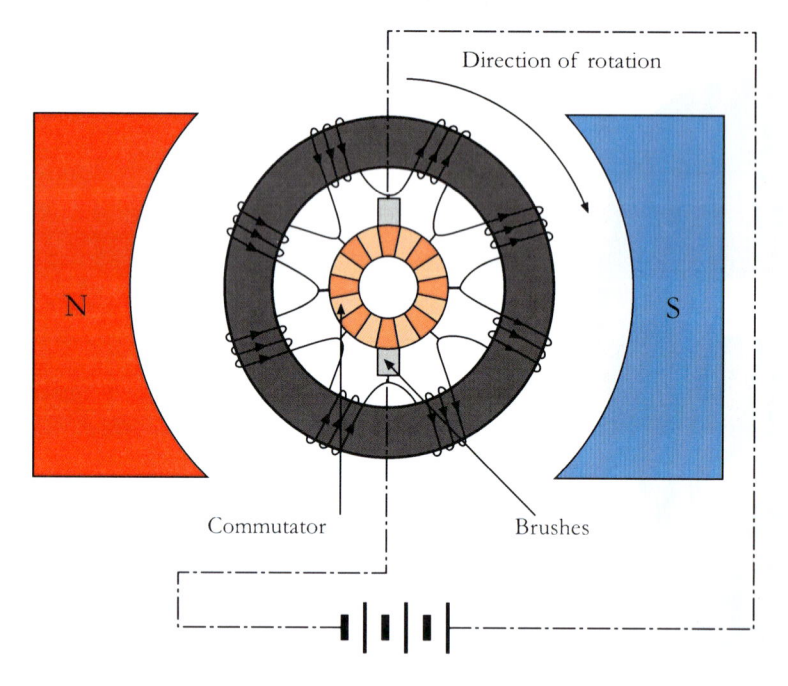

Figure 12.28 – Gramme Ring Armature

There are two parallel paths for current to follow, using conventional current flow, from the positive side of the battery; one down the left side and one down the right side of the armature windings. These paths are completed through the bottom brush back to the battery's negative lead.

We can check the direction of rotation of this armature by using the left-hand rule for motors i.e. holding the thumb, forefinger, and middle finger at right angles.

If the forefinger is in the direction of field flux; in this case, from left to right then turning the wrist so that the middle finger points in the direction that the current flows in the winding on the outside of the ring, the thumb points in the direction the winding will move. Note that conventional current flows out of the page in the left-hand windings and into the page in the right-hand windings.

The Gramme-ring armature is seldom used in modern DC motors as the windings on the inside of the ring are shielded from magnetic flux, which causes this type of armature to be inefficient. However, I have discussed it here primarily to help you better understand the drum-wound armature.

Drum-Wound Armature

The drum-wound armature is generally used in ac motors and is identical to the drum winding discussed in the chapter on DC generators.

If the drum-wound armature was cut in half, an end view at the cut would resemble the drawing in figure 12.29, view 'A', while view 'B' is a side view of the armature and pole pieces, the conductors in the armature slots are shown as small circles.

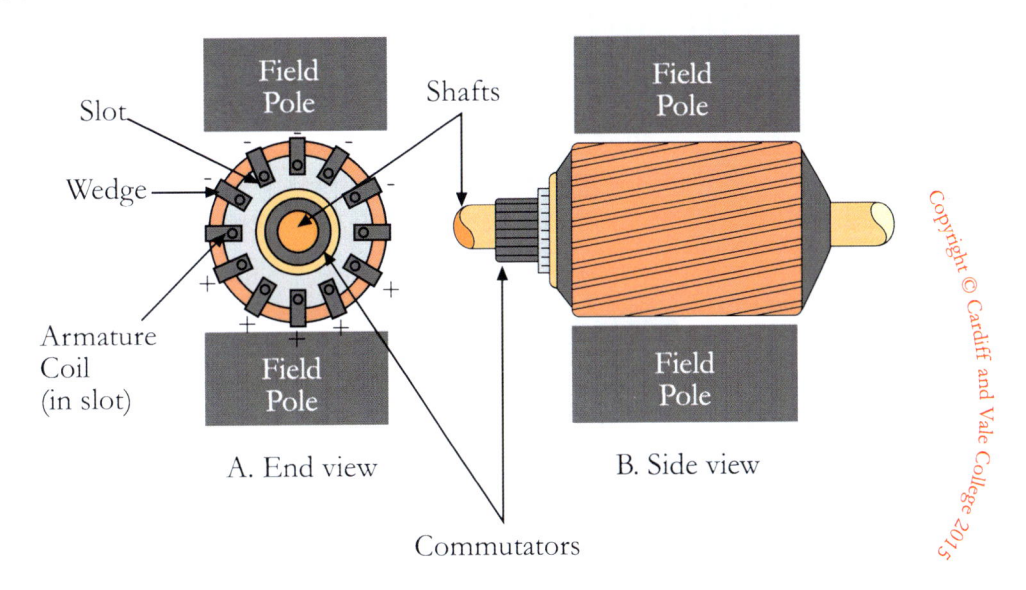

Figure 12.29 - Drum-Type Armature

Notice that the length of each conductor is positioned parallel to the faces of the pole pieces and so each conductor can cut the maximum flux of the motor field. The inefficiency of the Gramme-ring armature is overcome by positioning the conductors in this way.

By using the left-hand motor rule, the direction of the torque on the armature can be determined. With reference to Figure 12.30, the cross in the small circle indicates current flowing into the page and the dot indicates current flowing ot of the page.

In this example, the current on the left of the armature is flowing into the page, and the field flux is from right to left. The left-hand rule then indicates that the left hand side of the conductor would move up through the field. On the opposite side of the armature, the right hand side of the conductor would move down through the field, and the motor should run in a clockwise direction.

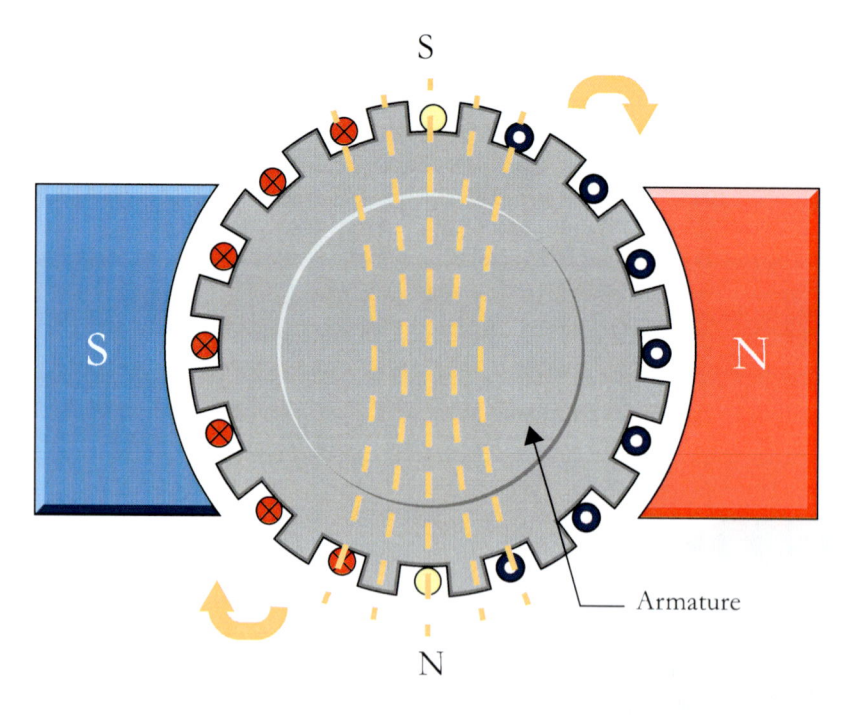

Figure 12.30 - Determining the Direction of Rotation

Strips of insulation are inserted in the slots to keep windings in place when the armature spins, shown as wedges in figure 12.29.

Direction of Rotation

The direction of rotation of a DC motor depends on the direction of the magnetic field and the direction of current flow in the armature. If either is reversed, the rotation of the motor will reverse, but if both are reversed at the same time, the motor will continue rotating in the same direction. In practice, the field excitation voltage is reversed in order to reverse motor direction.

Ordinarily, a motor is set up to do a particular job that requires a fixed direction of rotation, but there are times when it is necessary to change the direction of rotation, such as a drive motor for an aircraft's control surface, which must be able to move in both directions.

Motor Speed

A motor whose speed can be controlled is called a variable-speed motor; which is what many DC motors are. The speed of a DC motor is altered by changing the current in the field or by changing the current in the armature.

When the field current is decreased, the field flux is reduced, and the counter emf decreases, allowing more armature current, and so the motor speeds up. When the field current is increased, the field flux is increased, causing an increase in counter emf, which opposes the armature current, and the armature current decreases, and slows the motor down.

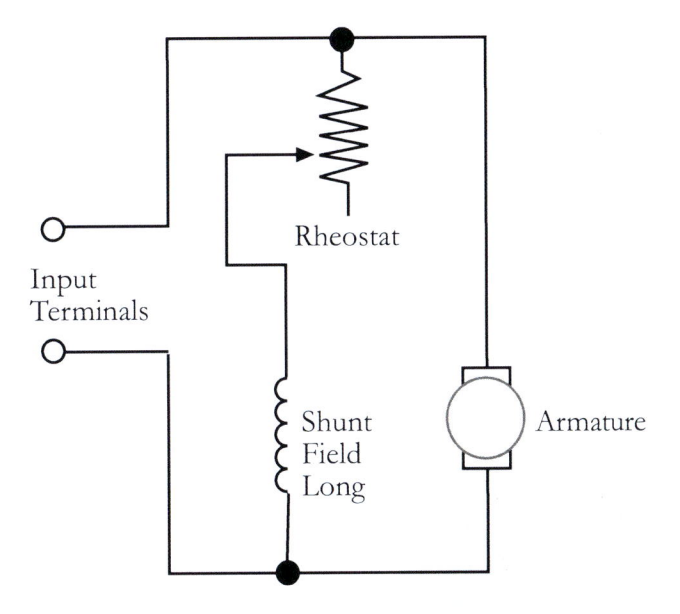

Figure 12.31 - Controlling Motor Speed

When the voltage applied to the armature is decreased, the armature current is decreased, and the motor again slows down. When the armature voltage and current are both increased, the motor speeds up.

In a shunt motor, a rheostat, connected in series with the field windings, usually controls the speed, as shown in figure 12.31. When the resistance of the rheostat is increased, the current through the field winding is decreased and the decreased flux momentarily decreases the counter emf and the motor then speeds up with the increase in counter emf keeping the armature current the same. In a similar way, a decrease in rheostat resistance increases the current flow through the field windings and causes the motor to slow down.

In a series motor, the rheostat speed control may be connected either in parallel or in series with the armature windings. In either case, moving the rheostat in a direction that lowers the voltage across the armature lowers the current through the armature and slows the motor and the opposite increases motor speed.

Armature Reaction

The reasons for armature reaction and the methods of compensating for its effects are basically the same for DC motors as for DC generators. Figure 12.32 reiterates for you the distorting effect that the armature field has on the flux between the pole pieces. Notice, however, that the effect has shifted the

CHAPTER TWELVE
DC GENERATOR & MOTOR THEORY

neutral plane backward, against the direction of rotation. This is different from a DC generator, where the neutral plane shifted forward in the direction of rotation.

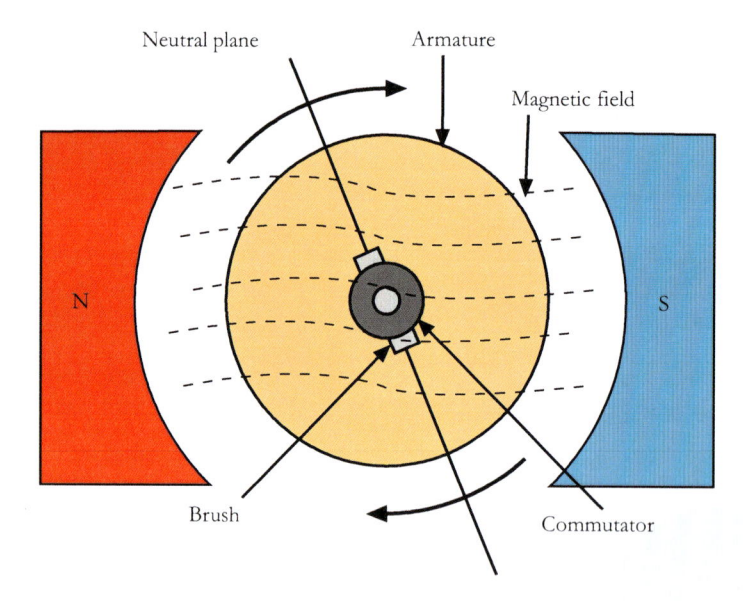

Figure 12.32 - Armature Reaction

As before, the brushes must be shifted to the new neutral plane, in this case counter-clockwise and again, the proper location is reached when there is no sparking from the brushes.

Compensating windings and interpoles, as discussed earlier, cancel armature reaction in DC motors. Shifting brushes reduces sparking, but it also makes the field less effective and cancelling armature reaction eliminates the need to shift brushes in the first place.

Compensating windings and interpoles are as important in motors as they are in generators. However, compensating windings are relatively expensive and so most large DC motors depend on interpoles to correct armature reaction. Compensating windings are the same in motors as they are in generators, but interpoles, however, are slightly different. The difference is that in a generator the interpole has the same polarity as the main pole **ahead** of it in the direction of rotation. In a motor the interpole has the same polarity as the main pole *following* it.

The interpole coil in a motor is connected so as to carry the armature current the same as in a generator. As the load varies, the interpole flux varies, and commutation is automatically corrected as the load changes so eliminating the need to shift the brushes when there is an increase or decrease in load. Therefore, in a practical motor, the brushes are located on the no-load neutral plane and they remain in that position for all conditions of load.

Manual & Automatic Starter/Generators

Because the DC resistance of most motor armatures is low, typically 0.05Ω to 0.5Ω, and because the counter emf does not exist until the armature begins to turn, it is necessary to use an external starting resistance in series with the armature of a DC motor to keep the initial armature current to a safe value. As the armature begins to turn, counter emf increases; and, since the counter emf opposes the applied voltage, the armature current is reduced. The external resistance in series with the armature is then decreased or eliminated as the motor comes up to normal speed and full voltage is applied across the armature.

Controlling the starting resistance in a DC motor is accomplished manually, by an operator, or by automatic devices. The manual systems use switching sequences controlled by the flight crew while automatic devices are usually just switches controlled by motor speed sensors.

The starter generator is connected to the engine and is usually powered directly off a battery or in more sophisticated aircraft, the ***starter busbar***. The aircraft start sequence is initiated by switches, usually on the aircraft's overhead panel, and the starter generator, usually connected to one of the compressor stages will start to rotate the engine and continue to do so until the engine becomes self sustaining.

Most modern aircraft that use DC as their main power source obtain primary electrical power from dual-purpose ***28 Volt DC Starter/Generators***. A starter/generator is installed on each engine gearbox and provides torque for engine starting, and in the generating mode supplies 28 volts DC, regulated, usually at approximately 300A continuous. Speed sensors and sequential relay control accomplish the transition from engine starting to generation automatically. A control unit associated with each starter/generator incorporates the functions of regulation, sensing and protection of the DC power system, and also provides field weakening for starter control and automatic, speed sensing cut-off. Typically, a starter/generator is self-exciting and cooled by airflow from its integral fan and also from in-flight ram air through an intake on the engine cowling and exhausted to atmosphere via a duct on the lower engine cowling.

Typical Start Sequence

This section is not part of the syllabus but has been included for your information as a typical example of a start sequence.

An aircraft's engine start control switches, as mentioned above, are usually found on its cockpit roof panel, figure 12.33 shows part of a typical example.

CHAPTER TWELVE
DC GENERATOR & MOTOR THEORY

Figure 12.33 – Part of a Typical Aircraft Overhead Panel

As shown in figure 12.33, the starter section consists of three annunciator type push button switches, which we will look at now.

Start PWR Switch

Functions as a start master switch for external or internal power starts. It is a 'push-on/push-off' type switch, and combines two functions indicated by 'PWR ON' and 'PUSH FOR ABORT'. The annunciator illuminates when the switch is initially pushed 'ON' to indicate 'PWR ON' is available. A second push of the illuminated switch will terminate an engine start cycle during the start sequence and extinguish the switch annunciation.

Eng. Starter Switches (2)

The engine start switches are *push-on* spring return switches, and labelled **ENG 1** and **ENG 2**, each switch with an integral *OPERATING* annunciator.

When the switch is operated, it causes power to be fed to the starter motor, the switch annunciator lights, within 2 seconds and may then be released. Usually, especially at well-equipped airfields, to conserve the aircraft batteries, external ground power should be used whenever possible for engine starting and testing of electrical services.

Note: Most DC aircraft require an external ground power unit capable of supplying 28 volts DC at loads up to 1500 amps with negligible voltage drop. In addition, the power unit should be fitted with a current limiter operating at, typically, 1100A.

290

CHAPTER TWELVE
DC GENERATOR & MOTOR THEORY

Once the start sequence is initiated, the starter motor rotates the engine, and when the engine N2 rpm reaches 10%, at least for this aircraft the *Hawker 800*, the HP cock lever is opened, initiating the engine fuel system and ignition.

For this aircraft, the engine start termination is automatically effected by a speed sensing circuit at *46% N2* speed, which is the self-sustaining speed of its engine. This causes the start contactor to de-energise, disconnects the ignition circuit and extinguishes the starter operating annunciator.

Power Generation

The transition from engine start to generating mode is achieved automatically by the *Generator Control Unit (GCU)* and the generator will come 'on line' without selection, when the quality of the electrical power as monitored by the GCU is correct, provided the engine start and ground power inhibit circuits are de-energised.

The GCU then connects the generator output to the aircraft's primary and emergency busbars by energising and closing the *Generator Line Contactor (GLC)* ; the aircraft's master warning *GEN FAIL* annunciator extinguishes and the associated power contactors is energised. This extinguishes the relevant warning annunciators and completes the charging circuit for the aircraft's main and standby batteries.

Usually, in a multi-engine aircraft, when a generator is 'on line' manual selection of a *BUSTIE* switch closes a bustie contactor or relay linking the main power busbars together. However, this will vary from aircraft type to aircraft type as some use a split busbar system and others use a tied busbar system; more on electrical systems later in the course.

With all the aircraft's generators operating and *on line*, all system failure annunciators are extinguished.

The GCUs now regulate the generator outputs at 28 volts and ensure that the generator busbar loads are equalised when operated as a tied busbar system. The GCUs also provide engine start control and protection circuits for reverse current, over excitation, over voltage and generator feeder earth faults by opening the GLC if a fault is detected. In addition to disconnecting generator output from the busbars when a fault is detected opening the GLC also *kills* the generator output, by interruption of the generator field excitation supply.

Fault detection that causes a GLC to open will light the appropriate *GEN FAIL* annunciator. After fault condition is isolated the generator may sometimes be re-instated by using a 3 position *GEN CLOSE TRIP* switch on overhead roof panel, as shown in figure 12.33 above. The switch is sprung loaded to its centre position, and when selected to *CLOSE* momentarily will ensure that the generator field circuit is reset. This completes the excitation supply circuit that promotes generator output build up, and when the switch is released, it may now come back *on line*. In the *TRIP* position, the GCU opens the GLC and *kills* the generator output.

Revision

DC Generator & Motor Theory

Questions

1. **A DC generator is a machine that converts:**

    a) Electrical energy into mechanical energy

    b) Mechanical Energy into Electrical energy

    c) AC to DC

2. **For a DC generator to function there must be:**

    a) Relative motion between the flux and conductor

    b) The magnetic field must rotate

    c) The conductor must rotate

3. **The amount of voltage generated by a DC generator depends on:**

    a) The speed at which the conductor moves

    b) The direction of the magnetic flux

    c) An odd number of poles

4. **The amount of voltage generated by a DC generator depends on:**

    a) The direction of the magnetic field

    b) The shape of the magnetic field

    c) The strength of the magnetic field

5. The polarity of the generated voltage depends on:

    a) The direction of the magnetic field

    b) The shape of the magnetic field

    c) The strength of the magnetic field

6. The rotating wire in a DC generator is called the:

    a) Inductor

    b) Armature

    c) Loop

7. The position where no voltage is induced in a DC generator's rotating wire is called the:

    a) Horizontal plane

    b) Vertical plane

    c) Neutral Plane

8. In a DC generator, the purpose of the commutator is to:

    a) Mechanically reverse the loop connections

    b) Electrically reverse the loop connections

    c) Maintain the ac flow

9. As a result of using a commutator, the output of a DC generator is:

    a) Pulsating DC at the frequency of rotation

    b) Pulsating DC at twice the frequency of rotation

    c) Pulsating DC at three times the frequency of rotation

10. When the output of a DC generator is varying it is said to be

    a) A flowing voltage

    b) A pulsating ac voltage

    c) A ripple voltage

11. Electromagnetic poles are used instead of permanent magnets as they increase the field strength and

    a) Make it easily reversible

    b) Provide a means of controlling it

    c) Prevent it from decaying

12. Armature reaction can be overcome by:

    a) Moving the commutator brushes

    b) Moving the armature

    c) Moving the magnetic poles

13. Armature reaction can be overcome by:

    a) Compensation windings

    b) Additional armature windings

    c) Reversing the armature connections

14. In the armature core, eddy current losses are improved by:

    a) Increasing its cross-sectional area

    b) Reducing the number of laminations

    c) Increasing the number of laminations

15. A generator that supplies its own field excitation is called a Self-Excited Generator, but this is only possible if the field pole pieces have

   a) Residual magnetism

   b) Low retentivity

   c) Residual current

16. Self-excited generators are classed according to the type of field connection they use and there are three general types, series wound, shunt wound and:

   a) Serial Wound

   b) Parallel Wound

   c) Compound wound

17. A series wound DC generator's characteristic is:

   a) That the output voltage varies with input voltage

   b) That the output voltage varies with load current

   c) That the output does not vary

18. The output voltage in a DC shunt wound generator increases:

   a) As load current increases

   b) As load current decreases

   c) Does not vary

19. The output of a compound wound generator:

    a) As load current increases

    b) As load current decreases

    c) Does not vary

20. If the no-load voltage of a generator is 110 volts and its full-load voltage of is 100 volts, its % of regulation is:

    a) 20%

    b) 10%

    c) 5%

21. A series wound motor develops:

    a) Large amounts of starting torque

    b) Small amounts of starting torque

    c) No starting torque

22. With a series motor, before it is turned on

    a) There must always be some load connected

    b) There must never be a load connected

    c) It doesn't matter if a load is connected

23. When the shunt field is connected in parallel with the series field and armature, it is called a

    a) Long shunt

    b) Short shunt

    c) Parallel shunt

24. With a DC motor if both the direction of the magnetic field and the direction of current flow in the armature are reversed the motor's direction will:

   a) Reverse

   b) Remain unchanged

   c) Remain unchanged but slow down

25. When the field current of a DC motor decreases the motor's speed:

   a) Decreases

   b) Increases

   c) Reverses

26. In a DC motor, when the voltage applied to the armature is decreased, motor:

   a) Slows down

   b) Speeds up

   c) Reverses

27. In a DC motor, due to armature reaction its neutral plane moves:

   a) Anti-clockwise

   b) Clockwise

   c) Either direction depending on the original motor rotation

28. With a starter/generator, the engine start sequence can be terminated automatically by:

   a) A timer circuit

   b) An inertia switch

   c) A speed sensing circuit

29. With a starter/generator, the engine start sequence can be terminated automatically by:

a) A timer circuit

b) A centrifugal switch

c) Voltage build up in the output

Revision

DC Generator & Motor Theory

Answers

1.	B	20.	B
2.	A	21.	A
3.	A	22.	A
4.	C	23.	A
5.	A	24.	B
6.	B	25.	B
7.	C	26.	A
8.	A	27.	C
9.	B	28.	C
10.	C	29.	B
11.	B		
12.	A		
13.	A		
14.	C		
15.	A		
16.	C		
17.	B		
18.	A		
19.	C		

AC Theory

Introduction

Alternating quantqities are quantities that change direction. On mathematical graphs, alternating quantities are shown with positive and negative parts. The negative part indicates a change in direction, but a signal that is varying just in amplitude is not always an alternating signal. For example, figure 13.1 shows a signal that is varying in amplitude but at no time goes negative.

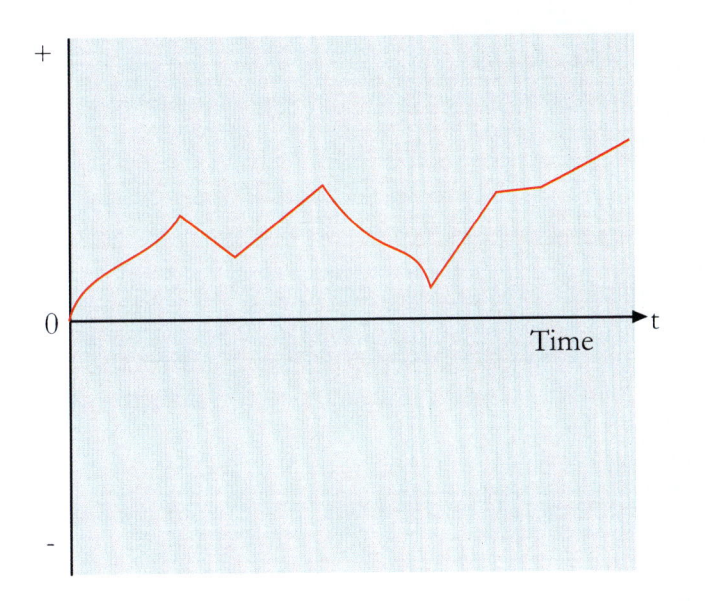

Figure 13.1 - Unidirectional Signal

As the signal does not go negative at any time it cannot be thought of as an alternating signal, so we can confer from this that an alternating indication, must have a ***positive*** and ***negative*** element. Alternating signals obviously come in all shapes and sizes, let's start by looking at one in its purest form.

CHAPTER THIRTEEN
AC THEORY

Sinusoid or Sine Wave

A *sinusoid*, more commonly called a *sine wave*, is the purest form of alternating signal in that its quantity goes positive and negative in equal amounts. This is perhaps better explained diagrammatically, so let us look at figure 13.2.

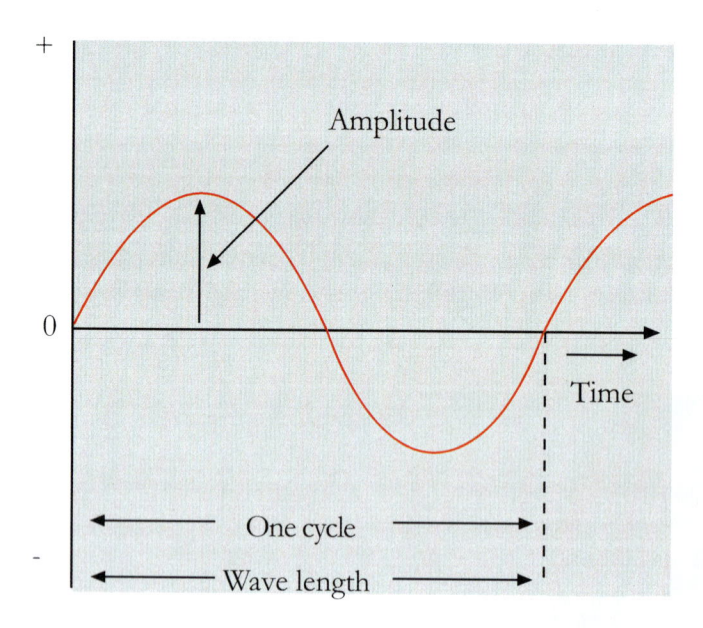

Figure 13.2 – Simple Sine Wave

This simple sine wave shows a positive half cycle and a negative half cycle of equal size or amplitude. A sine wave completes one cycle each time the wave goes negative and positive, usually over a period of time, but can also be explained in terms of degrees, radians or time.

Typically, when a sign wave is looked upon as an angular shift, measured in degrees, with a full cycle being 360°, it is best explained by looking at a point on a circle plotted on a graph as shown that in Figure 13.3.

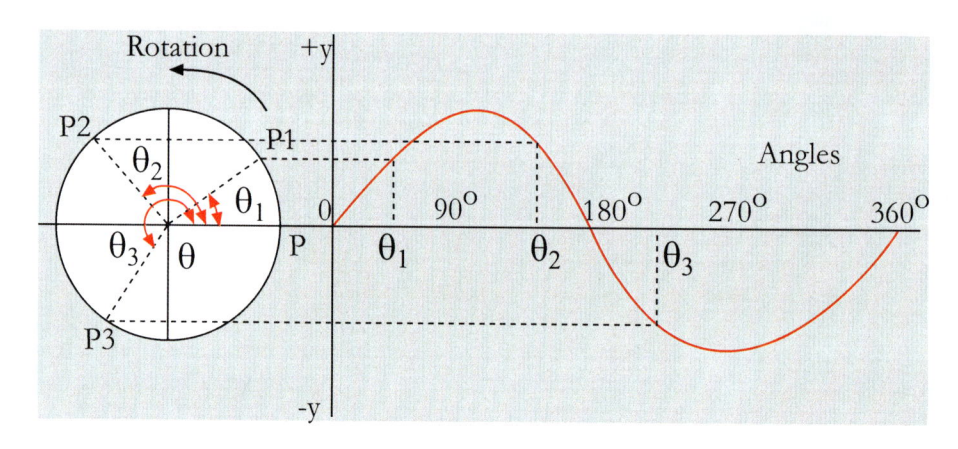

Figure 13.3 – Graph of a Point 'P' Rotated Through 360°

Figure 13.3, shows a point '**P**' on the circumference of a circle that is rotated anti clockwise through 360°. If its vertical height and angle are plotted on the graph as it rotates we get the resultant sine wave as shown. In electrical terms, the main alternating quantities are of current and voltage measured against time. However, before we go any further we perhaps need to define the terms that apply to all of the alternating quantities, Figure 13.4.

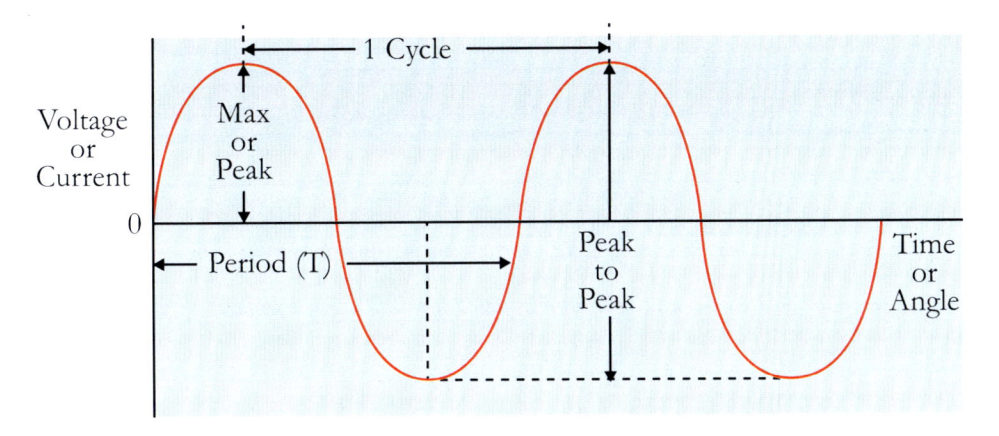

Figure 13.4 – Sinusoidal Waveform

Amplitude

This is defined as the vertical height of the signal at its maximum level above or below the horizontal or zero axis. As already mentioned , sine waves have both positive and negative amplitude.

Instantaneous Amplitude

This is the amplitude of the waveform at any point or instant along its path.

Cycle

When a signal completes one revolution, for example from 0° to 360° or one positive and negative waveform, it is known as a *Cycle*. It can all also be expressed as the distance between a point on the waveform and the next point where the waveform starts to repeat itself and this is also known as the *Period* of the sine wave.

Periodic Time

This is the time taken for an alternating quantity to complete one cycle, usually measured in seconds. Periodic Time is given the symbol T.

Frequency

This is the number of cycles that occurs in one second and is measured in cycles/second or Hertz. There is a direct relationship between frequency and periodic time as shown below:

$$f = \frac{1}{T} \text{ and } T = \frac{1}{f}$$

Wavelength

The distance between the beginning of one cycle, and the beginning of the next is called one *wavelength*, and the symbol used for wavelength in electronics is the Greek letter lambda (λ).

Average or Mean Value

The average value of the sine wave is zero. Therefore, a sine wave's average value is only taken over ½ a cycle. Average value is calculated using the following formula.

$$\text{Average or Mean Value} = \frac{\text{Area Under Curve}}{\text{Length of Base}}$$

To get the actual value of this requires calculus, which for the uninitiated is a branch of mathematics. However, thankfully we do not have to go through the mathematics and the answer is *0.637 × the maximum value*.

Whenever an average value is written, be it Voltage Current or power, it is always written as V_{av}, I_{av} or P_{av} as appropriate.

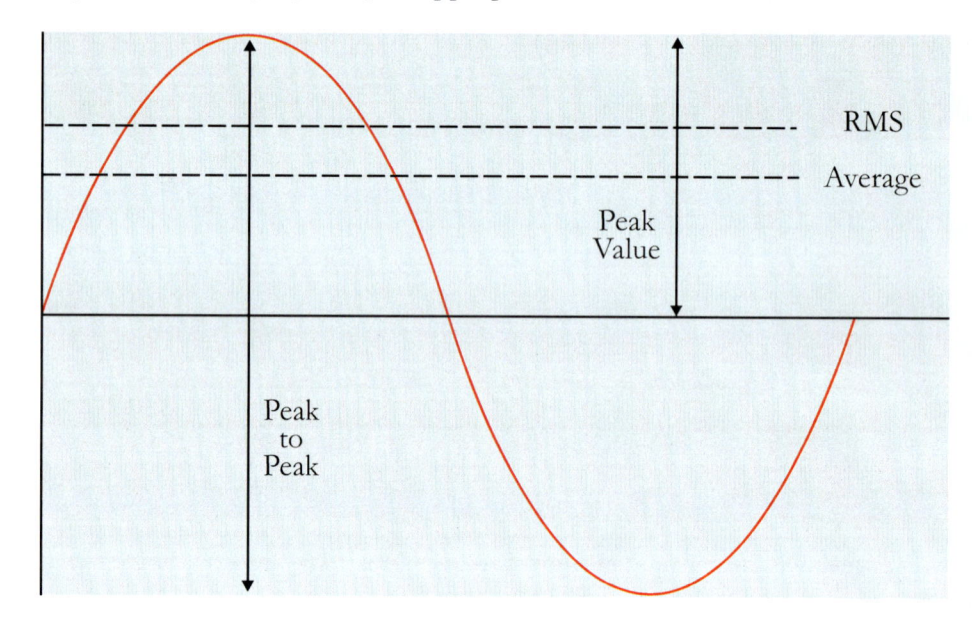

Figure 13.5 – Average and RMS values

Root Mean Square (RMS) Value

The value of an AC voltage is continually changing from zero up to the positive peak, through zero to the negative peak and back to zero again. Clearly, for most of the time it is less than the peak voltage, so this is not a good measure of its real effect. Instead we use the Root Mean Square Voltage

The RMS value is the square root of the average of the squares of all the instantaneous values during one half cycle. The RMS value of a sine wave is given by multiplying its maximum value by 0.707. However, this is sometimes expressed as $\frac{1}{\sqrt{2}}$.

Alternating currents and voltages are usually expressed in RMS values. The *RMS* value is the effective value of a varying voltage or current. It is the equivalent steady DC (constant) value which gives the same effect. For example, a lamp connected to a 12V RMS AC supply will shine with the same brightness when connected to a steady 12V DC supply. However, the lamp will be dimmer if connected to a 12V peak AC supply because the RMS value of this is only 8.5V (it is equivalent to a steady 8.5V DC).

Peak Amplitude

This is the maximum height or value of the waveform, which has a positive and negative value. Peak amplitude is usually called the maximum value of the signal or component, or element. However, alternating values are usually given as its RMS value. To calculate the peak value from RMS, it is multiplied by 1.414, sometimes written as √2.

Peak to Peak Amplitude

The difference in amplitude between the highest positive value, and highest negative value is called the *Peak-To-Peak* value, which is equal to twice the peak.

Angular Frequency (ω)

This is the rate of increase of the angle θ/second. Note that $\omega = 2\pi f$ radians/second. One radian per second is approximately 57.29578 degrees per second.

Instantaneous Values of a Sine Wave

In order to obtain the instantaneous value of a sign wave we need to do a little mathematics again I'm afraid. Let us take a closer look at the sine wave in figure 13.2 by looking at an individual point on its travels, figure 13.6.

CHAPTER THIRTEEN
AC THEORY

Figure 13.6 – Instantaneous Value of a Sine Wave

Figure 13.6 shows an alternating voltage waveform of peak value V_{max} that rotates around the circle with a radius of 0 to P. The radius has been rotated about an angle θ_1 to position OP_1. The line AP_1 now represents the instantaneous value, i.e. vertical height, of the voltage V_1 at θ_1 on the graph. This then forms a right-angled triangle θ to A to P_1 to 0. Using this triangle and Pythagoras's Theorem, we can calculate this instantaneous value using the formula:

$$Sin\theta = \frac{AP_1}{OP_1} \text{ and so } AP_1 = OP_1 Sin\theta$$

But from the graph in figure 13.6 we can see that $AP_1 = V_1$, which is the instantaneous value of the voltage waveform at θ_1; and we can also see that $OP_1 = V_{max}$. Therefore, we can rearrange the formula as:

$$V_1 = V_{max} Sin\theta$$

Therefore, for any instantaneous value of a sine wave, be it current (I) or Voltage (V), we can use the general expressions:

$$v = V_{max} Sin\theta$$

Where the lower case letter, 'v', 'i' is the instantaneous value of the sine wave and θ is its angle at any point along its wavelength. The horizontal axis of the sine wave can be in degrees, i.e. 360° or 2π radians, or time.

Note: 2π Radians = 360°, and so 1 radian = 57.3°.

With electrical signals that vary as a sine wave, the angle θ increases uniformly with time and therefore it is usual to write θ as a constant multiplied by time as θ = ωt, where ω is the constant and is called the angular frequency of the sine wave. Now using basic mathematics we can see that ω = 2πf Radians/seconds and t is measured in seconds. Therefore, ωt is a value in Radians and the instantaneous equation can be written, for all components as in table 13.1.

Quantity	Angular Frequency	Degrees
Voltage	$v = V_{max} \sin \omega t$	$v = V_{max} \sin \theta$
Current	$i = I_{max} \sin \omega t$	$i = I_{max} \sin \theta$
EMF	$e = E_{max} \sin \omega t$	$e = E_{max} \sin \theta$

Table 13.1 – Sine Wave Values

Sine Wave Phase Relationships

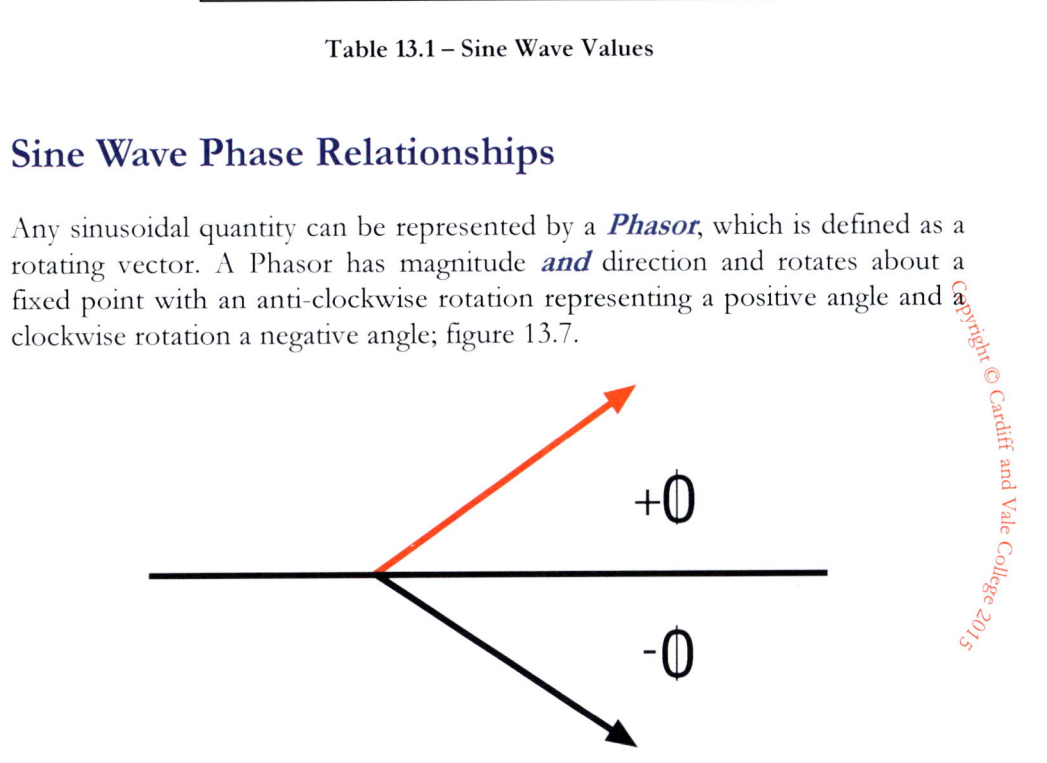

Any sinusoidal quantity can be represented by a **Phasor**, which is defined as a rotating vector. A Phasor has magnitude **and** direction and rotates about a fixed point with an anti-clockwise rotation representing a positive angle and a clockwise rotation a negative angle; figure 13.7.

Figure 13.7 – Positive & Negative Phasors

Let us look at a practical example of phasors. Figure 13.8 shows an alternating voltage with phasor 0A representing its peak value.

Figure 13.8 – Voltage Sine Wave Represented by a Phasor

As the voltage sine wave in figure 13.8 starts at time '0', the phasor is also shown as starting at '0'.

Signal Phase Relationships

When two or more signals are at the same frequency we can compare their *phase relationship* or more commonly their *phase angle*. The phase angle is defined as that angle between the phasor used to represent a sinusoidal quantity and a reference phasor. With *series* circuits, it is customary to use *current* as the reference phase, while in *parallel* circuits it is customary to use *voltage* as the reference.

Signals can only be in one of two states with respect to each other, in phase and out of phase.

Signals are in phase when they have the following characteristics:

- They pass through zero (0) at the *same* time and in the *same* direction
- They reach their positive maximum values at the *same* time
- They reach their negative maximum values at the *same* time

Consider the two sine waves in figure 13.9; an alternating voltage and an alternating current that are in phase.

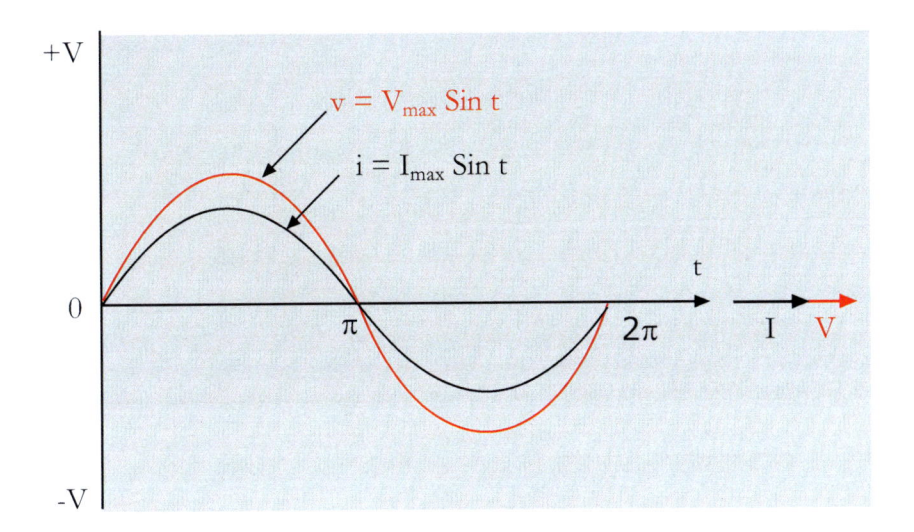

Figure 13.9 – Alternating Voltage & Alternating Current

Notice that they are in phase as they meet the criteria mentioned earlier, but their amplitudes are different. Signal amplitude ***does not*** affect the phase relationship and they can still be represented as phasor values as also shown in figure 13.8.

To consider the other option now, two or more signals are out of phase when they have the following characteristics:

- They pass through zero (0) at ***different*** times

- They reach their maximum positive values at ***different*** times

- They reach their maximum negative values at ***different*** times

When signals are out of phase they can be leading, lagging or anti-phase, ie out of phase by 180°. Let us consider the situation in figure 13.10 below:

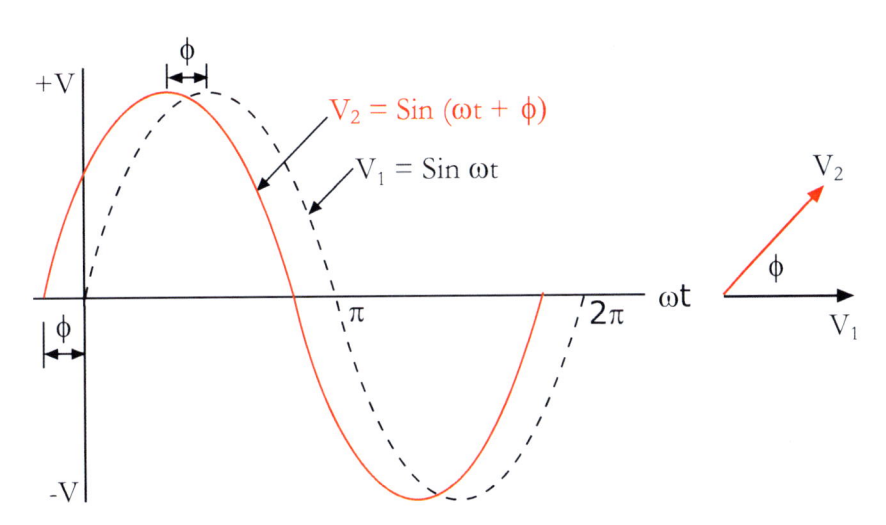

Figure 13.10 – Out of Phase Signals, Leading

CHAPTER THIRTEEN
AC THEORY

In figure 13.10, V_1 is the reference signal as it starts at time zero (0). V_2 starts before time zero (0) and so is said to **lead** V_1 by an angle ϕ.

When writing the general expressions for the instantaneous values of v_1 and v_2, the phase difference must also be brought into the formula as follows:

$$v_1 = V_{max}\mathrm{Sin}(\omega t) \text{ [Reference Signal]}$$

$$v_2 = V_{max}\mathrm{Sin}(\omega t + \phi) \text{ [Leading]}$$

The phase angle ϕ is obviously the difference between the signals and the '+' indicates which one is leading. These can also be represented as phasor values as also shown in figure 13.10. Now let us look at the opposite case in figure 13.11.

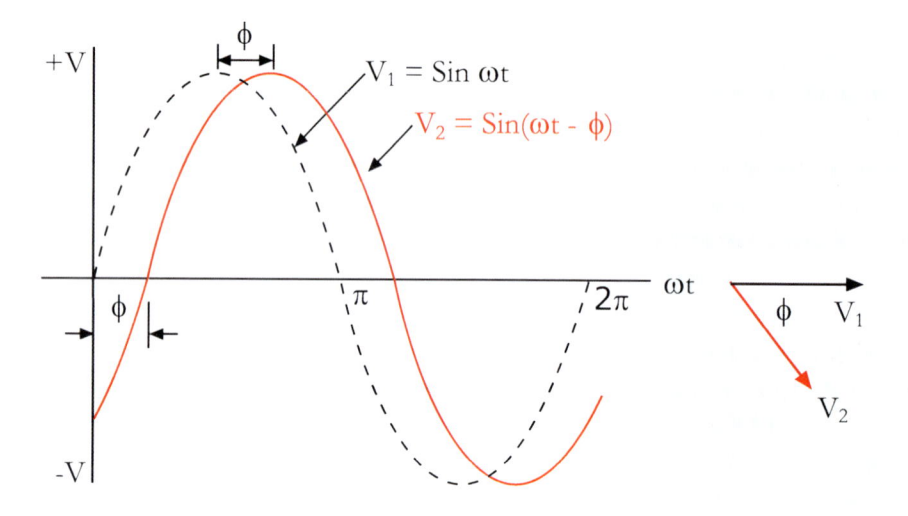

Figure 13.11 – Out of Phase Signals, Lagging

These two voltages are again out of phase but now V_2 starts **after** V_1 and so is now a lagging signal, and is said to lag by an angle ϕ.

Once again the general expressions for the instantaneous values of V_1 and V_2 are:

$$v_1 = V_{max}\mathrm{Sin}(\omega t) \text{ [Reference Signal]}$$

$$v_2 = V_{max}\mathrm{Sin}(\omega t - \phi) \text{ [Lagging]}$$

The phase angle ϕ is obviously the difference between the signals and the '-' indicates which one is lagging. These can also be represented as phasor values as also shown in figure 13.11. Notice that in the phasor diagram, the reference signal is **always** the horizontal value as this is the zero (0) start time.

Adding Phasors

Phasors, like all vector diagrams, can be added and subtracted using the parallelogram method as illustrated in figure 13.12, but *only if the phasors are drawn to scale* as arrow-headed lines.

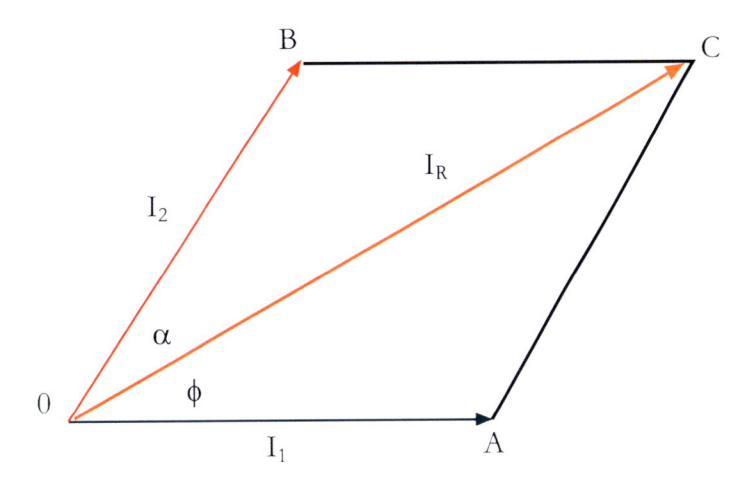

Figure 13.12 – Phasor Addition

Figure 13.12 shows two (2) current phasors I_1 and I_2 represented by the lines 0A and 0B respectively. To form the parallelogram, lines BC, AC are drawn parallel to their respective originals, and the resultant 0C is the addition of the currents I_1 and I_2. The resultant, whatever the component, is usually identified by the suffix 'R', and as shown in figure 13.12, its resultant is I_R. In figure 13.12, the resultant I_R is leading the reference value I_1 by the angle ϕ and lags I_2 by the angle α. However, it is important to note that phasor addition is **not** the same as mathematical addition, except at one point, as the angle between phasors must be included. Where is this one point? This point is where the individual components are in phase with each other.

If the phasors can be drawn as in figure 13.12 then a standard mathematical protractor can be used to find the angles ϕ and α.

The phase difference can also be measured against the amplitude/time graph as long as the individual components, i.e. current, voltage, etc, are at the same frequency. Figure 13.13 shows two voltage sine waves, V_1 and V_2, at a frequency of 12.5 kHz.

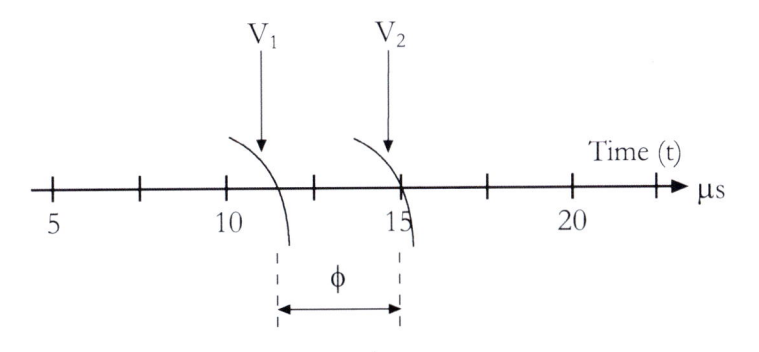

Figure 13.13 – Graph Measurement of Phase Difference

If the scale in figure 13.13 is represented by 1cm = 2.5μS and the difference between V_1 and V_2 is 1.6cm, then the time difference between these two voltages is 4μS. Now we know from the definitions given above that:

$$T = \frac{1}{f} = \frac{1}{12500} = 80\mu S$$

However, this represents 360° and so 1μS represents 4.5°. Therefore, as the time difference between these voltages is 4μS, the phase difference is 18°. If the horizontal scale were already in radians or degrees it would make the mathematics simpler, but the basic steps are the same.

Power in AC Circuits

When alternating current systems were first introduced, some scientists claimed that it was impossible to achieve energy in this way. Their argument was that power transfer would take place during the first half of the cycle, and would then transfer back during the second half.

Curiously, there is some truth in what they claimed but they had overlooked the relationship of **$P = I^2R$**. From basic mathematics, you should know that any number squared is always a positive resultant, and therefore I^2 means that the power is *always* positive no matter what the current is, ie positive or negative.

However, in an AC circuit it is only the resistive element that dissipates energy from the circuit as Inductors and Capacitors do not dissipate energy, which supports the theory of impossible energy, i.e. energy cannot be created or destroyed.

In this chapter, I only intend to look at power in a purely resistive circuit. Therefore, let us examine the circuit and waveforms in figure 13.14 below.

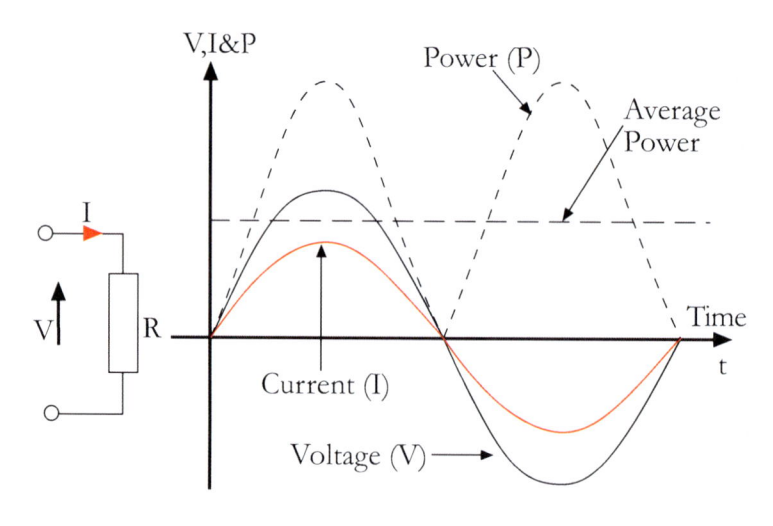

Figure 13.14 - V, I & P Waveforms of a Resistive Circuit

In figure 13.14, the circuit is purely resistive and when current flows through such a circuit with a resistance of 'R' ohms, the average power over a complete cycle is given by I^2R, where 'I' is the rms value of the current. If 'V' volts is also given as the rms value of the applied voltage, then V = IR, also in accordance with Ohm's Law. However, because the power is continually fluctuating, as in figure 13.14, the power of a purely resistive ac circuit is taken as the average power. This can then be obtained from the usual formula:

$$P = I^2R$$

Which we know can also be expressed as P=VI. Therefore, in a purely resistive circuit, the power can be obtained by multiplying the voltage and current read off the multimeter, exactly as in a dc circuit.

Other Alternating Waveforms

Earlier on in this chapter, I mentioned that to be an alternating signal, it must have a negative and positive element. So far we have only discussed the sine wave, but in practice, an alternating signal can be in any format. However, analysis of these formats requires detail knowledge of a branch of mathematics called Fourier's Analysis, which, thank heavens, is too deep for us on this short course. Therefore, we will only look at two (2) other types of alternating signal, the square and triangular wave.

Square Wave

The square wave is used throughout the aircraft as a signal format for the many digital databuses found throughout the aircraft, e.g. ARINC 429, ARINC 571; figure 13.15 shows its typical shape.

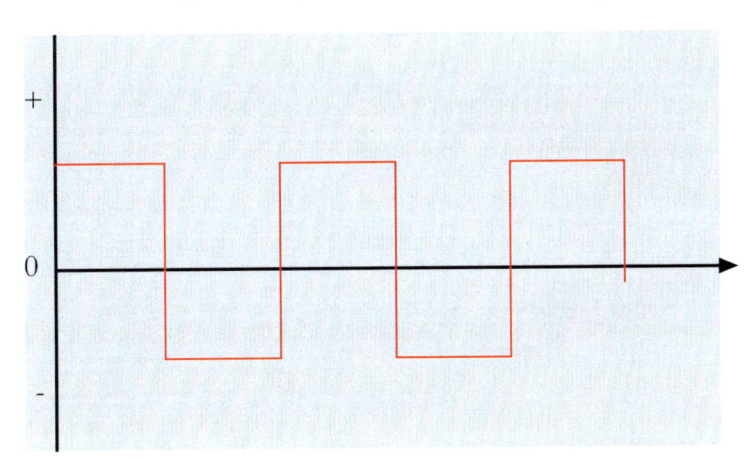

Figure 13.15 – Typical Alternating Square Wave

CHAPTER THIRTEEN
AC THEORY

As with the sine wave, the average value over a full cycle would be '0' and therefore, the average value is taken over a ½ cycle and so is the maximum value, ie V_{max} or I_{max}. In addition, because of its shape, the rms value of this waveform is also its maximum value.

Figure 13.16 shows a triangular waveform as an alternating signal.

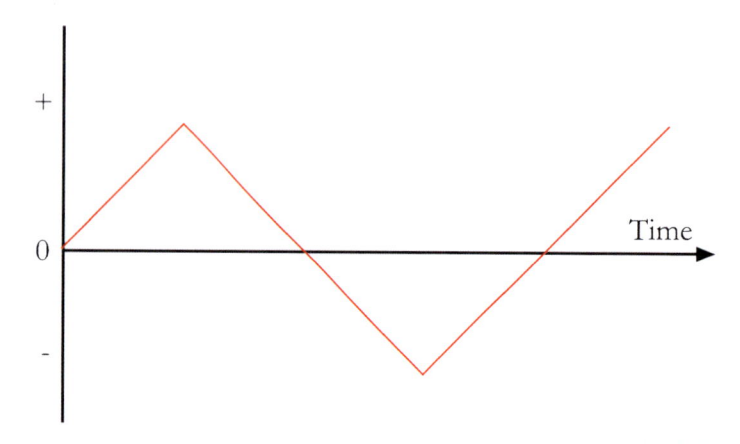

Figure 13.16 – Typical Alternating Triangular Wave

As with the sine wave, the average value over a full cycle would be '0' and therefore, the average value is taken over a ½ cycle and for a triangular waveform is 0.5 its maximum value. In addition, the rms value of this waveform is 0.577 its maximum value.

Form Factor

The ratio of a waveform's rms value to its average value over a ½ cycle is called its *form factor*, which gives an indication of the signal's shape. To summarise, I have tabulated the various values for the sine, square and triangular waveforms in table 13.2 below.

Wave Shape	Average Value	RMS Value	Form Factor
Sine	0.637 V_{max} or I_{max}	0.707 V_{max} or I_{max}	1.11
Square	V_{max} or I_{max}	V_{max} or I_{max}	1.00
Triangular	0.5 V_{max} or I_{max}	0.577 V_{max} or I_{max}	1.15

Table 13.2 – Form Factor

Single Phase

Up to now, in this chapter we have looked at single-phase circuits that have an alternating signal or signals. However, although single-phase systems are appropriate for smaller aircraft and primary dc aircraft that only require limited AC, eg for windscreen heating, reference voltages, etc, but this is inefficient the larger the load requirement. We therefore need to look at multi-phase systems, which we will do when we look at AC generators. However, we can still look at the basics of 3-phase in this chapter.

Three-Phase AC

When it is necessary to get the maximum amount of power from an ac circuit, we can use three (3) sets of windings that are physically 120° apart, although we still only use one rotating field; more on this later. When rotated, the outputs of each phase will also be at 120° with each other as illustrated in figure 13.17.

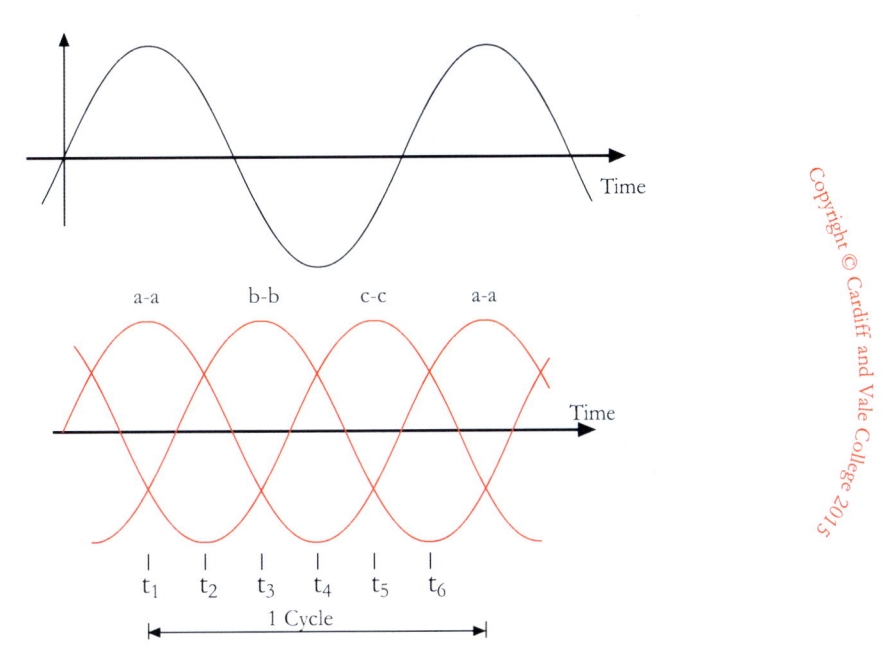

Figure 13.17 – Single Versus 3-Phase Signals

This approach has many advantages and one of the main ones is that if the ac is rectified, there are already three times as many pulses of rectified dc and these overlap so that the current never reaches zero.

There are two ways of connecting ac circuits; however, these will be discussed in a later chapter when we look in detail at AC generators and generation.

CHAPTER THIRTEEN
AC THEORY

Revision

AC Theory

Questions

1. In the figure below, the item highlighted by the arrow marked (1) is the sine wave's

   a) Amplitude

   b) Peak to Peak value

   c) Average value

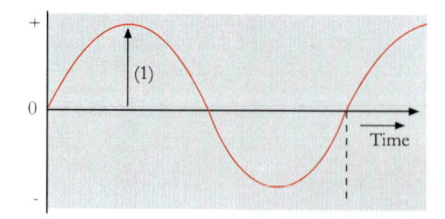

2. In the figure below, the item highlighted by the arrow marked (2) is the sine wave's:

   a) Wavelength

   b) Frequency

   c) Amplitude

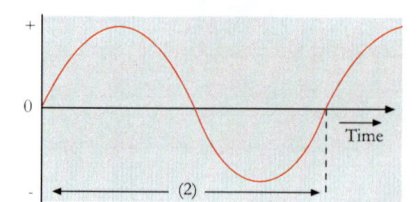

3. If the time for one cycle of a sine wave is 10 milli-seconds, its frequency is:

   a) 1000 Hz

   b) 10 Hz

   c) 100 Hz

4. If the maximum voltage of a sine wave is 10 V, what is the instantaneous voltage when Sin θ is 0.707:

   a) 70.7 V

   b) 0.707 V

   c) 7.07 V

5. If the maximum voltage of a sine wave is 20 V, what is its average value:

   a) 12.74 V

   b) 14.14 V

   c) 10 V

6. If the rms value of a sine wave is 30 V, what is its peak value?

   a) 42.42 V

   b) 21.21 V

   c) 19.11 V

7. If two signals pass through zero (0) at the *same* time, and reach their maximum values at different times, the signals are:

   a) In phase

   b) Out of phase

   c) Un-related

Revision

AC Theory

Answers

1. **A**
2. **A**
3. **C**
4. **C**
5. **A**
6. **A**
7. **B**

Resistive, Capacitive & Inductive Circuits

Introduction

You have already learned how resistance, inductance and capacitance behave individually in a direct current (DC) circuit. In this chapter we are going to look at how inductance, capacitance, and resistance affect alternating current.

Inductance & Alternating Current

When two things are in step, going through a cycle together, falling together and rising together, they are in phase. When they are out of phase, the angle of lead or lag, i.e. the number of electrical degrees by which one of the values leads or lags the other, is a measure of the amount they are out of step. The time it takes the current in an inductor to build up to maximum and to fall to zero is important for another reason. It helps illustrate a very useful characteristic of inductive circuits, i.e. *the current through the inductor always lags the voltage across the inductor*. Let us take a look at the voltage and current characteristics of some typical circuits.

A circuit having pure resistance, if such a thing were possible, would have the alternating current through it and the voltage across it rising and falling together. This is illustrated in figure 14.1 part (1), which shows the sine waves for *current* and *voltage* in a purely resistive circuit having an ac source. The current and voltage do not have the same amplitude, but they are in phase.

In the case of a circuit having inductance, the opposing force of the counter emf would be enough to keep the current from remaining in phase with the applied voltage. You learned that in a dc circuit containing pure inductance the current took time to rise to maximum even though the voltage was fully applied immediately at maximum.

Figure 14.1 parts (2) to (5) shows the waveforms for a purely inductive ac circuit in ¼ cycle steps.

CHAPTER FOURTEEN
RESISTIVE, CAPACITIVE & INDUCTIVE CIRCUITS

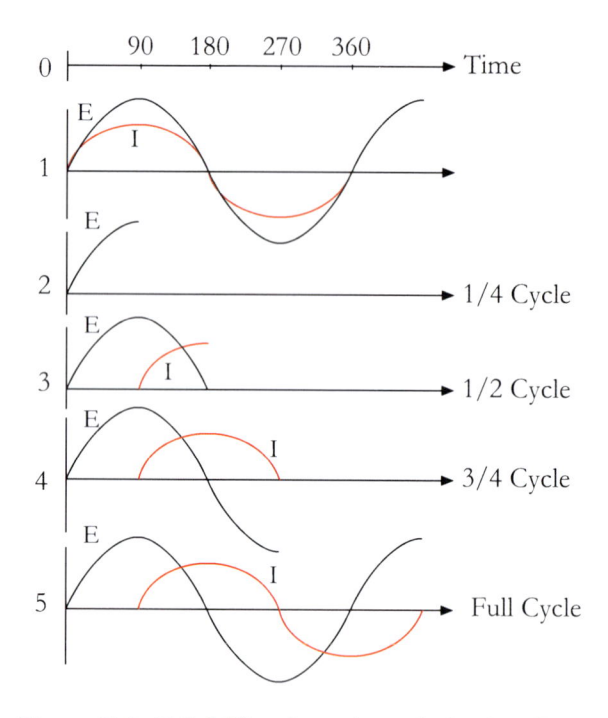

Figure 14.1 - V & I Waveforms in an Inductive Circuit

With an inductive circuit, in the first quarter-cycle, i.e. 0° to 90°, the applied ac voltage is continually increasing. If there were no inductance in the circuit, the current would also increase during this first quarter-cycle. However, since inductance opposes any change in current flow, no current flows during this first quarter-cycle.

In the next quarter-cycle, 90° to 180°, the voltage decreases back to zero; current begins to flow in the circuit and reaches a maximum value at the same instant the voltage reaches zero. The applied voltage now begins to build up to maximum in the other direction, to be followed by the resulting current. When the voltage again reaches its maximum at the end of the third quarter-cycle, i.e. at 270°, all values are exactly opposite to what they were during the first half-cycle. **The applied voltage leads the resulting current by 1/4-cycle or 90°.** To complete the full 360° cycle of the voltage, the voltage again decreases to zero and the current builds to a maximum value.

You must not get the idea that any of these values stops cold at a particular instant; **until the applied voltage is removed, both current and voltage are always changing in amplitude and direction**.

As you know from earlier chapters, the sine wave can be compared to a circle; just as you mark off a circle into 360°, you can mark off the *time* of one cycle of a sine wave into 360°. This relationship is shown in figure 14.2.

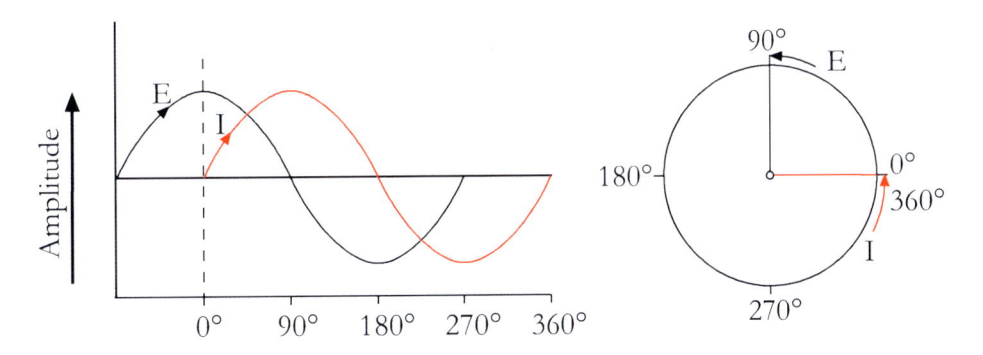

Figure 14.2 - Comparison of a Sine Wave & Circle in an Inductive Circuit

By looking at this figure you can see why the **current** is said to **lag** the **voltage**, in a purely inductive circuit, by 90°. Furthermore, by looking at figures 14.2 and 14.1, you can see why the current and voltage are said to be in phase in a purely resistive circuit.

In a circuit having both resistance and inductance then, as you perhaps would expect, *the current lags the voltage* by an amount somewhere between 0° to 90°.

A simple memory aid to help you remember the relationship of voltage and current in an inductive circuit is the word **ELI**. Since E is the symbol for voltage, L is the symbol for inductance, and I is the symbol for current; the word ELI demonstrates that current comes after, i.e. Lags, voltage in an inductor. There is an even better acronym, but I will leave this until we have also looked at the capacitive case.

Inductive Reactance

When an ac current flowing through an inductor continuously reverses itself, the inertia effect of the counter emf is greater than with DC. The greater the amount of inductance (L), the greater the opposition from this inertia effect and in addition, the faster the reversal of current, the greater the inertial opposition. This opposing force that an inductor presents to the flow of alternating current cannot be called resistance, since it does not act in the same way as a resistor to the flow of current. Therefore, an inductor's resistance to current flow is called **Inductive Reactance** because it is the '*reaction*' of the inductor to the changing value of alternating current. Inductive reactance is measured in ohms (Ω) and has the symbol 'X_L'.

As you know from our earlier work, the induced voltage in a conductor is proportional to the rate at which magnetic lines of force cut the conductor. The greater the rate, i.e. the higher the frequency, the greater the counter emf and in addition, the induced voltage increases with an increase in inductance, the more ampere-turns, the greater the counter emf. Reactance, then, increases with an increase of frequency and with an increase of inductance.

The formula to calculate inductive reactance is:

$$X_L = 2\pi f L$$

Where:

X_L is the Inductive Reactance

f is the frequency of the alternating voltage, measured in Hertz (z)

L is the inductance in Henrys

Using Pi = 3.14

Let us look at a typical example; L = 20H, f = 400 Hz, find X_L.

$$X_L = 2\pi \times 400 \times 20$$
$$X_L = 50,240 \Omega$$

However, rather than leave this as such a large number, it is usually written as 50.24 kΩ.

Capacitors & Alternating Current

The four parts of figure 14.3 show the variation of alternating voltage and current in a capacitive circuit, for each ¼ of a cycle. The black line represents the voltage across the capacitor, and the red line represents the current. The line running through the centre is the zero, or reference point, for both the voltage and the current and the bottom line marks off the time of the cycle in terms of degrees. I have also elongated the vertical scale to make the visual representation easier to view.

In a theoretical circuit of pure capacitance, the voltage across the capacitor exists only after current flows to charge the plates. At the instant a capacitor starts to charge, the voltage across its plates is zero and the current flow is maximum. The charging current, will continue to flow until the voltage across both plates (and hence the capacitor) is equal to the applied voltage. At this point the capacitor is said to be "fully charged" with electrons. The strength or rate of this charging current is at its maximum value when the plates are fully discharged (initial condition) and slowly reduces in value to zero as the plates charge up towards the supply voltage.

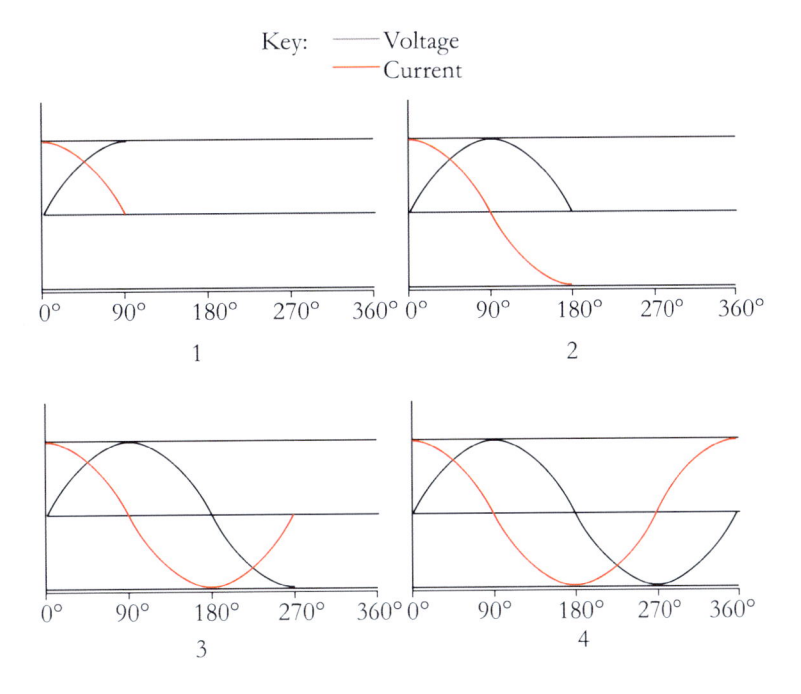

CHAPTER FOURTEEN
RESISTIVE, CAPACITIVE & INDUCTIVE CIRCUITS

Figure 14.3 - Phase Relationship of Voltage & Current in a Capacitive Circuit

To explain this function, first assume that the ac voltage has been acting on the capacitor for some time before the time represented by the starting point of the sine wave in figure 14.3. At the beginning of the first ¼ cycle, i.e. 0° to 90°, the voltage has just passed through zero and is increasing in a positive direction. Note that as the zero point is the steepest part of the sine wave, the voltage is changing at its greatest rate. The charge on a capacitor varies directly with the voltage, and therefore the charge on the capacitor is also changing at its greatest rate at the beginning of the first ¼ -cycle.

In other words, the greatest numbers of electrons are moving off one plate and onto the other and so the capacitor current is at its maximum value, as part 1 of the figure 14.3 shows.

As the voltage proceeds toward maximum at 90°, its rate of change becomes less and less, and so the current decreases towards zero. At 90°, the voltage across the capacitor is at a maximum, the capacitor is fully charged and there is no further movement of electrons from plate to plate. This is why the current at 90° is zero.

At the end of this first ¼ cycle, the alternating voltage stops increasing in the positive direction and starts to decrease. However, it is still a positive voltage, but to the capacitor the decrease in voltage means that the plate that has just accumulated an excess of electrons must lose some electrons. The current flow, therefore, must reverse its direction and part 2 of figure 14.3 shows the current curve to be below the zero line, i.e. negative current direction, during the second ¼ cycle, i.e. 90° to 180°.

At 180°, the voltage has dropped to zero and this means that for a brief instant the electrons are equally distributed between the two plates; the current is at a

CHAPTER FOURTEEN
RESISTIVE, CAPACITIVE & INDUCTIVE CIRCUITS

maximum because the rate of change of voltage is at a maximum. Just after 180° the voltage has reversed polarity and starts building up its maximum negative peak, which is reached at the end of the third ¼ cycle, i.e. 180° to 270°. During this third ¼ cycle the rate of voltage change gradually decreases as the charge builds to a maximum at 270°. At this point the capacitor is fully charged and it carries the full-applied voltage. Because the capacitor is fully charged there is no further exchange of electrons; therefore, the current flow is zero at this point. The conditions are exactly the same as at the end of the first ¼ cycle, i.e. 90°, but the polarity is reversed.

Just after 270°, the applied voltage once again starts to decrease, and the capacitor must lose electrons from the negative plate. It must discharge, starting at a minimum rate of flow and rising to a maximum and this discharging action continues through the last ¼ cycle, i.e. 270° to 360°, until the applied-voltage has reached zero. At 360° it is back at the beginning of the entire cycle, and everything starts over again.

If we now look at the complete voltage and current curves in part 4, you will see that the current always arrives at a certain point in the cycle 90° ahead of the voltage, because of the charging and discharging action. You should remember that this time and place relationship between the current and voltage is called the ***phase relationship***.

Note that the voltage-current phase relationship in a capacitive circuit is exactly opposite to that in an inductive circuit. ***The current of a capacitor leads the voltage across the capacitor by 90°.***

You should obviously realise that the current and voltage are both going through their individual cycles at the same time during the time the ac voltage is applied. The current does not go through part of its cycle, i.e. charging or discharging, stop, and wait for the voltage to catch up. The amplitude and polarity of the voltage and the amplitude and direction of the current are continually changing. Their positions with respect to each other and to the zero line at any instant between 0° to 360°, can be seen by reading upwards from the time/degree line. The current swing from the positive peak at 0° to the negative peak at 180° is ***not*** a measure of the number of electrons, or the charge on the plates but a picture of the direction and strength of the current in relation to the polarity and strength of the voltage appearing across the plates.

At times it is convenient to use the word '***ICE***' to recall that the phase relationship of the current and voltage in capacitive circuits. 'I' is the symbol for current, and in the word ICE it leads, or comes before, the symbol for voltage, i.e. E. This memory aid is similar to the '***ELI***' used to remember the current and voltage relationship in an inductor. The phrase '***ELI the ICE man***' is helpful in remembering the phase relationship in both the inductor and capacitor. Another more common acronym is ***CIVIL***; i.e. in a Capacitor current 'I' leads voltage 'V' which leads current 'I' in an Inductor 'L'.

Since the plates of the capacitor are changing polarity at the same rate as the ac voltage, the capacitor ***seems*** to pass an alternating current. Actually, the electrons do not pass through the dielectric, but their rushing back and forth from plate to plate does cause a current flow in the circuit. It is convenient, however, to say that the alternating current flows *through* the capacitor. You know this is not true, but the expression avoids a lot of trouble when speaking of current flow in a circuit containing a capacitor. By the same short cut, you may say that the capacitor does not pass a direct current because if both plates are connected to a dc source, current will flow only long enough to charge the capacitor. With a capacitor in a circuit containing both ac and dc, only the ac will be *passed* onto another circuit.

Capacitive Reactance

So far we have looking at the capacitor as a device that passes ac and in which the only opposition current has been the normal circuit resistance present in any conductor. However, capacitors themselves offer a very real opposition to current flow and this arises from the fact that, at a given voltage and frequency, the number of electrons that go back and for from plate to plate is limited by the storage ability, i.e. the capacitance, of the capacitor. As the capacitance is increased, a greater number of electrons change plates every cycle, and, since current is a measure of the number of electrons passing a given point in a given time, the current is increased.

Increasing the frequency will also *decrease* the opposition offered by a capacitor. This occurs because the number of electrons that the capacitor is capable of handling at a given voltage will change plates more often. As a result, more electrons will pass a given point in a given time, i.e. greater current flow. The opposition that a capacitor offers to ac is therefore *inversely* proportional to frequency and to capacitance and is called *Capacitive Reactance*. The symbol for capacitive reactance is X_C and the formula for calculating it is:

$$X_C = \frac{1}{2\pi f C}$$

Where:

X_C is capacitive reactance in ohms

f is frequency in Hertz

C is capacitance in farads

Using Pi = 3.14

$$X_C = \frac{1}{2\pi \times 100 \times 50 \times 10^{-6}}$$
$$X_C = \frac{1}{0.0314}$$
$$X_C = 31.85 \Omega$$

Let us look at a typical example; C = 50 µF, f = 100 Hz, find X_C.

$$X_C = \frac{1}{2\pi \times 100 \times 50 \times 10^{-6}}$$

$$X_C = \frac{1}{0.03142}$$

$$X_C = 31.83 \Omega$$

Reactance, Impedance & Power Relationships in AC Circuits

Up to this point inductance and capacitance have been explained individually in ac circuits. Now we need to look at the combination of inductance, capacitance, and resistance in ac circuits.

Series RLC Circuits

To explain the various properties that exist within ac circuits, let us look at the series RLC circuit illustrated in figure 14.4.

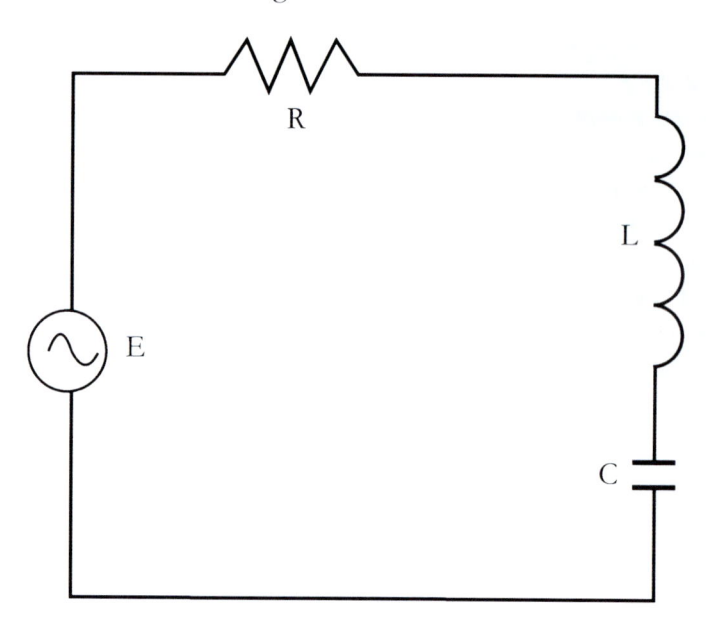

Figure 14.4 - Series RLC Circuit

The symbol marked 'E' in figure 14.4 is the general symbol used to indicate an AC voltage source.

Reactance

The effect of inductive reactance is to cause the current to lag the voltage, while that of capacitive reactance is to cause the current to lead the voltage. Therefore, since inductive reactance and capacitive reactance are exactly opposite in their effects, what will be the result when the two are combined? It is not hard to see that the net effect is a tendency to cancel each other out, with the combined effect then equal to the difference between their values. This resultant is called *Reactance* ; which is represented by the symbol 'X'; and expressed by the equation $X = X_L - X_C$ or $X = X_C - X_L$, whichever is the larger.

Therefore, if a circuit contains 50Ω of inductive reactance and 25Ω of capacitive reactance in series, the net reactance, or X, is 25Ω of inductive reactance.

For a practical example, suppose we look at a circuit with an inductor of 100μH in series with a capacitor of 0.001 μF, operating at a frequency of 4MHz. What is the value of net reactance, or X?

Solution:

$$X_L = 2\pi \times 4 \times 10^6 \times 100 \times 10^{-6}$$
$$X_L = 2512 \Omega$$

and

$$X_C = \frac{1}{2\pi \times 4 \times 10^6 \times 0.001 \times 10^{-6}}$$
$$X_C = 40 \Omega$$

Therefore, this circuit has a net reactance, 'X' of 2512Ω - 40Ω = 2.472 kΩ, which is *inductive*.

Note that when capacitive and inductive reactances are combined in series, the smaller value is always subtracted from the larger and the resultant reactance *always* takes the characteristics of the larger value.

Impedance

From the work we have done on inductance and capacitance you should now know how inductive reactance and capacitive reactance act in opposition to the flow of current in an ac circuit. However, there is another factor, the resistance, which also opposes the flow of the current. Since in practice ac circuits containing *reactance* also contain *resistance*, the two combine to oppose the flow of current. This combined opposition by resistance and reactance is called the circuit's *Impedance*, and is represented by the symbol 'Z'.

CHAPTER FOURTEEN
RESISTIVE, CAPACITIVE & INDUCTIVE CIRCUITS

Since the values of resistance and reactance are both given in ohms, it might at first glance seem possible to determine the value of the impedance by simply adding them together. However, this is not the case as in an ac circuit that contains only resistance, the current and the voltage will be in step, ie in phase, and will reach their maximum values at the same instant. In addition, in an AC circuit containing only reactance the current will either lead or lag the voltage by one ¼ of a cycle or 90°.

Therefore, the voltage in a purely reactive circuit will differ in phase by 90° degrees from that in a purely resistive circuit and for this reason reactance and resistance cannot be combined by simply adding them.

When reactance and resistance are combined, the value of the impedance will be greater than either on its own. It is also true that the current will not be in phase with the voltage, nor will it differ in phase by *exactly* 90°, but will be somewhere *between* the in phase and the 90°. In practice, the *larger* the reactance compared with the resistance, the *more* the phase difference will approach 90° and conversely, the *larger* the resistance compared to the reactance, the *more* the phase difference will approach 0°.

If the value of resistance and reactance cannot simply be added together to find the impedance, or Z, how is it determined? Well, because the current through a resistor is in phase with the voltage across it and the current in a reactance differs by 90° from the voltage across it, the two are at right angles to each other. Therefore they can be combined by using the same method used in the construction of a right-angle triangle.

Assume you want to find the impedance of a series combination of 8Ω resistance and 5Ω inductive reactance. Start by drawing a horizontal line to scale to represent the 8Ω resistance, '**R**', as the base of the triangle; then, since the effect of the reactance is always at right angles, i.e. 90°, to that of the resistance, draw to scale a vertical line, representing the 5Ω inductive reactance, 'X_L', as the height of the triangle, figure 14.5. Now, complete the hypotenuse, i.e. longest side, of the triangle and this then represents the impedance of the circuit, '**Z**'.

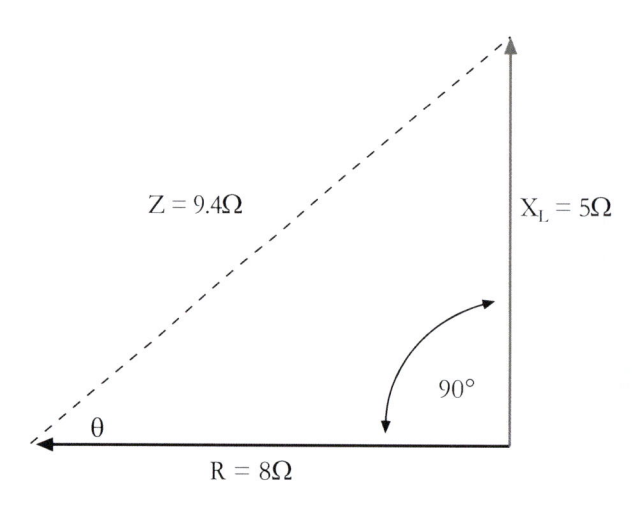

Figure 14.5 - Vector Diagram of R, XL & Z in a Series Circuit

CHAPTER FOURTEEN
RESISTIVE, CAPACITIVE & INDUCTIVE CIRCUITS

As shown in figure 14.5, representing these values as vectors form a right-angled triangle, which can also be represented mathematically as:

$$\text{Hypotenuse}^2 = \text{Base}^2 + \text{Vertical}^2$$

or

$$\text{Hypotenuse} = \sqrt{\text{Base}^2 + \text{Vertical}^2}$$

This is Pythagoras's theorem for a right-angled triangle and can be applied to the various values of resistance, etc as:

$$\text{Impedance(Z)} = \sqrt{\text{Resistance(R)}^2 + \text{Reactance(X)}^2}$$

i.e. $Z = \sqrt{R^2 + X^2}$

Now let us apply this equation to the values in figure 14.5:

$$Z = \sqrt{8^2 + 5^2}$$
$$Z = \sqrt{64 + 25}$$
$$Z = 9.4\,\Omega$$

When there is a capacitive reactance to deal with instead of inductive reactance, it is customary to draw the line representing the capacitive reactance in a downward direction, illustrated in figure 14.6.

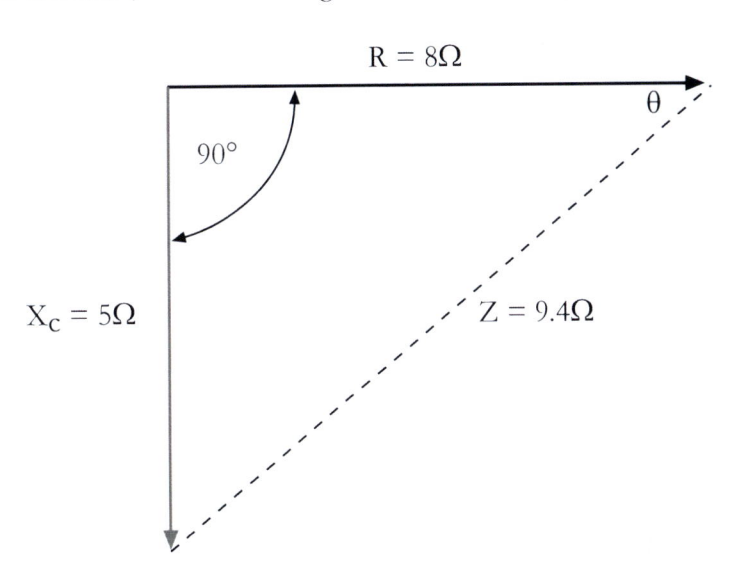

Figure 14.6 - Vector Diagram of R, XC & Z in a Series Circuit

CHAPTER FOURTEEN
RESISTIVE, CAPACITIVE & INDUCTIVE CIRCUITS

The line is drawn downward for capacitive reactance to indicate that it acts in a direction opposite to inductive reactance, which is drawn upward.

In a series circuit containing capacitive reactance the equation for finding the impedance becomes:

$$Z = \sqrt{R^2 + X^2}$$

In many series circuits you will find resistance combined with both inductive and capacitive reactance. Since you know that the value of the reactance, X, is equal to the difference between the values of the inductive reactance, X_L, and the capacitive reactance, X_C, the equation for the impedance in a series circuit containing R, X_L, and X_C then becomes:

$$Z = \sqrt{R^2 + (X_L - X_C)^2}$$

However, this formula can only be used when the resistance, capacitance and inductance are in series.

Figure 14.7 below shows the vector addition method that may be used to determine the impedance in a series circuit consisting of resistance, inductance, and capacitance.

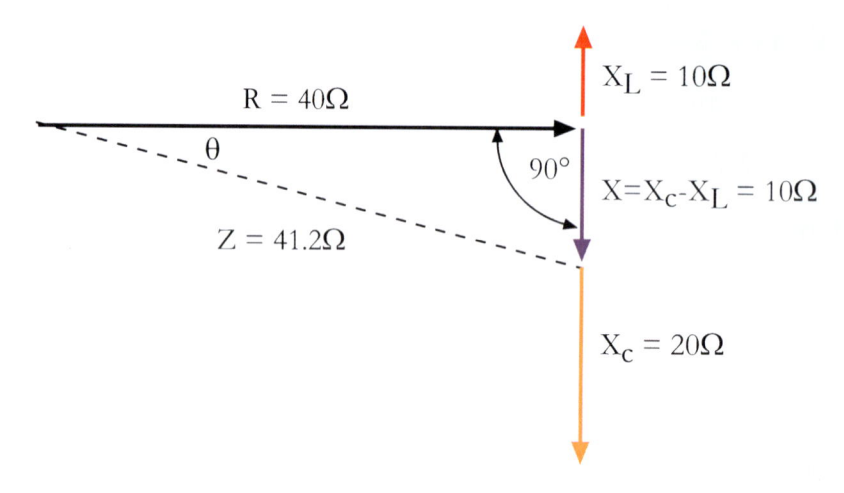

Figure 14.7 - Vector Diagram Showing Relationship of R, X & Z in a Series Circuit

If a 10Ω inductive reactance and 20Ω capacitive reactance are connected in series with 40Ω ohms resistance then let the horizontal line represent the resistance 'R'. The line drawn upward from the end of R, represents the inductive reactance, X_L, and the capacitive reactance, X_C, is represented by a line drawn downward at right angles from the same end of R. The resultant of X_L and X_C is found by subtracting X_L from X_C and this resultant represents the value of X. The subsequent hypotenuse is then the impedance 'Z', i.e. 41.2Ω.

Ohms Law for AC

In general, Ohm's law cannot be applied to ac circuits, as it does not consider the reactance that is always present in such circuits. However, by a modification of Ohm's law that does take into consideration the effect of reactance we obtain a general law, which is applicable to ac circuits. As the impedance Z, represents the combined opposition of all the reactance and resistances, this general law for ac is:

$$I = \frac{E}{Z}$$

This general modification applies to ac flowing in any circuit, and any one of the values may be found from the equation if the other two are known.

In a purely capacitive circuit, the only opposition to current flow is X_C, whereas in a purely inductive circuit the only opposition to current flow is X_L. These types of circuits are theoretical as there will always be some form of resistance in the circuit from conductors and connectors.

It should also be noted that if $X_L = X_C$ which cancel each other out, making the circuit appear to be purely resistive, the circuit is then deemed to be at **resonance**. More on this later.

Power in AC Circuits

We know that in a DC circuit the power is equal to the voltage times the current, or $P = V \times I$. Therefore, if a voltage of 100 V applied to a circuit produces a current of 10 A, the power is 1 kW.

This is also true in an ac circuit when the current and voltage are in phase; i.e. when the circuit is effectively resistive. However, if the ac circuit contains reactance, the current will lead or lag the voltage by a certain amount, i.e. the phase angle. When the current is out of phase with the voltage, the power indicated by the product of the applied voltage and the total current gives only what is known as the *Apparent Power*. The *True Power* depends upon the phase angle between the current and voltage and this is given the symbol 'θ', i.e. the Greek letter Theta.

When an alternating voltage is applied across a capacitor, power is taken from the source and stored in the capacitor as the voltage increases from zero to its maximum value. Then, as the applied voltage decreases from its maximum value to zero, the capacitor discharges and returns the power to the source.

Likewise, as the current through an inductor increases from its zero value to its maximum value the field around the inductor builds up to a maximum, and when the current decreases from maximum to zero the field collapses and returns the power to the source. You can see therefore that no power is used up in either case, since the power alternately flows to and from the source.

CHAPTER FOURTEEN
RESISTIVE, CAPACITIVE & INDUCTIVE CIRCUITS

This power that is returned to the source by the reactive components in the circuit is called **Reactive Power**.

In a purely *resistive* circuit **all of the power is consumed and none is returned to the source.** When the instantaneous values of voltage and current are multiplied, the resultant power waveform is as shown in Figure 14.8

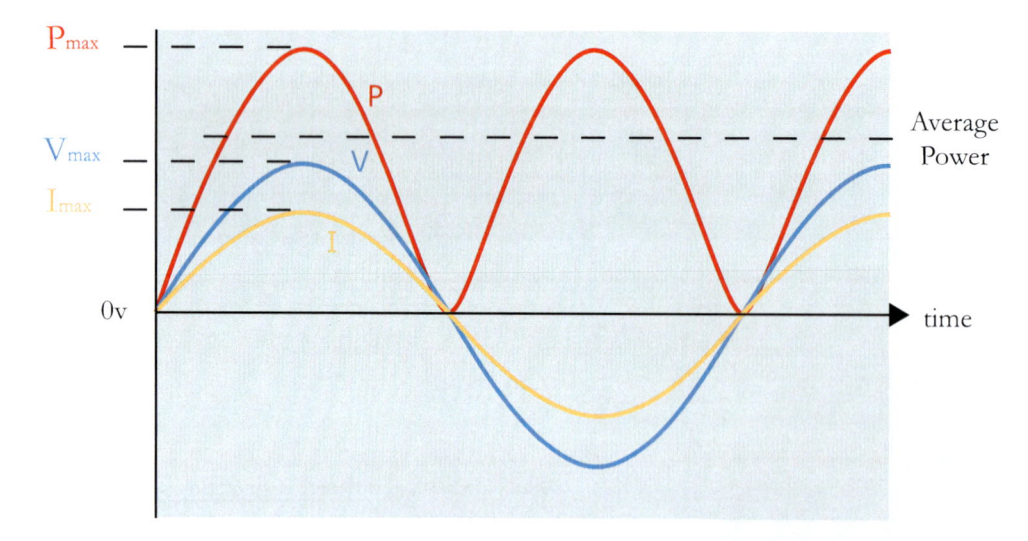

Figure 14.8 – Power in a Purely Resistive Circuit

In a purely *reactive* circuit **no power is consumed and all of the power is returned to the source**. Figure 14.9 details the resultant power waveform in a purely inductive circuit.

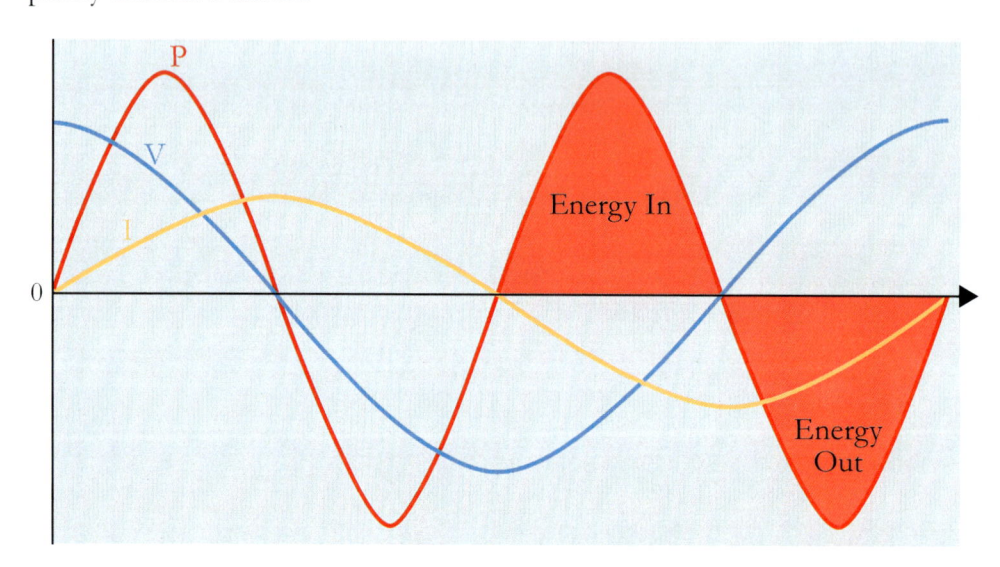

Figure 14.9 – Power in a Purely Inductive Circuit

As we can see, when the power curve is 'positive', the inductor takes power from the supply source. When the power curve is 'negative', the inductor returns power to the supply source.

CHAPTER FOURTEEN
RESISTIVE, CAPACITIVE & INDUCTIVE CIRCUITS

In the same way as a pure inductive circuit, a pure capacitive circuit also produces a current flow which effectively does 'no work'. On one half-cycle, power is delivered to the capacitor (charging) from the supply source but the on the next half-cycle the capacitor returns power to the supply source (discharging). Figure 14.10.

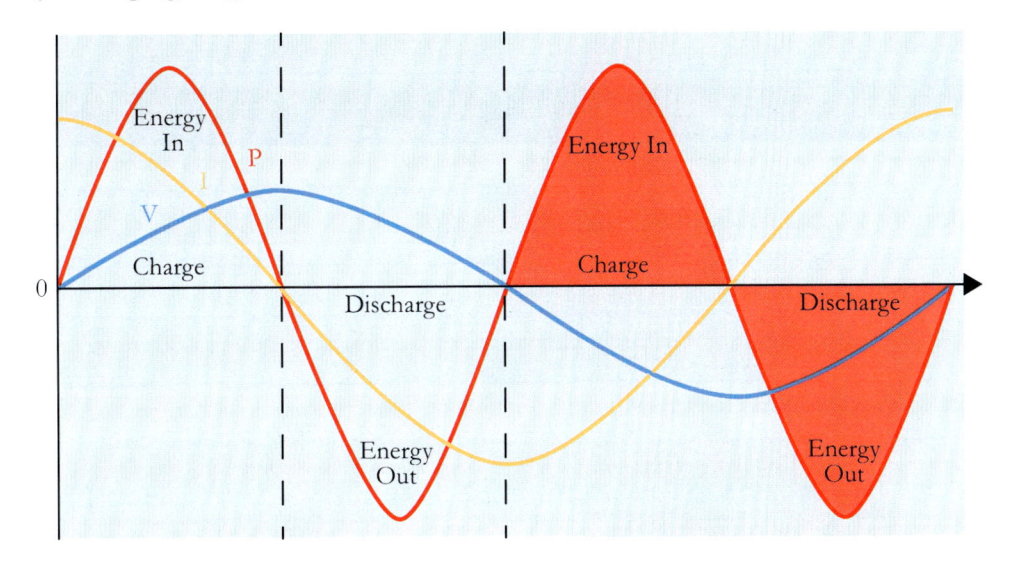

Figure 14.10 – Power in a Purely Capacitive Circuit

It follows then that in a circuit that contains both resistance and reactance, there must be some power dissipated in the resistance as well as some returned to the source by the reactance. In figure 14.11 we can see the relationship between the voltage, current, and power in such a circuit.

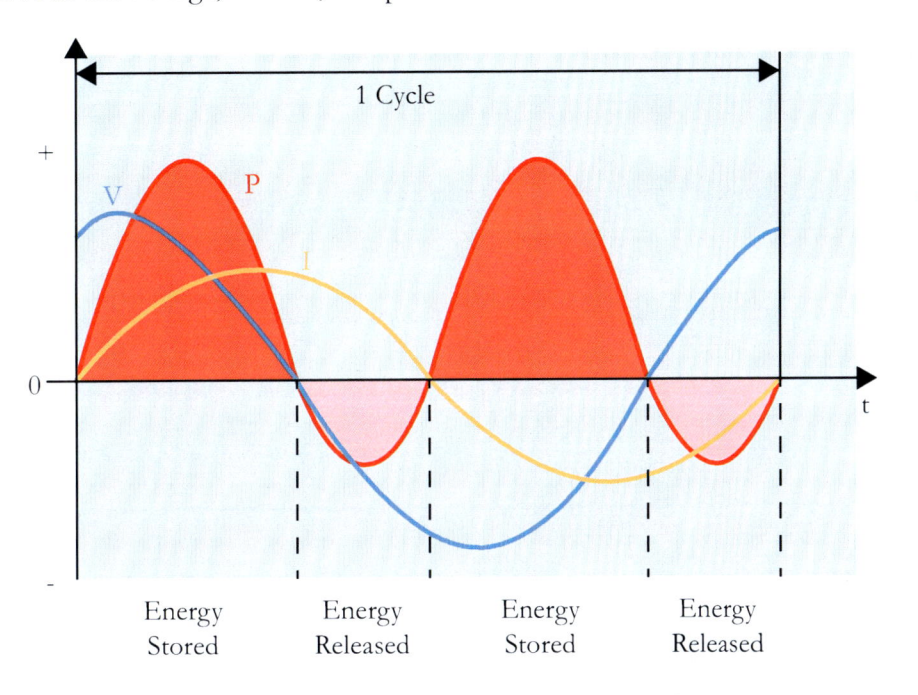

Figure 14.11 - Instantaneous Power When V & I are Out of Phase

The part of the power curve that is shown below the horizontal reference line is the result of multiplying a *positive* instantaneous value of current by a

333

negative instantaneous value of the voltage, or vice versa. As I am sure you know, the product obtained by multiplying a positive value by a negative value will always be negative. Therefore, the power at that instant must be considered as negative power; or in other words, during this time the reactance is returning power to the source.

The instantaneous power in the circuit is equal to the product of the applied voltage and current through the circuit. When the voltage and current are of the same polarity they are acting together and taking power from the source. When the polarities are unlike they are acting in opposition and power is being returned to the source. Briefly then, in an ac circuit that contains reactance as well as resistance, the apparent power is reduced by the power returned to the source, so that in such a circuit the net power, or *true power*, is always less than the apparent power.

Calculating True Power in AC Circuits

As mentioned before, the *true* power of a circuit is the power *actually used* in the circuit. This power, measured in watts, is the power associated with the total resistance in the circuit. To calculate the circuit's true power, only the voltage and current associated with the resistance must be used. Since the voltage drop across the resistance is equal to the resistance multiplied by the current through the resistance, true power can be calculated by the formula:

$$\text{True Power} = (I_R)^2 R$$

Where:

I_R is the resistive Current

R is the overall circuit resistance

Let us look at an example; find the true power of the circuit shown in figure 14.12.

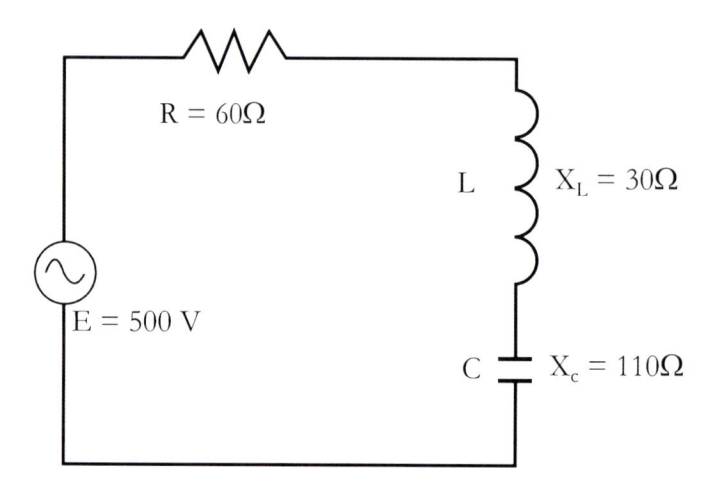

Figure 14.12 - Example Circuit

Solution:

$$X = X_C - X_L$$
$$X = 110 - 30 \, \Omega$$
$$X = 80 \, \Omega$$

Now

$$Z = \sqrt{R^2 + X^2}$$
$$Z = \sqrt{60^2 + 80^2}$$
$$Z = 100 \, \Omega$$

Now that we have Z and are given E, we can work out the current using the formula:

$$I = \frac{E}{Z}$$
$$I = \frac{500}{100}$$
$$I = 5A$$

Since the current in a series circuit is the same in all parts of the circuit, we can say that:

$$\text{True Power} = (I_R)^2 R$$
$$\text{True Power} = 5^2 \times 60$$
$$\text{True Power} = 1500W$$

Calculating Reactive Power in AC Circuits

Reactive power is the power returned to the source by the circuit's reactive components. This type of power is measured in *Volt-Amperes-Reactive*, usually abbreviated to *VAR* or *var*.

Reactive power is calculated by using the voltage and current associated with the circuit reactance. However, since the voltage of the reactance is equal to the reactance multiplied by the reactive current, reactive power can be calculated by the formula:

$$\text{Reactive Power} = (I_X)^2 X$$

Where:

Reactive Power is measured in VAR

I_X is the reactive current in Amperes

X is the total reactance measured in ohms

Another way to calculate reactive power is to calculate the inductive and capacitive power and subtract the smaller from the larger. Either way will work; which one you use will depend on the circuit values you are given.

Calculating Apparent Power in AC Circuits

Apparent power is the power that appears to the source because of the circuit impedance, since the impedance is the total opposition to the ac, and is measured in **Volts-Amperes**, abbreviated to **VA**. Apparent power is the combination of true power and reactive power, but is not found by simply adding true power and reactive power just as impedance is not found by adding resistance and reactance.

To calculate apparent power, we need to use either of the following formulas:

$$\text{Apparent Power} = (I_Z)^2 Z$$

or

$$\text{Apparent Power} = \sqrt{(\text{True Power})^2 + (\text{Reactive Power})^2}$$

Power Factor

Power Factor (PF) is a number, represented as a decimal or a percentage, which represents the portion of the apparent power dissipated in a circuit. The easiest way to find the power factor is to use trigonometry and find the cosine of the phase angle θ as this is equal to the power factor. However, we do not need to use trigonometry to find the power factor, as the power dissipated in a circuit is true power, then:

$$\text{True Power} = \text{Apparent Power} \times \text{PF}$$

$$\text{PF} = \frac{\text{True Power}}{\text{Apparent Power}}$$

If true power and apparent power are known you can use the formula shown above. Going one step further, another formula for power factor can be developed. By substituting the equations for true power and apparent power in the formula for power factor, we get:

$$\text{PF} = \frac{(I_R)^2 R}{(I_Z)^2 Z}$$

However, since current in a series circuit is the same in all parts of the circuit, $I_R = I_Z$ and therefore, in a series circuit:

$$\text{PF} = \frac{R}{Z}$$

Note: As stated earlier the power factor can be expressed as a decimal or percentage, e.g. a PF of 0.6 and 60% are the same. In a purely resistive circuit, true power is always equal to the apparent power so the power factor will always be 1 or unity. Because of considerations of automatic control over varying conditions, the power factor in aircraft systems is kept well away from unity. It is usual to operate at power factors in the order of 0.75 or 0.8 on aircraft.

Power Factor Correction

The apparent power in an ac circuit has been described as the power the source '*sees*'. As far as the source is concerned the apparent power is the power that must be provided to the circuit. However, from our work above, you also know that the true power is the power *actually* used in the circuit. The difference between apparent power and true power is wasted because, in reality, only true power is consumed and so the ideal situation would be for apparent and true power to be equal. If this were the case the power factor would be one (1), i.e. unity or 100%.

There are two ways in which this condition can exist.

1. If the circuit is purely resistive
2. If the circuit '*appears*' purely resistive to the source

To make the circuit appear purely resistive there must obviously be no reactance, but this will only be possible when the inductive reactance (X_L) and capacitive reactance (X_C) are equal. Remember that $X = X_L - X_C$ and so when $X_L = X_C$, $X = 0$.

The expression '*correcting the power factor*' refers to reducing the reactance in a circuit, and as we have just discussed, the ideal situation is to have no reactance in the circuit. This is accomplished by *adding* capacitive reactance to a circuit that is inductive and inductive reactance to a circuit that is capacitive.

Series RLC Resonant Circuits

In a series RLC circuit there is always a specific frequency were the inductive reactance of the inductor becomes equal in value to the capacitive reactance of the capacitor. In other words, $X_L = X_C$. The point at which this occurs is called the **Resonant Frequency** (f_r) of the circuit, Figure 14.13.

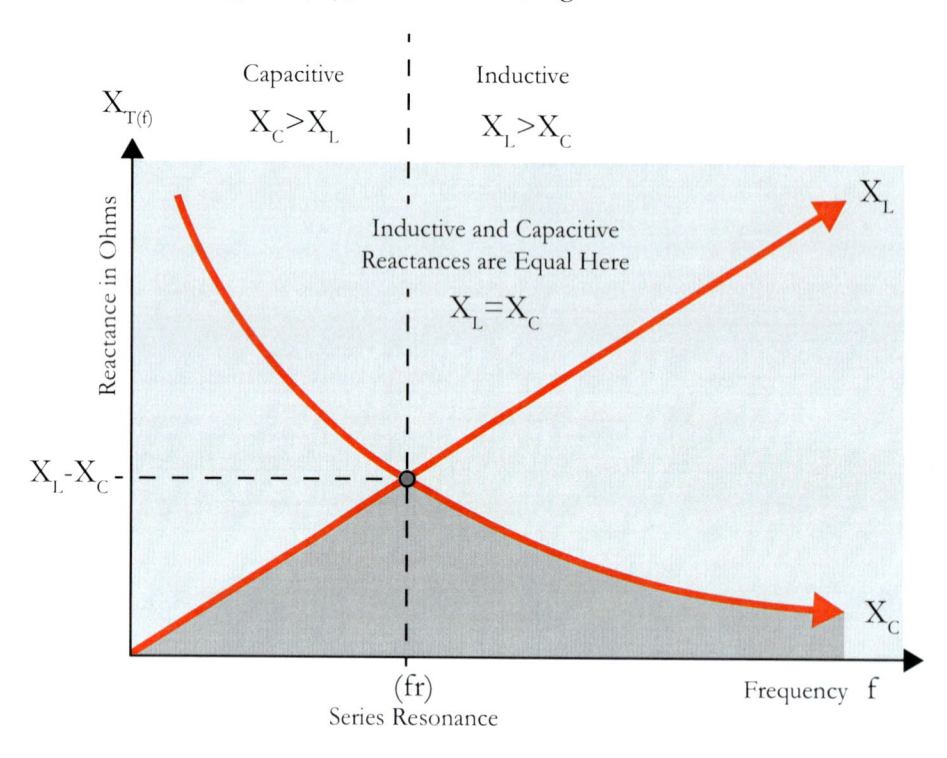

Figure 14.13 – Resonant Frequency

At the resonant frequency the voltages developed by the inductive and capacitive components cancel each other out. As a result a series circuit at resonance acts like a pure effective resistance, meaning as $X_L = X_C$ the impedance at resonance is equal to the effective resistance.

At frequencies below the resonant frequency X_L is lower than X_C and the capacitive reactance is predominant. At frequencies above the resonant frequency XL is higher than XC and the inductive reactance is predominant.

Without going into the mathematics of how this is achieved, a circuit's resonant frequency is found using the formula:

$$f_r = \frac{1}{2\pi\sqrt{LC}}$$

Where f_r is the resonant frequency in Hertz

L is the circuit inductance in Henries

C is the circuit capacitance in Farads

At resonance, the circuit reactance 'Z' must equal 'R' as $X_C = X_L$ and are 180° anti-phase and so cancel each other out. At this point, the circuit current will be at a maximum, as only 'R' is opposing it, and in phase with the applied voltage, Figure 14.14.

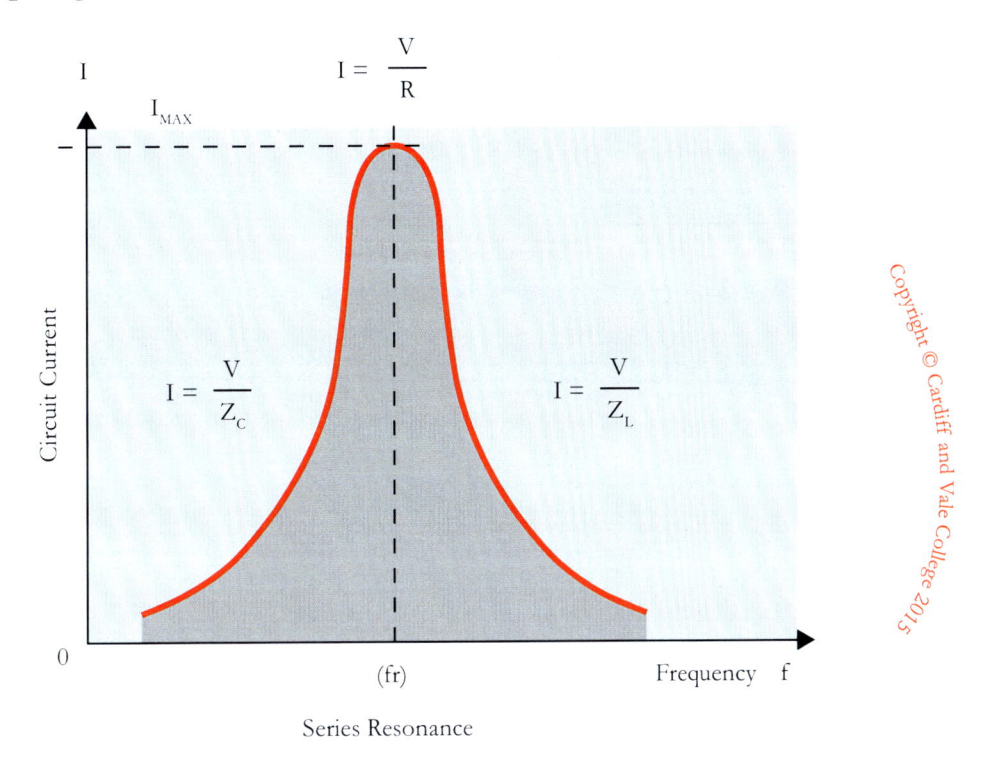

Figure 14.14 – Series Resonance Current Flow

Since the current flowing through a series resonance circuit is the product of voltage divided by impedance and, as we have seen, at resonance the impedance, Z is at its minimum value R, since X_L and X_C cancel each other out. The circuit current at this frequency will be at its maximum value (V/R)

A series resonance circuit is also known as an ***Acceptor Circuit*** because at resonance, the impedance of the circuit will be minimum so it easily accepts the current whose frequency is equal to the resonant frequency.

With reference to the simple series RLC circuit shown in Figure 14.15, the voltage across the series combination is the phasor sum of V_R, V_L and V_C.

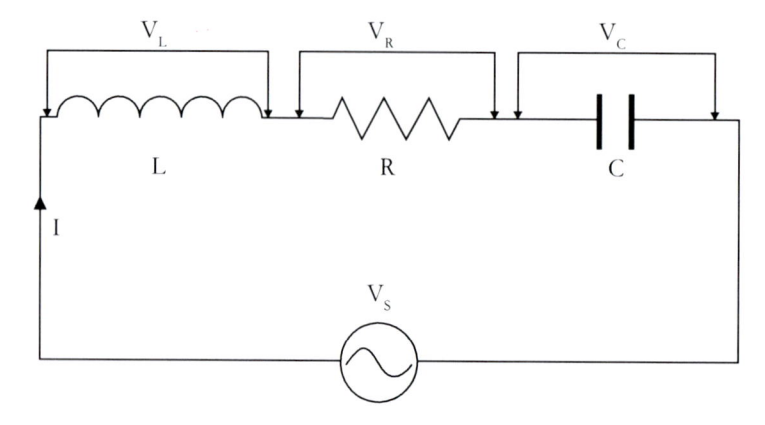

Figure 14.15 – Series RLC Circuit

Therefore if the circuit is at resonance the two reactances are equal and cancelling, then the two voltages representing V_L and V_C must also be opposite and equal in value thereby cancelling each other out. Hopefully we should remember that with pure components the phasor voltages are drawn at +90° and -90° respectively. As such, in a ***series resonance*** circuit $V_L = -V_C$ therefore, $V_S = V_R$, meaning that all of the supply voltage is dropped across the resistor. Please note that this will only occur in a series circuit at resonance.

Parallel R, L, & C Circuits

When dealing with a parallel ac circuit, the theories we have already discussed in this chapter for series ac circuits still apply. However, there is one major difference between a series circuit and a parallel circuit that must be considered; i.e. current is the same in all parts of a series circuit, whereas voltage is the same across all branches of a parallel circuit.

Because of this difference, the total impedance of a parallel circuit must be computed on the basis of the current in the circuit. Therefore, when working with a parallel circuit, we must use the following formulas instead:

$$I_X = I_L - I_C$$

$$I_Z = \sqrt{(I_R)^2 + (I_X)^2}$$

$$PF = \frac{I_R}{I_Z}$$

Note: If no value for E is given in a circuit, any value of E can be assumed to find the values of I_L, I_C, I_X, I_R, and I_Z as long as the same value of voltage is then used to find the impedance.

CHAPTER FOURTEEN
RESISTIVE, CAPACITIVE & INDUCTIVE CIRCUITS

Let us have a look at an example and find the value of Z in the circuit shown in figure 14.16.

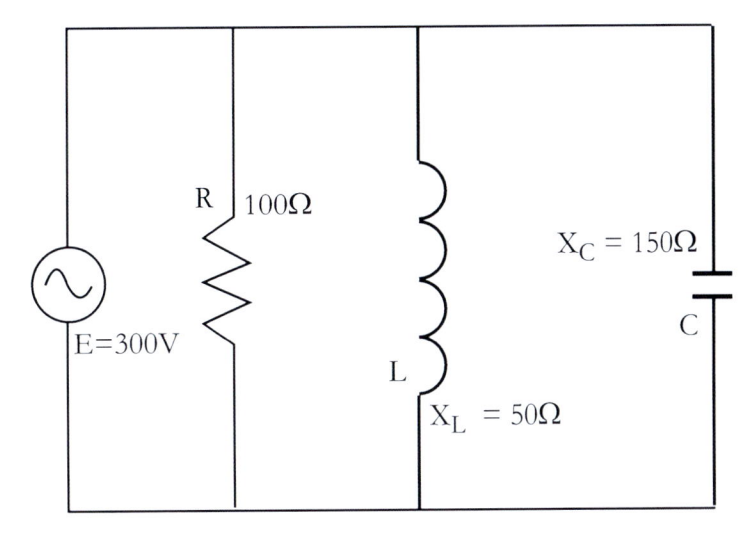

Figure 14.16 - Parallel R, L & C Circuit

The first step in solving for Z is to calculate the individual branch currents.

$$I_R = \frac{300}{100} = 3A$$

$$I_L = \frac{300}{50} = 6A$$

$$I_C = \frac{300}{150} = 2A$$

Using the values for I_R, I_L, and I_C, we can now solve for I_X and I_Z.

$$I_X = I_L - I_C$$
$$I_X = 6A - 2A$$
$$I_X = 4A \text{ Inductive}$$

Now

$$I_Z = \sqrt{3^2 + 4^2}$$
$$I_Z = 5A$$

Using this value of I_Z, we can now solve for Z as:

$$Z = \frac{E}{I_Z} = \frac{300}{5}$$
$$Z = 60\Omega$$

CHAPTER FOURTEEN
RESISTIVE, CAPACITIVE & INDUCTIVE CIRCUITS

When the circuit voltage is also given, you can use the values of currents, I_R, I_X, and I_Z, to calculate the true power, reactive power, apparent power, and power factor using the formulas given above.

A parallel resonance circuit is also influenced by the currents flowing through each parallel branch within the parallel LC tank circuit. A **tank circuit** is a parallel combination of L and C that is commonly used in filter networks to either select or reject AC frequencies, more on this later in Chapter 16. A parallel tank circuit is shown in Figure 14.17.

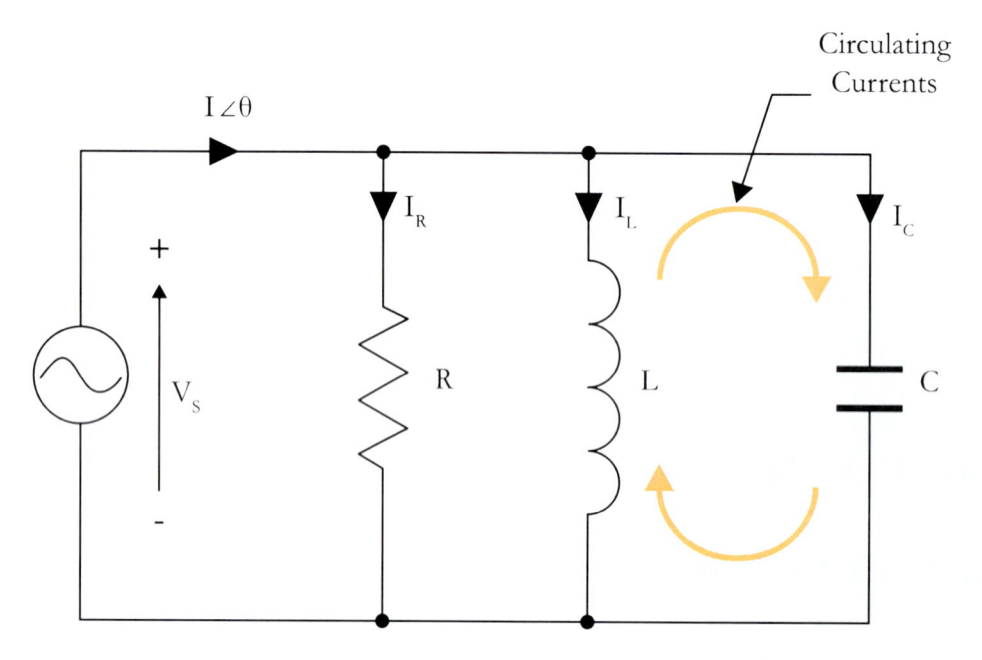

Figure 14.17 – Parallel RLC Circuit

At resonance there will be a large circulating current between the inductor and the capacitor due to the energy of the oscillations. This is because at resonance the parallel circuit stores energy in the magnetic field of the inductor and the electric field of the capacitor. This energy is constantly being transferred back and forth between the inductor and the capacitor which results in zero current and energy being drawn from the supply. This is because the corresponding instantaneous values of I_L and I_C will always be equal and opposite and cancel each other out. Therefore the total current drawn from the supply is the vector addition of these two currents and the current flowing in the Resistor. In this way, at resonance the parallel LC tank circuit acts like an open circuit with the circuit current being determined by the resistor, R only, therefore at resonance, the impedance of the parallel circuit is at its maximum value, Figure 14.18.

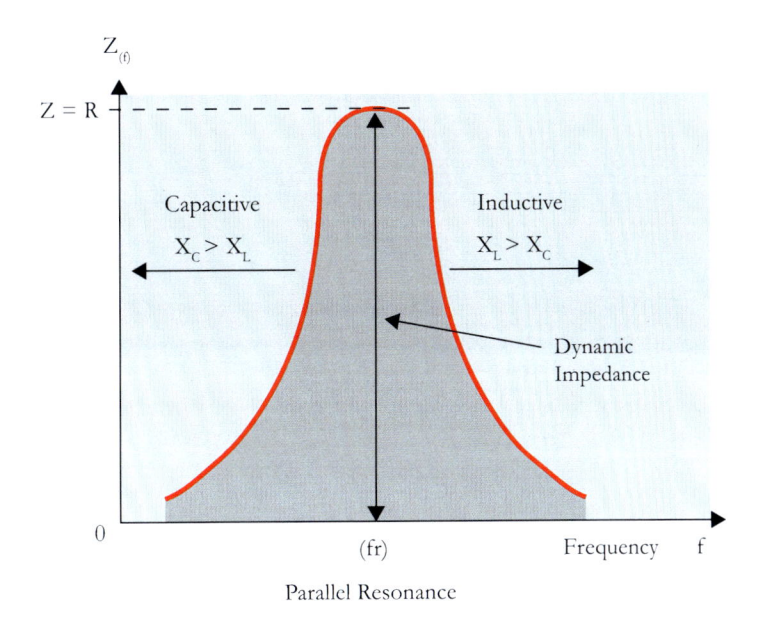

Figure 14.18 – Parallel Resonance Impedance

A parallel resonance circuit is also known as a **Rejector Circuit** because at resonance, the impedance of the circuit is at its maximum thus suppressing or rejecting the current whose frequency is equal to its resonant frequency.

Summary

With the completion of this chapter you now have all the building blocks for electrical circuits. The following summary is a brief review of the subjects covered in this chapter.

Inductance in AC Circuits

An inductor in an ac circuit opposes any change in current flow just as it does in a dc circuit.

Phase Relationships of an Inductor

The current lags the voltage by 90° in an inductor (ELI).

Inductive Reactance

The opposition an inductor offers to ac is called inductive reactance. It will increase if there is an increase in frequency or an increase in inductance. The symbol is X_L, and the formula is $X_L = 2\pi fL$.

Capacitance in AC Circuits

A capacitor in an ac circuit opposes any change in voltage just as it does in a dc circuit.

Phase Relationships of a Capacitor

The current leads the voltage by 90° in a capacitor (ICE).

CHAPTER FOURTEEN
RESISTIVE, CAPACITIVE & INDUCTIVE CIRCUITS

Capacitive Reactance

The opposition a capacitor offers to ac is called capacitive reactance. Capacitive reactance will decrease if there is an increase in frequency or an increase in capacitance. The symbol is X_C and the formula is:

$$X_C = \frac{1}{2\pi f C}$$

Total Reactance

The total reactance of a series ac circuit is determined by the formula $X = X_L - X_C$ or $X = X_C - X_L$. The total reactance in a series circuit is either capacitive or inductive depending upon the largest value of X_C and X_L. In a parallel circuit the reactance is determined by E/I_X where $I_X = I_C - I_L$ or $I_X = I_L - I_C$.

The reactance in a parallel circuit is either capacitive or inductive depending upon the largest value of I_L and I_C.

Phase Angle

The number of degrees that current leads or lags voltage in an ac circuit is called the phase angle and has the symbol θ.

Ohm's Law Formulas for AC

The formulas derived for Ohm's law used in ac are: $E = IZ$ and $I = E/Z$.

True Power

The power dissipated across the resistance in an ac circuit is called true power. It is measured in watts and the formula is: True Power = $(I_R)^2 R$.

Reactive Power

The power returned to the source by the reactive elements of the circuit is called reactive power. It is measured in volt-amperes reactive (var) and the formula is: Reactive Power = $(I_X)^2 X$.

Apparent Power

The power that appears to the source because of circuit impedance is called apparent power. It is the combination of true power and reactive power and is measured in volt-amperes (VA).

Power Factor

The portion of the apparent power dissipated in a circuit is called the power factor of the circuit. It can be expressed as a decimal or a percentage.

Power Factor Correction

To reduce losses in a circuit the power factor should be as close to unity or 100% as possible, achieved by adding capacitive reactance to a circuit when the total reactance is inductive or if the total reactance is capacitive, adding inductive reactance.

Resonance

Only occurs at one frequency when $X_L = X_C$.

Revision

Resistive, Capacitive & Inductive Circuits

Questions

1. With a capacitive ac circuit:

    a) The current leads the voltage

    b) The current lags the voltage

    c) The current and the voltage are in phase

2. In an ac inductive circuit, the elements have the following values; L = 20H, f = 500 Hz, find X_L taking π to be 3.14

    a) 62800 Ω

    b) 6280 Ω

    c) 1.6 Ω

3. In an ac capacitive circuit, the elements have the following values; C = 40 µF, f = 500 Hz, find X_C taking π to be 3.14:

    a) 125600 Ω

    b) 8 Ω

    c) 80 Ω

4. If X_L = 10 Ω, X_C = 7 Ω, R = 4 Ω, what is the circuit's impedance?

    a) 21 Ω Neutral

    b) 5 Ω Capacitive

    c) 5 Ω Inductive

5. If a voltage of 150 V ac is applied to a series circuit where: $X_L = 10\ \Omega$, $X_C = 7\ \Omega$ and $R = 4\ \Omega$, what is its current?

   a) Approximately 7.2 A

   b) 30 A

   c) 750 A

6. A circuit is at resonance when:

   a) $X_C = X_L$

   b) $R = X_L$

   c) $X_L + X_C = R$

CHAPTER FOURTEEN
RESISTIVE, CAPACITIVE & INDUCTIVE CIRCUITS

Revision

Resistive, Capacitive & Inductive Circuits

Answers

1. **A**

2. **A** $X_L = 2\pi fL = 2 \times \pi \times 500 \times 20 = 20,000\,\pi = 62,800$

3. **B** $X_C = 1 \div 2\pi fC = 1 \div (2 \times \pi \times 500 \times 40 \times 10^{-6}) = 8$

4. **C**

5. **B** $I = E \div Z = 150 \div 5 = 30$

6. **A**

Transformers

Introduction

A *Transformer* is a device that transfers electrical energy from one circuit to another by *electromagnetic induction*, sometimes referred to as the *transformer action*. The electrical energy is *always* transferred *without* a change in frequency, but may involve changes in magnitudes of voltage and current. Because a transformer works on the principle of electromagnetic induction, it *must* be used with an input source voltage that varies in amplitude, i.e. an alternating source. There are many types of power that fit this description, but for ease of explanation and understanding, in this chapter, transformer action will be explained using an ac voltage as the input source.

In a preceding chapter we discussed how alternating current has certain advantages over direct current and one important advantage is that when ac is used, the voltage and current levels can be *increased* or *decreased* by using a transformer.

As we have already discussed, the amount of power used by the load of an electrical circuit is equal to the current in the load times the voltage across the load, or P = VI.

Therefore, if the load in an electrical circuit requires an input of two (2) amperes at ten (10) volts, i.e. 20 watts, and the source is capable of delivering only one (1) ampere at twenty (20) volts, the circuit could not normally be used with this particular source. However, if a transformer is connected between the source and the load, the voltage can be *decreased*, commonly called *stepped down*, to ten (10) volts and the current *increased*, commonly called *stepped up*, to two (2) amperes.

Notice in the above case that the power remains the same, i.e. 20 volts × 1A is the same as 10 V × 2A, which is 20 Watts.

CHAPTER FIFTEEN
TRANSFORMERS

Basic Operation of a Transformer

In its most basic form a transformer consists of:

- A Primary Coil or Winding

- A Secondary Coil or Winding

- A Core that supports the coils or windings

Let us look at a simple schematic diagram of a transformer in figure 15.1.

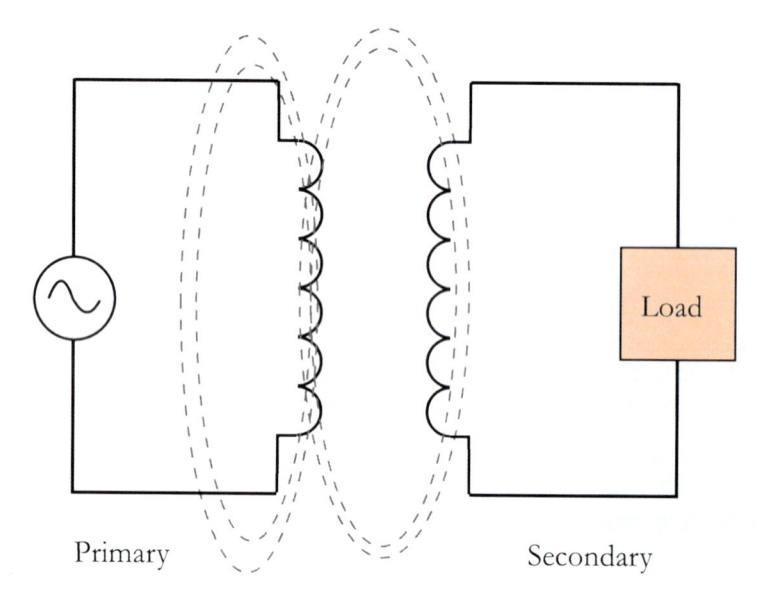

Figure 15.1 – Schematic of Simple Transformer

The primary winding is connected to an ac voltage source and as it alternates, the magnetic field, ie flux, builds up and collapses about the primary winding. This expanding and contracting magnetic field cuts the secondary winding and induces an alternating voltage into its winding, which causes alternating current to flow through the load. The voltage may be stepped up or down depending on the design of the primary and secondary windings; more on this later.

A Transformer's Components

Having just looked at the schematic of a simple transformer, let us now look in more detail at its construction. A typical transformer consists of the following component parts:

- The *Core* that provides a path for the magnetic lines of flux

- The *Primary Winding* that receives energy from the ac source

- The *Secondary Winding* that receives energy from the primary winding and delivers it to a load

- The *Enclosure* or *Case* that protects the above components from dirt, dust, moisture, and mechanical damage

Core Characteristics

The composition of a transformer core depends on what it is being used for and such factors as voltage, current, and frequency; size limitations and construction costs are also issues to be considered. Commonly used core materials are *air*, *soft iron*, and *steel*. Each of these materials, as perhaps you would expect, is suitable for particular applications and unsuitable for others. Generally, *air-core* transformers are used when the voltage source has a high frequency, typically above 20 kHz; *Iron-core* transformers are usually used when the source frequency is low, typically below 20 kHz; *soft-iron-core* transformers are very useful where the transformer must be physically small, yet efficient.

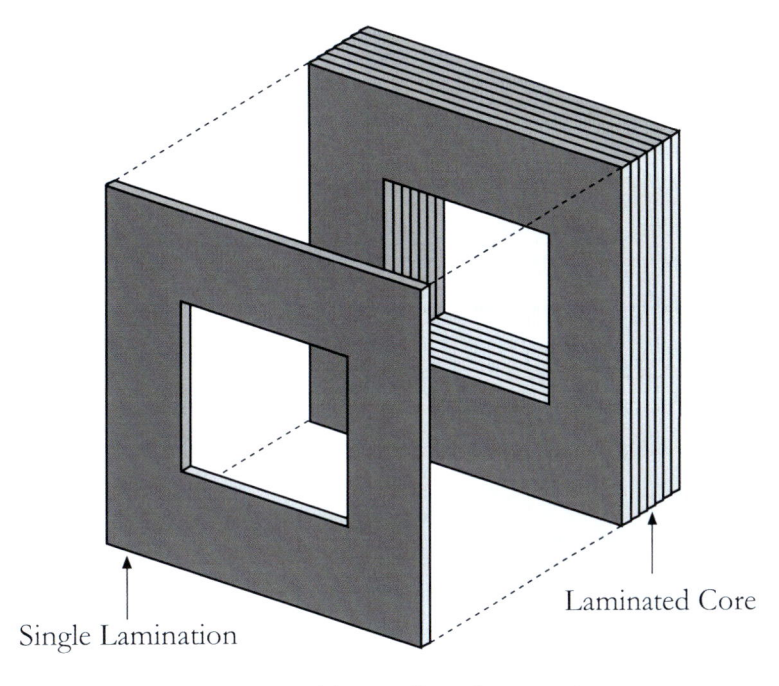

Figure 15.2 – Typical Core Construction

An iron-core transformer provides better power transfer than the air-core and a transformer whose core is constructed of laminated sheets of *soft iron* dissipates heat readily; thus it does provide for the efficient transfer of power. The majority of transformers you will encounter in aircraft equipment contain laminated-steel cores. These steel laminations, figure 15.2, are insulated with a non-conducting material, such as varnish, and then formed into a core that takes approximately 50 such laminations to make one an inch thick.

The purpose of the laminations is to reduce certain losses, which will be discussed later in this chapter. An important point to remember is that the most efficient transformer core is one that offers the best path for the lines of flux with the least loss in magnetic and electrical energy.

There are two main shapes of cores used in laminated-steel-core transformers; the *Hollow-Core*, so named because the core is shaped with a hollow square through the centre as in figure 15.2 , and the *Shell Core*.

Hollow-Core Transformers

Figure 15.2 illustrates this shape of core that is typically made up of many laminations of steel. Figure 15.3 shows how the transformer windings are wrapped around both sides of the core.

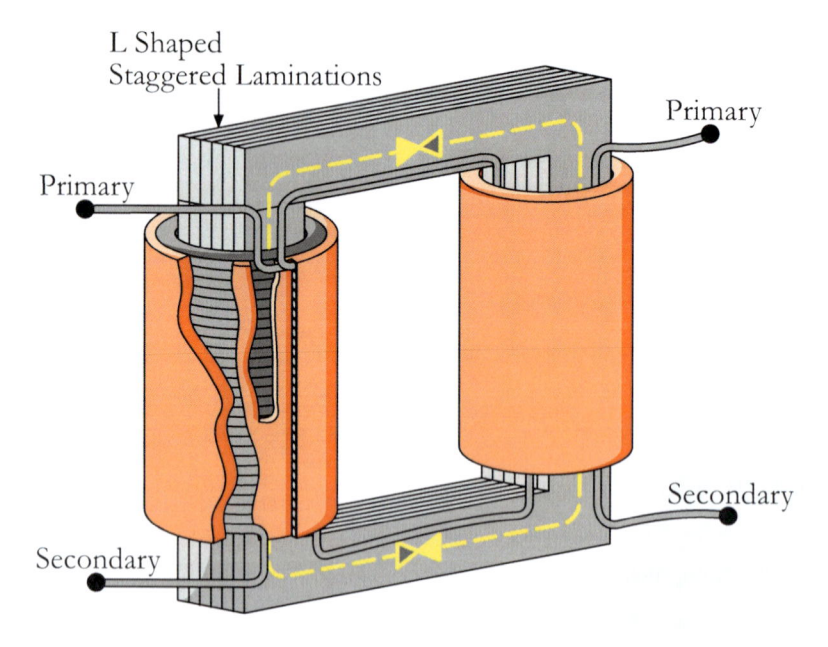

Figure 15.3 – Hollow Core Construction with Windings

Shell-Core Transformers

The shell core is the most popular and efficient transformer core, figure 15.4. Each layer of the core consists of 'E' and 'I' shaped sections of metal. These sections are butted together to form the laminations, which are insulated from each other and then pressed together to form the core.

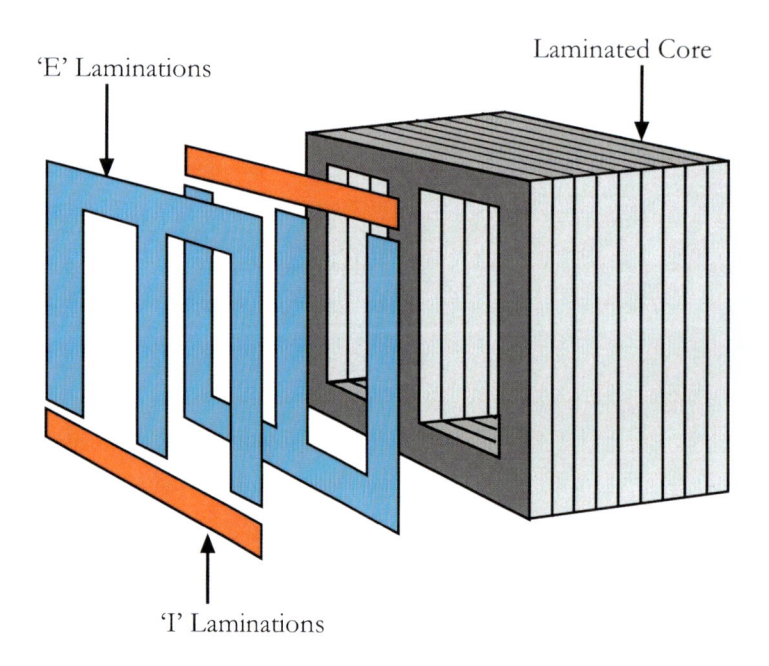

Figure 15.4 - Shell-Type Core Construction

In this transformer, the primary consists of many turns of relatively small wire, coated with varnish so that each turn of the winding is insulated from every other.

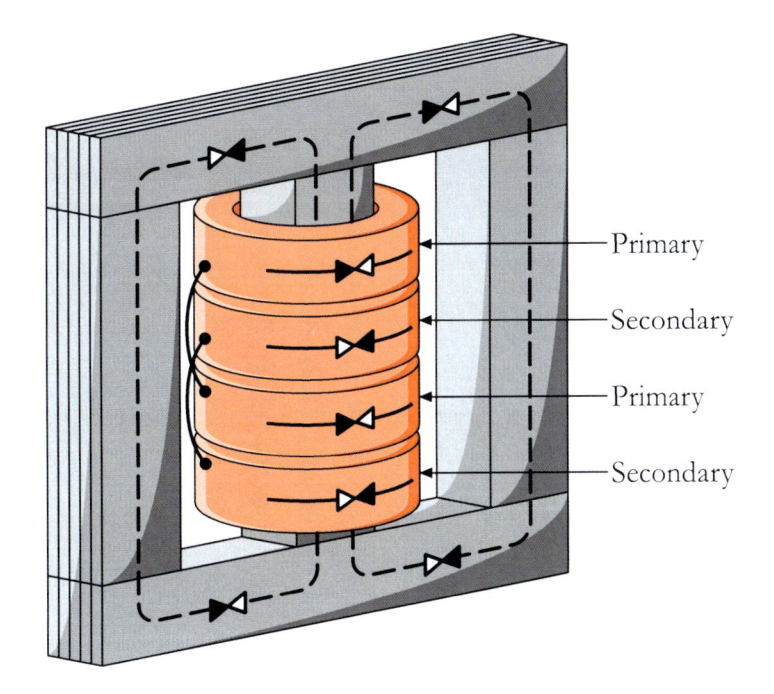

Figure 15.5 – Another Exploded View of a Transformer's Construction

Where the transformer is designed for high-voltage applications, sheets of insulating material, e.g. paper, are placed between the layers of windings to provide additional insulation.

When the primary winding is completely wound, it is wrapped in insulating paper or cloth and the secondary winding is then wound on top of the primary winding. After the secondary winding is complete, it too is covered with insulating paper and the 'E' and 'I' sections of the iron core are inserted into and around the windings as shown above in figure 15.5.

The leads from the windings are normally brought out through a hole in the enclosure of the transformer, but sometimes, terminals may be provided on the enclosure for connections to the windings. Figure 15.5 shows four leads, two from the primary and two from the secondary and these leads are connected to the source and load, respectively.

The shell core transformer has an advantage in that the magnetic flux has two closed magnetic paths to flow around external to the coils on both left and right hand sides before returning back to the central coils. This means that the magnetic flux has a closed path around the coils, thus decreasing core losses and increasing overall efficiency.

Enclosures

Transformer enclosures are usually made of metal, but this does vary, especially those miniature ones fitted on PCBs.

Transformers Schematic Symbols

Figure 15.6 shows typical schematic symbols used for transformers on electrical drawings.

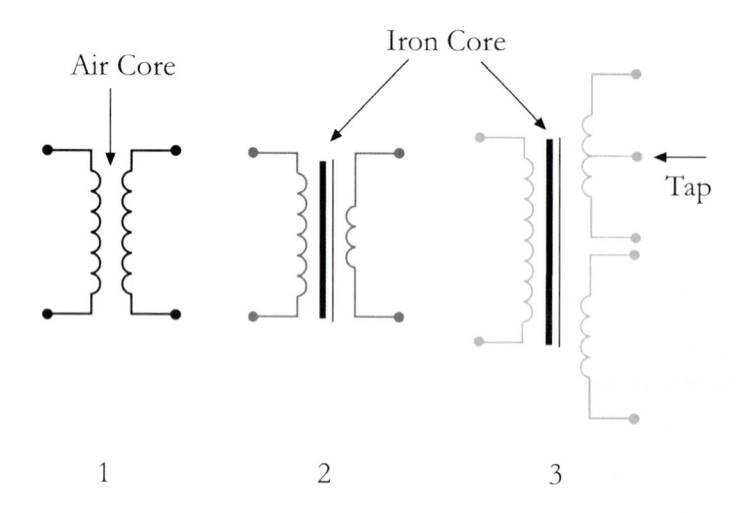

Figure 15.6 - Schematic Symbols for Various Transformers

The symbol for an air-core transformer is shown in figure 15.6 part 1 while part 2 and part 3 shows iron-core transformers with the bars between the coils being used to indicate an iron core. Frequently, additional connections, called *Taps*, are made to the transformer windings at points other than the ends and when one is connected to the centre of the winding, it is called a *Centre Tap*.

How Transformers Work

Up to this point this chapter has presented the basics of the transformer, including transformer action, the transformer's physical characteristics, and how it is constructed. Now you have the necessary knowledge to proceed into the theory of its operation.

No-Load Condition

Transformers, as we have already discussed, are capable of outputting voltages that are *usually* higher or lower than the source voltage, which is accomplished through mutual induction that takes place when the changing magnetic field produced by the primary voltage cuts the secondary winding. A no-load condition is said to exist when a voltage is applied to the primary, but no load is connected to the secondary, as illustrated by figure 15.7.

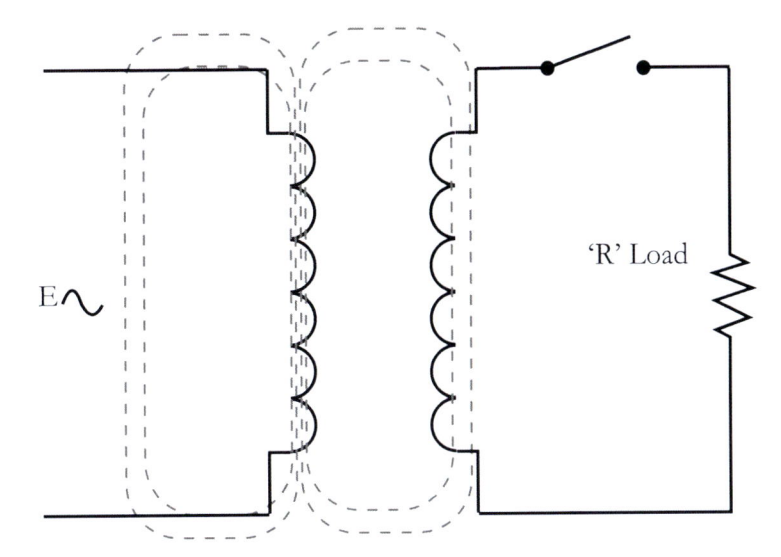

Figure 15.7 - Transformer Under No-Load Conditions

Because of the open switch, there is no current flowing in the secondary winding. With the switch open and an ac voltage applied to the primary, there is, however, a very small amount of current called the **Excitation Current** flowing in the primary.

Essentially, what the excitation current does is **excite** the coil of the primary to create a magnetic field. The amount of excitation current is determined by three factors:

- The amount of applied voltage (E_a)

- The resistance (R) of the primary coil's wire and core losses

- The Inductive reactance (X_l), which is dependent on the frequency of the excitation current

These last two factors are controlled by transformer design.

This very small amount of excitation current serves two functions:

- Most of the exciting energy is used to maintain the magnetic field of the primary

- A small amount of energy is used to overcome the resistance of the wire and core losses, which are dissipated in the form of heat, ie power loss

Excitation current will flow in the primary winding at all times to maintain this magnetic field, but no transfer of energy will take place as long as the secondary circuit is open, (under no-load conditions).

When an alternating current flows through a primary winding, as we have already established, a magnetic field is created around the winding. As the lines of flux expand outward, relative motion is present, and, due to the laws of magnetism already discussed, a *back emf* is induced in the winding. This is the same back emf that we have discussed earlier when learning about inductors.

As you know, by convention, flux leaves the primary at the North Pole and enters the primary at the South Pole and so the back emf induced in the primary has a polarity that opposes the applied voltage, thus opposing the flow of current in it. It is the back emf that limits excitation current to a very low value.

How a Voltage is induced in the Secondary

To visualise how a voltage is induced into the secondary winding of a transformer, let us look again at figure 15.7. As the exciting current flows through the primary, it generates magnetic lines of force, i.e. flux. During the time current is increasing in the primary, magnetic lines of force expand outward from the primary and cut the secondary. As you should remember, a voltage is induced into a coil when magnetic lines cut across it and therefore, the voltage across the primary cause a voltage to be induced across the secondary.

Primary and Secondary Phase Relationship

The secondary voltage of a simple transformer may be either in phase or out of phase with the primary voltage. This depends on the direction in which the windings are wound and the arrangement of the connections to the external load circuit. Simply, this means that the two voltages may rise and fall together or one may rise while the other is falling.

Transformers in which the secondary voltage is in phase with the primary are referred to as *like-wound* transformers, while those in which the voltages are 180° out of phase are called *unlike-wound* transformers, amazingly.

Dots are used to indicate points on a transformer schematic symbol that have the same instantaneous polarity, i.e. points that are in phase, figure 15.8.

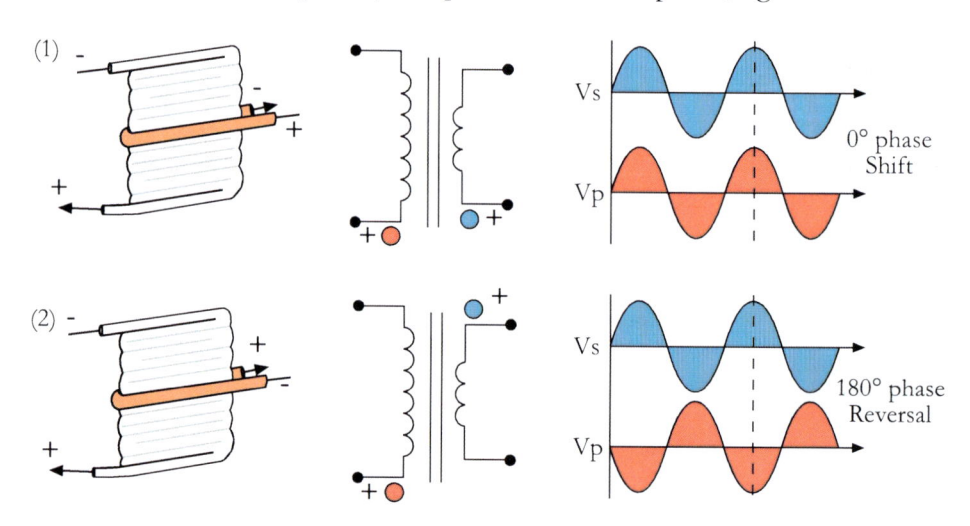

Figure 15.8 – Instantaneous Polarity Indications

The use of phase-indicating dots illustrated in figure 15.8 shows in case (1), both the primary and secondary windings are wound from top to bottom in a clockwise direction, as viewed from above the windings. When constructed in this manner, the top lead of the primary and the top lead of the secondary have the *same* polarity, indicated by the dots on the transformer symbol. A lack of phasing dots *always* indicates a reversal of polarity between primary and secondary.

Case (2) in figure 15.8, illustrates a transformer in which the primary and secondary are wound in opposite directions. As viewed from above, the primary is wound in a clockwise direction from top to bottom, while the secondary is wound in a counter-clockwise direction. Notice that the top leads of the primary and secondary have *opposite* polarities, indicated by the dots being placed on opposite ends of the transformer symbol. Consequently, the polarity of the voltage at the terminals of the secondary of a transformer depends on the direction in which the secondary is wound with respect to the primary.

Coefficient of Coupling (K)

The *Coefficient of Coupling*, given the symbol 'K', of a transformer is dependent on the portion of the total flux lines that cuts both primary and secondary windings. Ideally, *all* the flux lines generated by the primary should cut the secondary, and all the lines of the flux generated by the secondary should cut the primary and then the coefficient of coupling would then be one, i.e. unity, and maximum energy would be transferred from the primary to the secondary. Practical power transformers use high-permeability silicon steel cores and close spacing between the windings to provide a high coefficient of coupling.

Lines of flux generated by one winding that do not link with the other winding are called *Leakage Flux* and since leakage flux generated by the primary does not cut the secondary, it cannot induce a voltage into it. The voltage induced into the secondary is therefore less than it would be if the leakage flux did not exist. Since the influence of leakage flux is to lower the voltage induced into the secondary, the affect can be duplicated by assuming an inductor is connected in series with the primary. This series *Leakage Inductance* is assumed to drop part of the applied voltage, leaving less voltage across the primary.

Turns & Voltage Ratios

The total voltage induced into the secondary winding of a transformer is determined mainly by the *ratio* of the number of turns in the primary to the number of turns in the secondary, and by the amount of voltage applied at the primary. Let us look at two (2) examples in figure 15.9.

CHAPTER FIFTEEN
TRANSFORMERS

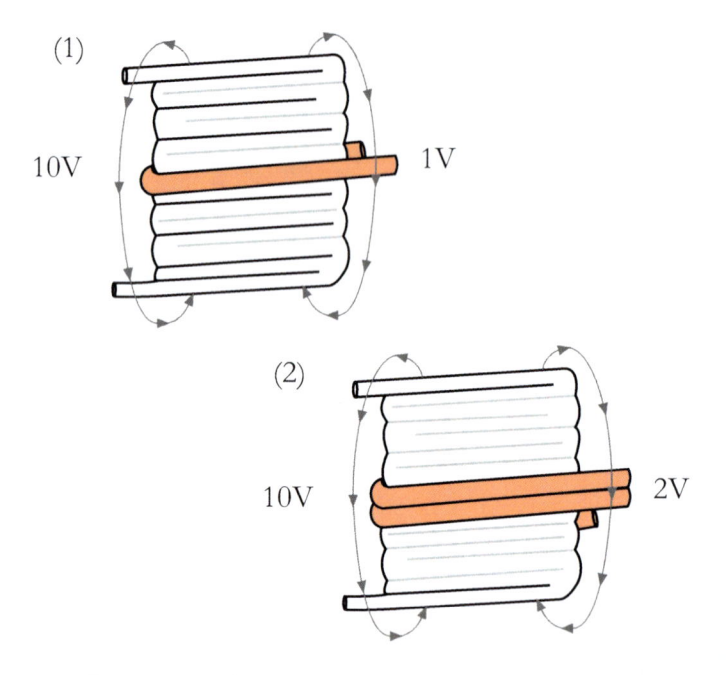

Figure 15.9 – Transformer Turns & Voltage Ratio

In case (1) of figure 15.9, it shows a transformer whose primary consists of ten turns of wire and whose secondary consists of a single turn of wire.

You know that as lines of flux generated by the primary expand and collapse, they cut **both** the ten turns of the primary and the single turn of the secondary. Since the length of the wire in the secondary is approximately the same as the length of the wire in each turn in the primary, **the emf induced into the secondary will be the same as the emf induced into each turn in the primary**. This means that if the voltage applied to the primary winding is 10 volts, the back emf in the primary is almost 10 volts and so each turn in the primary will have an induced back emf of approximately one-tenth of the total applied voltage, i.e. one volt. Since the same flux lines cut the turns in both the secondary and the primary, each turn will have an emf of one volt induced into it. However, the transformer in part (1) of figure 15.9 has only one turn in the secondary, and so the emf across the secondary is one volt.

The transformer depicted in case (2) of figure 15.9 has a ten-turn primary and a two-turn secondary. Since the flux induces one volt per turn, the total voltage across the secondary is now two volts. Notice that the volts/turn are the same for both primary and secondary windings.

Since the back emf in the primary is equal, or almost, to the applied voltage, we can calculate the value of the voltage induced in the secondary using the voltage applied to the primary and the number of turns in each winding.

358

This also shows the relationship between the number of turns in each winding and the voltage across each winding and is expressed by the equation:

$$E_S = \frac{N_S \times E_P}{N_P} = \frac{1 \times 10}{10} = 1V$$

Where:

N_P = Number of Turns in the primary

E_P = Voltage applied to the primary

N_S = Number of Turns in the secondary

E_S = Voltage induced in the secondary

Notice the equation shows that the ratio of secondary voltage to primary voltage is equal to the ratio of secondary turns to primary turns. Using transposition of formula, the equation can also be written as:

$$E_P N_S = E_S N_P$$

Now by transposing this formula, we can also derive:

$$E_S = \frac{E_P N_S}{N_P} \text{ and } E_P = \frac{E_S N_P}{N_S}$$

Therefore, if any three (3) of the quantities in the above formulas are known, the fourth quantity can be calculated.

Note: The ratio of the primary and secondary voltages is equal to the turns ratio. Sometimes, instead of specific values, only a turns or voltage ratio is given and in this case, you may assume any value for one of the voltages, or turns, and compute the other value from the ratio. For example, if a turn ratio is given as 6:1, you can assume a number of turns for the primary and compute the secondary number of turns, e.g. 60:10, 36:6, 30:5, etc.

As we have already discussed, a transformer in which the voltage across the secondary is less than the voltage across the primary is called a ***Step-Down*** transformer; e.g. the ratio of a four-to-one step-down transformer is written as 4:1. A transformer that has fewer turns in the primary than in the secondary will produce a greater voltage across the secondary than the voltage applied to the primary and as we have discussed earlier, is called a ***Step-Up*** transformer; e.g. the ratio of a one-to-four step-up transformer should be written as 1:4. Notice that whenever quoting the two ratios, the value of the primary winding is generally stated first.

Load Effects

When a load device is connected across the secondary winding of a transformer, current obviously flows through the secondary and the load. The magnetic field produced by the current in the secondary interacts with the magnetic field produced by the current in the primary resulting from the mutual inductance (M) between the primary and secondary windings.

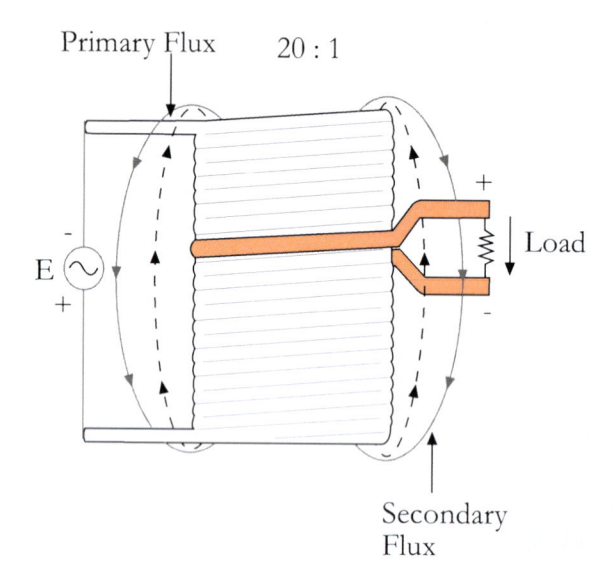

Figure 15.10 – Simple Transformer Primary & Secondary Flux Relationship

The total flux in the core of the transformer is common to both the primary and secondary windings and it is also the means by which energy is transferred from the primary winding to the secondary winding. Since this flux links both windings, it is called *Mutual Flux*. The inductance that produces this flux is also common to both windings and is called mutual inductance. Figure 15.10 shows the flux produced by the currents in the primary and secondary windings of a transformer when a source current is flowing in the primary winding.

When a load resistance is connected to the secondary winding, the voltage induced into it causes current to flow. This current produces a flux field about the secondary, shown as broken lines, which is in opposition to the flux field about the primary by virtue of Lenz's law. Therefore, the flux about the secondary cancels some of the flux about the primary. With less flux surrounding the primary, the back emf is reduced and more current is drawn from the source. The additional current in the primary generates more lines of flux, nearly re-establishing the original number of total flux lines. This sounds like a bit of a merry go round but explains why transformers are so efficient.

Turns & Current Ratios

The number of flux lines developed in a core is proportional to the magnetising force, in **Ampere-Turns**, of the primary and secondary windings. The ampere-turn, i.e. current (I) × number of turns (N), is a measure of magnetomotive force (mmf); and is defined as *the mmf developed by one ampere of current flowing in a coil of one turn*. The flux that exists in the core of a transformer surrounds both the primary and secondary windings and since the flux is the same for both windings, the ampere-turns in both the primary and secondary windings must be the same and therefore:

$$I_P N_P = I_S N_S$$

Where:

$I_P N_P$ = Ampere-turns in the primary

$I_S N_S$ = Ampere-turns in the secondary

Now by dividing both sides of the equation by $I_P N_S$ we get:

$$\frac{N_P}{N_S} = \frac{I_S}{I_P}$$

Since:

$$\frac{E_S}{E_P} = \frac{N_S}{N_P}$$

Then:

$$\frac{E_P}{E_S} = \frac{N_P}{N_S}$$

And:

$$\frac{E_P}{E_S} = \frac{I_S}{I_P}$$

Where:

E_P = Voltage applied to Primary

E_S = Voltage across the Secondary

I_P = Primary current

I_S = Secondary current

Notice the equations show the current ratio to be the inverse of the turns and voltage ratios. This means, a transformer having less turns in the secondary than in the primary would step down the voltage, but would step up the current.

The expression is called the transformer *Turns Ratio* and may be expressed as a single factor. Remember, the turns ratio indicates the amount by which the transformer increases or decreases the voltage applied to the primary. For example, if the secondary of a transformer has twice as many turns as the primary, the voltage induced into the secondary will be twice the voltage across the primary. If the secondary has ½ as many turns as the primary, the voltage across the secondary will be ½ the voltage across the primary. However, the turns ratio and the current ratio of a transformer have an inverse relationship. Therefore, a 1:2 step-up transformer will have ½ the current in the secondary as in the primary. A 2:1 step-down transformer will have twice the current in the secondary as in the primary.

Power Relationship between Primary & Secondary Windings

As we have seen, the turns ratio of a transformer affects current as well as voltage; for example, if voltage is doubled in the secondary, current is halved in the secondary; conversely, if voltage is halved in the secondary, current is doubled in the secondary. In this way, all the power delivered to the primary by the source is also delivered to the load by the secondary. However, this is the ideal case and the transfer must be minus whatever power is consumed by the transformer in the form of losses. Let us look again to the transformer illustrated in figure 15.10.

The turns ratio is 20:1 and so if the input to the primary is 0.1 A at 300 volts, the power in the primary is $P = V \times I = 30$ watts. If the transformer has no losses, 30 watts is delivered to the secondary. The secondary steps down the voltage to 15 volts and steps up the current to 2 amperes. Thus, the power delivered to the load by the secondary is $P = V \times I = 15\text{ V} \times 2\text{ A} = 30$ watts.

The reason for this is that when the number of turns in the secondary is decreased, the opposition to the flow of the current is also decreased and so more current will flow in the secondary. If the turns ratio of the transformer is increased to 1:2, the number of turns on the secondary is twice the number of turns on the primary, meaning the opposition to current is doubled and so the voltage doubles, but current is halved due to the increased opposition to current in the secondary.

The important thing to remember is that **with the exception of the power consumed within the transformer, all power delivered to the primary by the source will be delivered to the load**. The *form* of the power may change, but the power in the secondary almost equals the power in the primary. As a formula, this is:

$$P_S = P_P - P_L$$

Where:

P_S = Power delivered to the load by the secondary

P_P = Power delivered to the primary by the source

P_L = Power losses in the transformer

Transformer Losses

Practical power transformers, although highly efficient, are not perfect devices. Small power transformers used in electrical and avionic equipment have an 80% to 90% percent efficiency range, while large, commercial power-line transformers may have efficiencies exceeding 98%. The total power loss in a transformer is a combination of three types of losses:

- Due to the resistance in the primary and secondary windings, called *Copper loss* or I^2R *loss*

- Losses due to currents set up in the transformer's core, called *Eddy Current losses*

- *Hysteresis* in the transformer core

These losses result in the undesirable conversion of electrical energy into heat energy.

Copper Loss

Whenever current flows in a conductor, power is dissipated in the resistance of the conductor in the form of heat. The amount of power dissipated is directly proportional to the resistance of the wire, and to the square of the current flowing through it. The greater the value of either resistance or current, the greater the dissipated power. To help overcome this loss, a transformer's primary and secondary windings are usually made of low-resistance copper wire.

In addition, the resistance of a given winding is a function of the diameter of the wire and its length. Therefore, copper loss can be minimised by using the proper diameter wire; large diameter wire is required for high-current windings, whereas small diameter wire can be used for low-current windings.

Eddy-Current Loss

The core of a transformer is usually constructed of some type of ferromagnetic material because it is a good conductor of magnetic lines of flux. Whenever the primary of an iron-core transformer is energised by an alternating-current source, a fluctuating magnetic field is produced. This magnetic field cuts the conducting core material and induces a voltage into it. The induced voltage causes random currents to flow through the core, which dissipates power in the form of heat. These undesirable currents are called *Eddy Currents*.

To minimise the loss resulting from eddy currents, transformer cores are *laminated* as the thin, insulated laminations do not provide an easy path for current, and so eddy-current losses are greatly reduced.

Hysteresis Loss

When a magnetic field is passed through a core, the core material becomes magnetised. To become magnetised, the domains within the core must align themselves with the external field as we discussed in the chapter on magnetism. If the direction of the field is reversed, the domains must turn so that their poles are aligned with the new direction of the external field. Transformers normally operate from either 50 to 60 Hz, or 400 Hz AC and so each tiny domain must realign itself twice during each cycle, or a total of 100 times/second at 50Hz to 120 times/second when 60 Hz AC is used or 800 times/second with 400 Hz. The energy used to turn each domain is dissipated as heat within the iron core and this is called the *Hysteresis Loss*, which can be thought of as resulting from molecular friction. Hysteresis loss can be reduced by proper choice of core materials, as already discussed in the magnetism chapter.

Transformer Efficiency

To calculate the efficiency of a transformer, the input power to and the output power from the transformer must be known. The input power is equal to the product of the voltage and current applied to the primary. The output power is equal to the product of the voltage and current across the secondary. The difference between the input power and the output power represents a power loss as we have just discussed. The transformer's percentage of efficiency is easily calculated by using the formula:

$$\text{Efficiency (\%)} = \frac{P_{out}}{P_{in}} \times 100$$

Let us take a look at an example. If the input power to a transformer is 650 watts and the output power is 610 watts, what is the efficiency?

Solution:

$$\text{Efficiency (\%)} = \frac{610}{650} \times 100$$

$$\text{Efficiency} = 93.8\%$$

With an efficiency of 93.8%, 6.2%, or 40 watts is wasted due to heat losses.

Transformer Ratings

When a transformer is used in a circuit, the voltage, current, power handling and turns ratio must be considered as a whole. The maximum voltage that can safely be applied to any winding is determined by the type and thickness of the insulation; the thicker the insulation the higher operating voltage. While the maximum current is determined by the diameter of the wire used for the winding. If winding current is excessive, a higher than normal amount of power will be dissipated by the winding in the form of heat. This heat may be sufficiently high to cause the insulation around the wire to break down and if this happens, the transformer may be permanently damaged.

The power-handling capacity of a transformer is dependent upon its ability to dissipate heat. If the heat can safely be removed, e.g. by using heat sinks or cooling fan, the power-handling capacity of the transformer can be increased. This can also be achieved by immersing the transformer in oil, or by using cooling fins. A transformer's power-handling capacity is measured in either the volt-ampere unit, i.e. VA or kVA, or just the straightforward watt.

Operating frequency also has an influence on a transformer's performance. If the frequency applied to a transformer is increased, the inductive reactance of the windings also increases, causing a greater ac voltage drop across the windings and a lesser voltage drop across the load. However, an increase in the frequency applied to a transformer should not damage it.

However, if the frequency applied to the transformer is decreased, the reactance of the windings is decreased and the current through the transformer winding is increased. If the decrease in frequency is enough, the resulting increase in current will damage the transformer. For this reason a transformer may be used at frequencies above its normal operating frequency, but not below that frequency.

The vast majority of aircraft and avionic transformers operated at 400 Hz.

Types & Applications of Transformers

The transformer has many useful applications in an electrical circuit. A brief discussion of some of these applications will help you recognise the importance of the transformer in modern electricity, electronics and avionics.

Power Transformers

Power transformers are used to supply various voltages to the numerous electrical and avionic systems throughout an aircraft. These transformers have two or more windings wound on a laminated iron core with the number of windings and the turns per winding depending on the voltages that the transformer is to supply. Their coefficient of coupling, i.e. efficiency is typically 0.95, or more.

CHAPTER FIFTEEN
TRANSFORMERS

It is usually easy to distinguish between the high-voltage and low-voltage windings in a power transformer by measuring their resistance. The low-voltage winding usually carries the higher current and therefore has the larger diameter wire, meaning its resistance is less than the resistance of the high-voltage winding.

So far we have only looked at transformers that have just one secondary winding. The typical power transformer has multiple secondary windings, each providing a different voltage. Figure 15.11 shows the schematic symbol for a typical power-supply transformer with several different outputs.

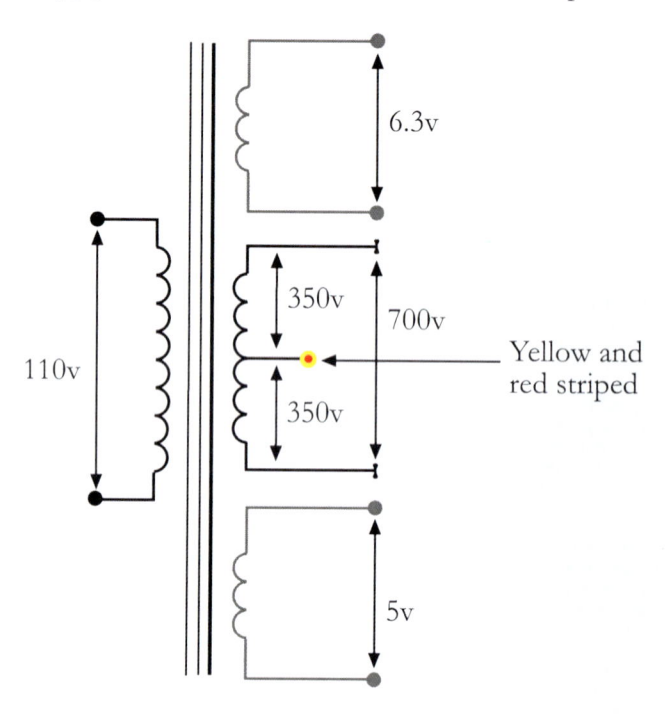

Figure 15.11 – Schematic of a Typical Power Transformer

For any given voltage across the primary, the voltage across each of the secondary windings is determined by the number of turns each one has.

A winding may be centre-tapped like the secondary 350-volt winding shown in figure 15.11. To centre tap a winding means to connect a wire to the centre of the coil, so that between this centre tap and either terminal of the winding there appears ½ of the voltage developed across the entire winding. Most power transformers have coloured leads so that it is easy to distinguish between the various windings. Usually, red is used to indicate the high-voltage leads, but it is possible for a manufacturer to use some other colour or colours.

There are many types of power transformers and they range in size from the huge transformers weighing several tons-used in power substations of commercial power companies-to very small ones weighing as little as a few ounces-used in electronic and avionic equipment.

Autotransformers

In order to achieve a voltage increase or decrease, it is not necessary for a transformer to have separate and distinct primary and secondary windings. This type of transformer is known as shows an **Autotransformer,** Figure 15.12.

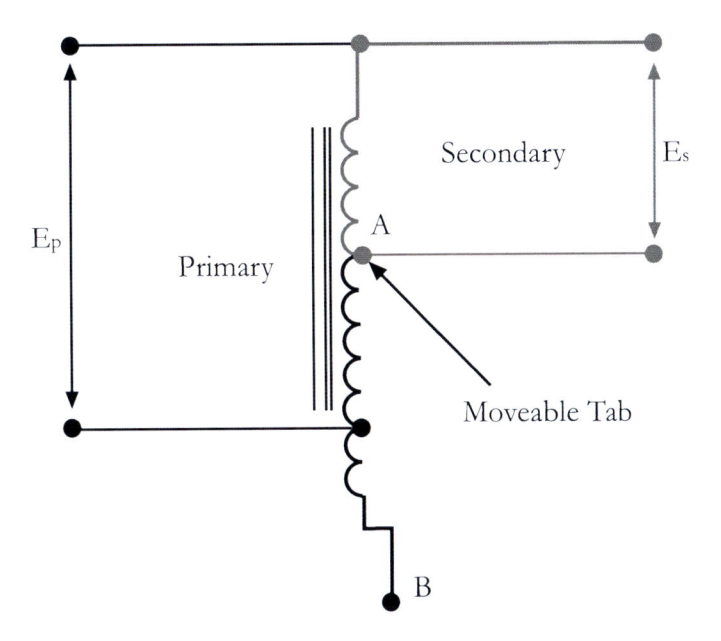

Figure 15.12 - Schematic Diagram of an Autotransformer

In an Autotransformer a single coil of wire is *tapped* to produce, what is electrically, a primary and secondary winding. The voltage across the secondary winding has the same relationship to the voltage across the primary that it would have if they were two distinct windings.

The movable tap in the secondary is used to select a value of output voltage, either higher or lower than E_p within the range of the transformer, e.g. when the tap is at point 'A', E_S is less than E_p but when the tap is at point 'B', E_S is greater than E_p.

Autotransformers are often referred to by the trademark name Variac, meaning Variable AC.

One major disadvantage of auto transformers, especially step down types, is that should the common portion of the winding go open circuit, the primary voltage is applied directly to the load on the secondary, Figure 15.13.

CHAPTER FIFTEEN
TRANSFORMERS

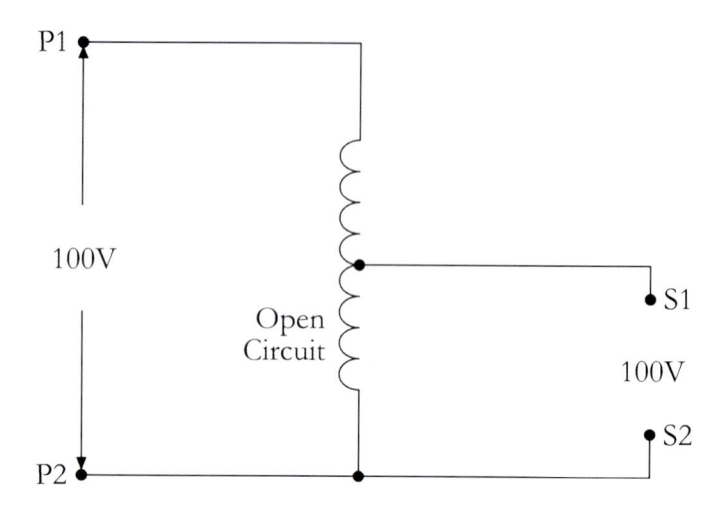

Figure 15.13 – Effects of Open Circuit on an Autotransformer

As no current flows in the circuit, there can be no voltage drop across the top half of the winding and all the supply voltage will be directly fed to the load. However improved reliability, through modern manufacturing methods, has increased their use on aircraft.

Three-phase Transformers

Modern large transformers are usually of the three-phase core type. Three similar limbs are connected by top and bottom yokes, each limb having primary and secondary windings, arranged concentrically. How the windings are connected will produce different outputs in terms of Voltage and Current. When transformers are used to provide three or more phases they are generally referred to as a **Polyphase Transformer**. The combinations of the three windings may be with the primary delta-connected and the secondary star-connected, or star-delta, star-star or delta-delta, depending on the transformers use.

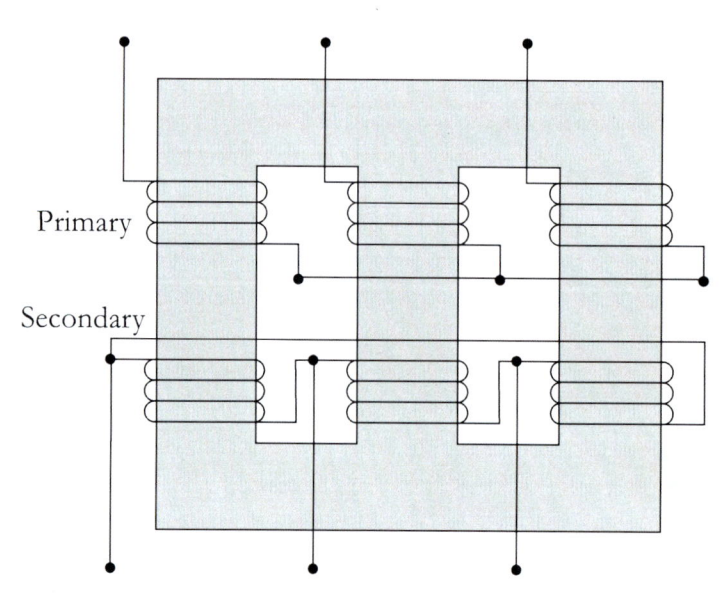

Figure 15.14 – Three-Phase Transformer

As we can see, the three-phase transformer actually has 6 windings (or coils) 3 primary and 3 secondary. How these sets of windings are interconnected, determines whether the connection is a star or delta configuration.

The three phase voltages are each displaced from the other by 120°. The flow of the transformers currents are also determined by the type of the electrical connection used on both the primary and secondary sides.

The standard method for marking three phase transformer windings is to label the three primary windings letters A, B and C, used to represent the three-phases of RED, YELLOW and BLUE.

Figure 15.15 shows a Delta configuration for the primary windings.

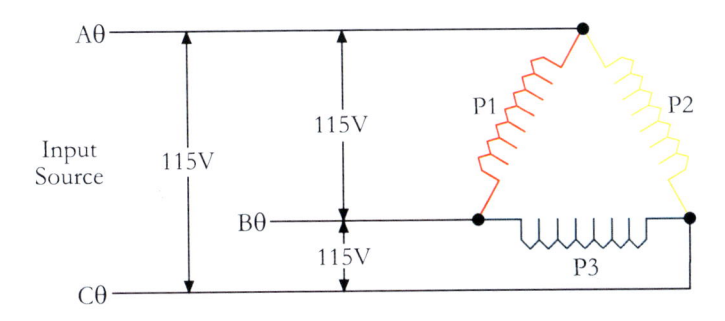

Figure 15.15 – Delta Configured Primary Winding

When connected to a three-phase source in a Delta configuration, each primary winding will have the same voltage across it. If the secondary windings are also connected Delta then they have equal voltages across each winding, Figure 15.16.

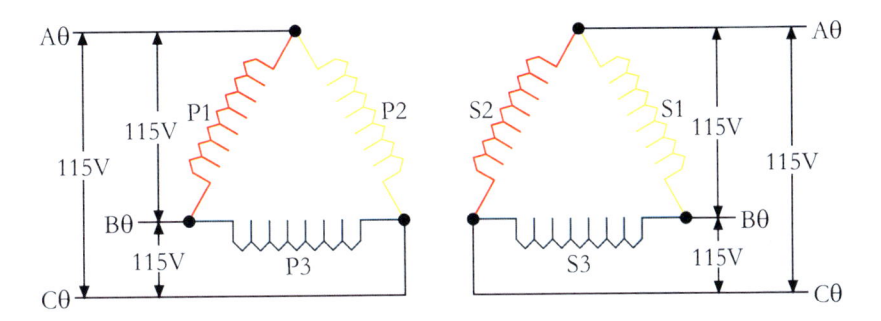

Figure 15.16 – Delta/Delta Configured Transformer

Of course, this voltage will be either higher or lower than the primary depending upon the "turns ratio".

Whilst we have seen that the voltages in a Delta connected system are the same, this is not the case for the current flow. In a Delta transformer, the line current does not equal the phase current because each line from the transformer is connected to two transformer phases. Therefore the line current from a 3Ø load will be greater than the phase current by √3, or put another way the line current will be 1.73 x the phase current.

A star connected system is considered to be a four wire system, where one end of each winding is connected to a common or Neutral" point and the other ends to the output terminals, Figure 15.17.

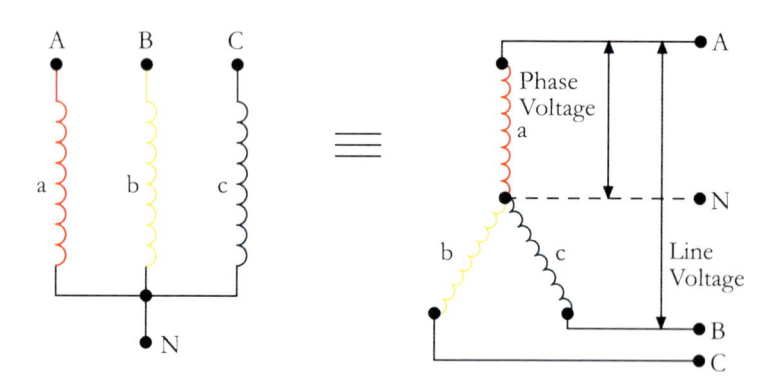

Figure 15.17 – Star connected Transformer

The voltage from the neutral line or star point to the other end of each phase winding is called the **phase voltage**; the voltage from one phase to another is called the **line voltage**. In a three-phase, Star-connected transformer, the total voltage, or line voltage, across any two of the three line leads is the vector sum of the individual phase voltages. As the phase voltages are displaced by 120° the line voltage cannot be double the phase voltage, as such each line voltage is 1.73 (√3) times one of the phase voltages. Because the windings form only one path for current flow between phases, the line and phase currents are the same.

The windings can be connected star/delta, delta/delta, star/star or delta/star, depending on the conditions under which the transformers are to be used. Figure 15.18.

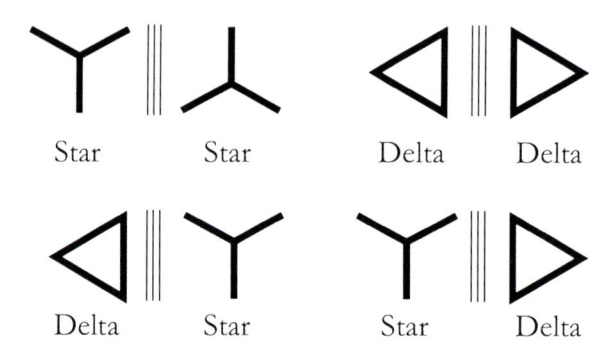

Figure 15.18 – Transformer Configurations

Three Phase Transformer Power Rating Calculations

As we have already seen, Power can be calculated using the formula:

$$P = VI$$

This formula still applies in regards to 3 Ø transformers, but it is important to remember that that we must use the highest values of current and voltage in our calculations.

For Example let us consider a star connected transformer with a phase voltage of 200 volts and a line current of 144 Amps. To calculate the Power we must convert the phase voltage into the higher value line voltage as shown:

$$P = \sqrt{3} \times V \times I$$
$$P = 1.73 \times 200 \times 144$$
$$P = 50kVA$$

Audio-Frequency Transformers

Audio-Frequency (AF) transformers are used in AF circuits as coupling devices, designed to operate at frequencies in the audio frequency spectrum, i.e. 15Hz to 20kHz. They consist of a primary and a secondary winding wound on a laminated iron or steel core. Because these transformers are subjected to higher frequencies than are power transformers, special grades of steel, such as silicon steel, or special alloys of iron that have a very low hysteresis loss must be used for the core material. These transformers usually have a greater number of turns in the secondary than in the primary; common step-up ratios being 1:2 or 1:4. With audio transformers the impedance of the primary and secondary windings is as important as the ratio of turns, since the selected transformer should have its impedance match the circuits to which it is connected.

Radio-Frequency Transformers

Radio Frequency (RF) transformers are used to couple circuits at frequencies above 20 kHz. Their windings are wound on a tube of nonmagnetic material, and they have a special powdered-iron core, or just an air core. In standard broadcast radio receivers, e.g. ADF, they operate in a frequency range of from 530kHz to 1550kHz; in HF from 2MHz to 30MHz; VHF from 108MHz to 152MHz and in Radar systems to 1GHz and beyond.

Current Transformers

Current transformers (CTs) are a special type of transformer which allows the measurement of AC currents without breaking into the circuit, Figure 15.19.

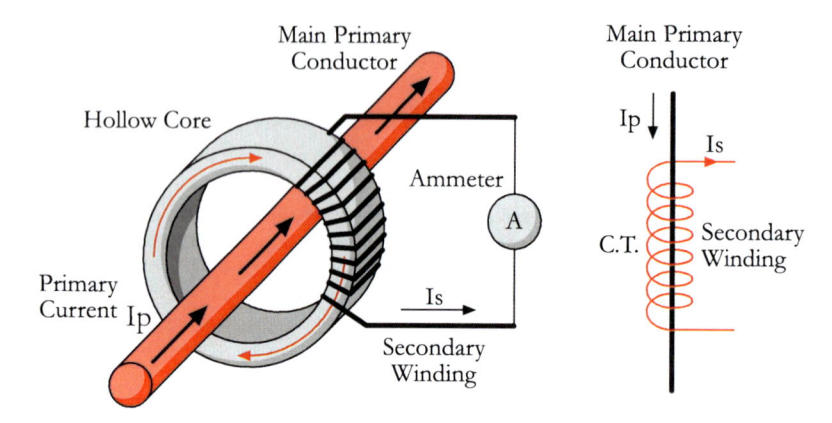

Figure 15.19 – Current Transformer

Current transformers are used to reduce high voltage currents to a much lower value but the principal of operation of a current transformer is no different from that of an ordinary transformer. CT`s provide a convenient way of safely monitoring the actual electrical current flowing in an AC transmission line using a standard ammeter. The secondary winding will typically have a large number of coil turns wound on a laminated core depending upon how much the current needs to be stepped down.

Generally, the primary of a current transformer is a single feeder cable from the aircraft generator, and the secondary is wound on a laminated magnetic core, placed around the conductor in which the current needs to be measured. Figure 15.20 shows a typical 3 phase configuration of current transformers on an aircraft.

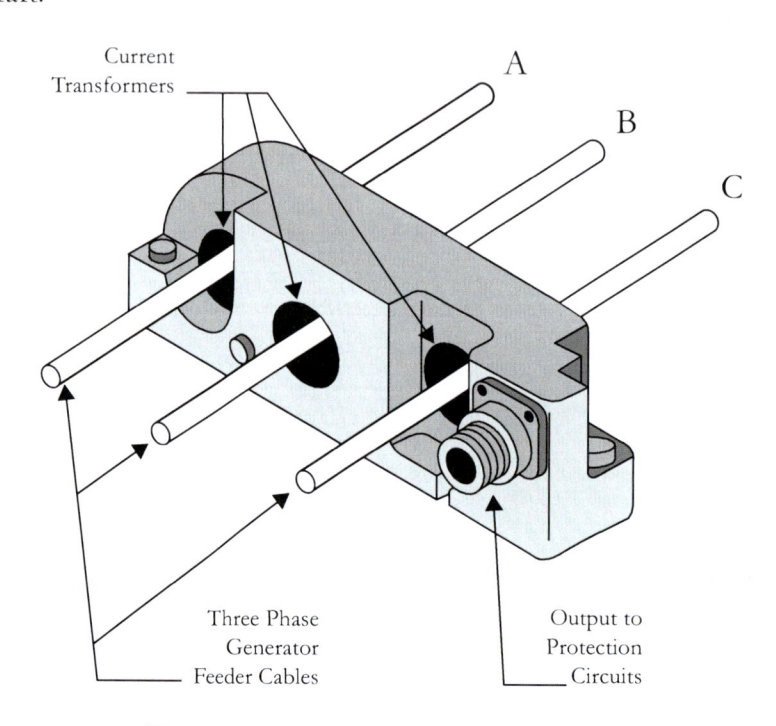

Figure 15.20 – 3 Phase Current Transformers

Impedance-Matching Transformers

For maximum or optimum transfer of power between two circuits, it is always necessary for the impedance of one circuit to match that of the other circuit as close as possible; one common method of impedance matching is the transformer.

To obtain proper matching, a transformer must have the correct turns ratio and the number of turns on the primary and secondary windings and the impedance of the transformer have the following mathematical relationship:

$$\frac{N_P}{N_S} = \sqrt{\frac{Z_P}{Z_S}}$$

Z_p represents the primary impedance, which is the output impedance of the power source. Z_s represents the secondary, or load, impedance (Z_l).

For example, a power source's 300-Ω output impedance is transformed into 75 Ω by a transformer to match the 75-Ω load with a turns ratio of 2:1:

$$N_p / N_s = \sqrt{(Z_p / Z_s)} = \sqrt{(300/75)} = \sqrt{4} = 2$$

Because of this ability to match impedances, the impedance-matching transformer is widely used in electronic and avionic equipment.

Summary

As a study aid and for future reference, the important points of this chapter have been summarised below.

Basic Transformer

The basic transformer is an electrical device that transfers alternating-current energy from one circuit to another circuit by magnetic coupling of the primary and secondary windings of the transformer. This is accomplished through mutual inductance (M). The coefficient of coupling (K) of a transformer is dependent upon the size and shape of the coils, their relative positions, and the characteristic of the core between the two coils. An ideal transformer is one where all the magnetic lines of flux produced by the primary cut the entire secondary. **The higher the K of the transformer, the higher the transfer of the energy.**

The voltage applied to the primary winding causes current to flow in the primary. This current generates a magnetic field, generating a back emf that has the opposite phase to that of the applied voltage. The magnetic field generated by the current in the primary also cuts the secondary winding and induces a voltage in this winding.

Transformer Construction

A transformer consists of two coils of insulated wire wound on a core. The primary winding is usually wound onto a form and is then wrapped with an insulating material such as paper or cloth. The secondary winding is then wound on top of the primary and both windings are wrapped with insulating material. The windings are then fitted onto the core of the transformer. Cores come in various shapes and materials. The most common materials are air, soft iron, and laminated steel.

The most common types of transformers are the shell and hollow-core. The type and shape of the core is dependent on the intended use of the transformer and the voltage applied to the current in the primary winding.

Excitation Current

When voltage is applied to the primary of a transformer, excitation current flows in the primary. The current causes a magnetic field to be set up around both the primary and the secondary windings. The moving flux causes a voltage to be induced into the secondary winding, countering the effects of the back emf in the primary.

Phase

When the secondary winding is connected to a load, causing current to flow in the secondary, the magnetic field decreases momentarily. The primary then draws more current, restoring the magnetic field to almost its original magnitude. The phase of the current flowing in the secondary circuit is dependent upon the phase of the voltage applied across the primary and the direction of the winding of the secondary. If the secondary were wound in the same direction as the primary, the phase would be the same; if wound opposite to the primary, the phase would be reversed. This is shown on a schematic drawing by the use of phasing dots. The lack of phasing dots on a schematic means a phase reversal.

Turns Ratio

The turns ratio of a transformer is the ratio of the number of turns of wire in the primary winding to the number of turns in the secondary winding. When the turns ratio is stated, the number representing turns on the primary is generally stated first. For example, a 1:2 turns ratio means the secondary has twice the number of turns as the primary. In this example, the voltage across the secondary is two times the voltage applied to the primary.

Power & Current Ratios

The power and current ratios of a transformer are dependent on the fact that power delivered to the secondary is always equal to the power delivered to the primary minus the losses of the transformer. This will always be true, regardless of the number of secondary windings. Using the law of power and current, it can be stated that current through the transformer is the inverse of the voltage or turns ratio, with power remaining the same or less.

Transformer Losses

Transformer losses have two sources-copper loss and magnetic loss. Copper losses are caused by the resistance of the wire, ie I^2R, and eddy currents and hysteresis in the core cause magnetic losses. Copper loss is a constant after the coil has been wound and therefore a measurable loss.

Hysteresis loss is constant for a particular voltage and current. Eddy-current loss, however, is different for each frequency passed through the transformer.

Transformer Efficiency

The amplitude of the voltage induced in the secondary is dependent upon the efficiency of the transformer and the turns ratio. The efficiency of a transformer is related to the power losses in the windings and core of the transformer. Efficiency in percentage terms is:

$$\frac{P_{out}}{P_{in}} \times 100$$

A perfect transformer would have an efficiency of 1.0 or 100%, but this is never possible; *no transformer has 100% efficiency transfer*.

Power Transformer

A transformer with two or more windings wound on a laminated iron core. The transformer is used to supply stepped up and stepped down values of voltage to the various circuits in electrical and avionic equipment.

Autotransformer

A transformer with a single winding in which the entire winding can be used as the primary and part of the winding as the secondary, or part of the winding can be used as the primary and the entire winding can be used as the secondary.

Audio-Frequency Transformer

A transformer used in audio-frequency circuits to transfer AF signals from one circuit to another.

Radio-Frequency Transformer

A transformer used in a radio-frequency circuit to transfer RF signals from one circuit to another.

Current Transformers

Current transformers (CTs) are a special type of transformer which allows the measurement of AC currents without breaking into the circuit

Impedance-Matching Transformer

A transformer used to match the impedance of the source and the impedance of the load. The mathematical relationship of the turns and impedance of the transformer is expressed by the equation:

$$\frac{N_P}{N_S} = \sqrt{\frac{Z_P}{Z_S}}$$

Revision

Transformers

Questions

1. If a transformer has a ratio of 4:1 and the primary voltage and current is 100 V and 2 A, what are the values of the secondary voltage and current?

    a) 25 V & 8 A

    b) 400 V & 0.5 A

    c) 100 V & 2 A

2. The symbol in the figure below represents:

    a) An iron core transformer

    b) A laminated core transformer

    c) An air core transformer

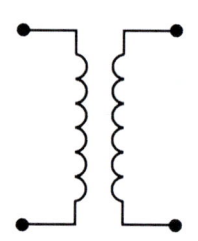

3. The symbol in the figure below represents:

   a) An unlike wound transformer

   b) A like wound transformer

   c) An air core transformer

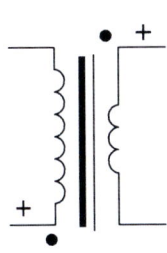

4. With the transformer illustrated in question 3 above, the primary and secondary voltages are:

   a) In phase

   b) The same value

   c) Anti-phase

5. If the distance between the primary & secondary of a transformer are brought closer together, it's coefficient of coupling 'K':

   a) Increases

   b) Decreases

   c) Remains the same

6. A transformer's hysteresis core loss is improved by altering its:

   a) Number of laminations

   b) Core material

   c) Primary/Secondary gap

7. If the power from the source of a transformer is 200 W, and the output power from the load is 180 W, what is the efficiency of the transformer?

   a) 90%

   b) 100%

   c) 80%

**CHAPTER FIFTEEN
TRANSFORMERS**

Revision

Transformers

Answers

1. **A**   6. **B**
2. **C**   7. **A**
3. **A**
4. **C**
5. **A**

Filters

Introduction

Real-world signals contain both wanted and unwanted information. Therefore, some kind of filtering technique must separate the two before processing and analysis can begin. In addition, it is sometimes desirable to have circuits capable of *selectively* filtering one frequency or range of frequencies out of a mix of different frequencies coming from a common signal source; these circuits are called *Filters*.

A common need for filter circuits that you may already know of is in high-performance stereo systems, where certain ranges of audio frequencies need to be amplified or suppressed for best sound quality and power efficiency. You may also be familiar with *Equalisers* that allow the amplitudes of several frequency ranges to be adjusted to suit the listener's taste and acoustic properties of the listening area. You may also be familiar with *Crossover Networks* that block certain ranges of frequencies from reaching speakers. A tweeter, (high-frequency speaker), would waste any signal power of low frequencies such as drum beats, so a crossover circuit is connected between the tweeter and the stereo's output terminals to block low-frequency signals, only passing high-frequency signals to the speaker's connection terminals. A capacitor connected in series with the tweeter speaker will impose a high impedance to low-frequency bass signals, thereby preventing that power from being wasted, Figure 16.1.

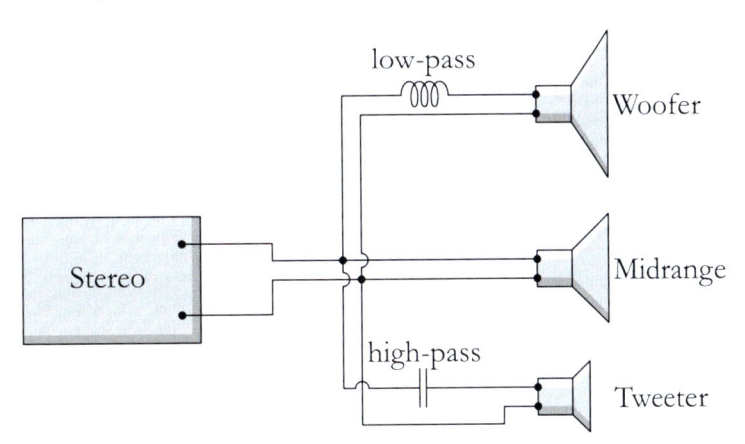

Figure 16.1 – Crossover Circuit

CHAPTER SIXTEEN
FILTERS

An inductor connected in series with the woofer (bass) speaker will serve as a low-pass filter for the low frequencies that particular speaker is designed to reproduce, whilst blocking the higher frequencies.

Both equalizers and crossover networks are examples of filters, designed to accomplish filtering of certain frequencies.

Another practical application of filter circuits is in the *conditioning* of non-sinusoidal voltage waveforms in power circuits. If a distorted sine-wave voltage behaves like a series of harmonic waveforms added to the fundamental frequency, then it is possible to construct a filter circuit that only allows the fundamental waveform frequency to pass through. A harmonic is a signal or wave whose frequency is a multiple of the fundamental frequency. For a signal whose fundamental frequency is f, the second harmonic has a frequency $2f$; the third harmonic has a frequency of $3f$, and so on.

Filters can be divided into two distinct types:

- Active Filters
- Passive Filters.

Active filters contain amplifying devices to increase signal strength while passive do not contain amplifying devices to strengthen the signal.

An important point to remember with all filter circuits is the formulae:

$$X_L = 2\pi f L$$

$$X_C = \frac{1}{2\pi f C}$$

$$\text{Resonant Frequency} = \frac{1}{2\pi\sqrt{LC}}$$

Filters are named according to the frequency range of signals that they allow to pass through them, whilst **blocking** or "**attenuating**" the rest. The most commonly used filters are:

- Low-Pass
- High-Pass
- Band-Pass
- Band-Stop

Low-Pass Filters

By definition, a *low-pass filter* is a circuit offering easy passage to low-frequency signals and difficult passage to high-frequency signals. There are two basic kinds of circuits capable of accomplishing this objective, and many variations of each one:

- An inductive low pass filter
- A capacitive low pass filter

Inductive Low Pass Filter

Figure 16.2 shows a simple inductive low-pass filter. With this arrangement, the inductor's impedance *increases* with *increasing frequency*. This high series impedance tends to block high frequency signals from getting to the load and therefore, the load voltage decreases with increasing frequency.

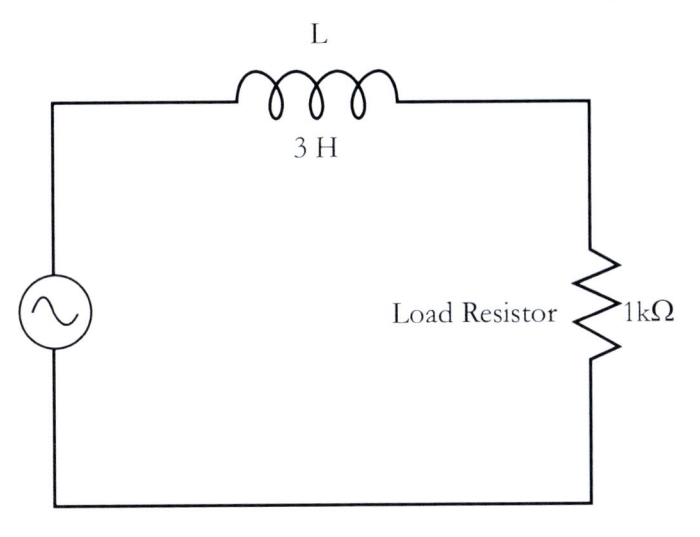

Figure 16.2 - Inductive Low Pass Filter

Capacitive Low Pass Filter

Figure 16.3 shows a simple capacitive low pass filter. With this arrangement, the capacitor's impedance *decreases* with *increasing frequency*. This low impedance in parallel with the load resistance tends to short out high-frequency signals, dropping most of the voltage gets across series resistor R1 and therefore, the load voltage decreases with increasing frequency.

CHAPTER SIXTEEN
FILTERS

Figure 16.3 - Capacitive Low Pass Filter

The inductive low-pass filter is the pinnacle of simplicity, with only one component making up the filter and the capacitive filter is not that much more complex, with only a resistor and capacitor needed for its operation. However, capacitive filter designs are generally preferred over inductive, because capacitors tend to be *purer* reactive components than inductors and therefore are more predictable in their behaviour. By *pure* I mean that capacitors possess little stray resistance, making them almost 100% reactive. Inductors, on the other hand, typically have significant levels of stray resistance, both in the long lengths of wire used to make them, and in the magnetic losses of the core material.

However, the inductive low-pass filter is quite frequently used in ac to dc power supplies to filter out the ac *ripple* waveform created when ac is rectified into dc, passing only the pure dc component. The primary reason for this is the requirement of low filter resistance for the output of such a power supply. A capacitive low-pass filter necessitates an extra resistance in series with the source, whereas the inductive low-pass filter does not. In the design of dc power supply, or similar circuit, where additional series resistance is probably undesirable, the inductive low-pass filter is the better choice.

On the other hand, if low weight and compact size are higher priorities than low internal supply resistance in a power supply design, the capacitive low-pass filter might make more sense. See, it really is a juggling act when it comes to electronic circuit design!

All low-pass filters are rated at a certain **cut-off frequency**, (the frequency above which the output voltage falls below 70.7% of the input voltage). This cut-off percentage of 70.7% is not arbitrary, although it may seem so at first glance. In a simple capacitive/resistive low-pass filter, it is the frequency at which capacitive reactance in ohms equals resistance in ohms. In a simple capacitive low-pass filter, i.e. one resistor, one capacitor, the cut-off frequency is given as:

$$f_{cut\text{-}off} = \frac{1}{2\pi RC}$$

In radio frequency applications, low pass filters are typically designed using inductors and capacitors, Figure 16.4. The inductors are in line (in series) and the capacitors are across (in parallel) with the conductors connecting the supply to the load.

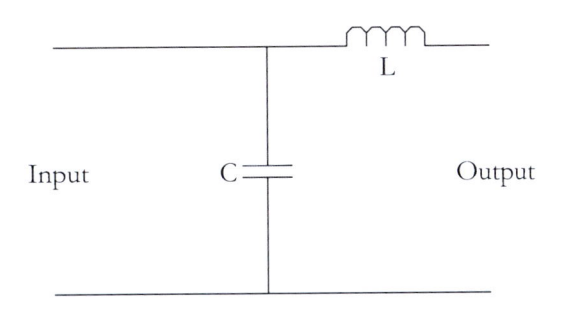

Figure 16.4 – LC Low Pass Filter

The inductor has a low reactance to the lower frequencies, allowing them to pass easily onto the output terminals, but has a high reactance to the higher frequencies.

In the same manner, the capacitor has a low reactance to the higher frequencies, so they are filtered off through it, but a high reactance to the required low frequencies and therefore does not attenuate them significantly.

A number of filter circuits are typically used to improve the attenuation of the higher frequencies, and ensure that the cut off region can be more sharply defined.

High-Pass Filters

A high-pass filter's task is just the opposite of a low-pass filter; i.e. to offer easy passage of a high-frequency signal and difficult passage to a low-frequency signal. As one might expect, the inductive and capacitive versions of the high-pass filter are just the opposite of their respective low-pass filter designs. Figure 16.5 shows a simple capacitive high pass filter.

The capacitor's impedance *increases* with *decreasing frequency* and this high impedance in series tends to block low-frequency signals from getting to the load and therefore the load voltage increases with increasing frequency.

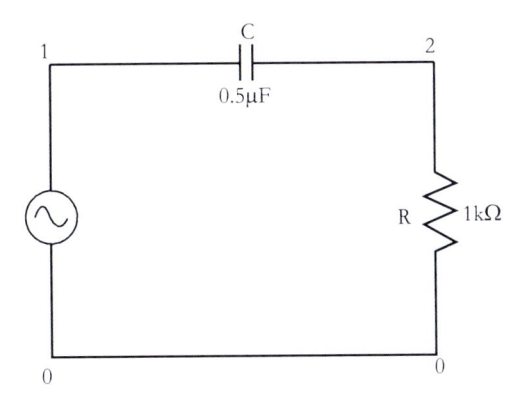

Figure 16.5 - Simple Capacitive High Pass Filter

Inductive High Pass Filter

The inductor's impedance *decreases* with *decreasing* frequency. This low impedance in parallel tends to short out low-frequency signals from getting to the load resistor and as a consequence, most of the voltage gets dropped across series resistor R1. As the frequency increases the inductor's impedance increases and therefore the voltage across the load increases. Figure 16.6 shows a typical circuit.

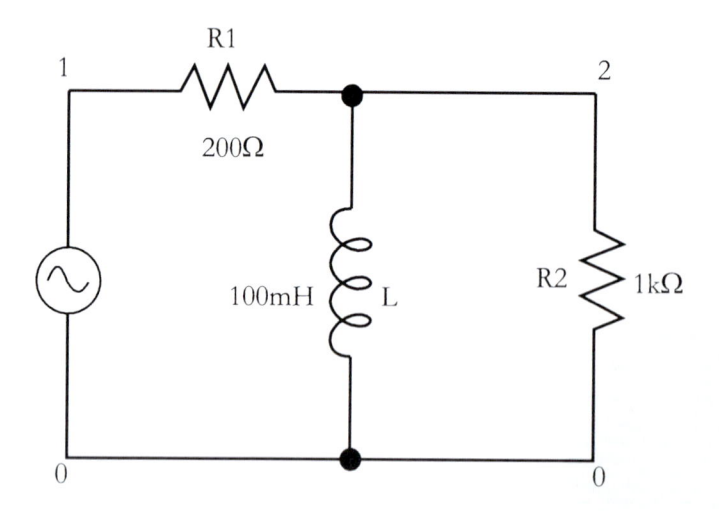

Figure 16.6 – Simple Inductive High Pass Filter

This time, the *capacitive* design is the simplest, requiring only one component above and beyond the load and, again, the reactive purity of capacitors over inductors tends to favour their use in filter design, especially with high-pass filters where high frequencies commonly cause inductors to behave strangely due to the skin effect and electromagnetic core losses we have discussed in earlier chapters.

High pass filters can also be designed using inductors and capacitors, Figure 16.7 The capacitors are in line (in series) and the inductors are across (in parallel) the conductors connecting the supply to the load.

The Inductor offers a low reactance to low frequencies, so they are filtered off through it, but offers a high reactance to the high frequencies.

The Capacitor allows the high frequencies to pass onto the output terminals, but offers a high reactance to the low frequencies.

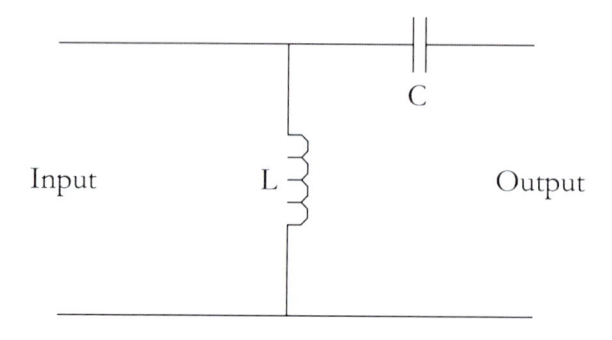

Figure 16.7 – LC High Pass Filter

As with low-pass filters, high-pass filters have a rated cut-off frequency, above which the output voltage increases above 70.7% of the input voltage and just as in the case of the capacitive low-pass filter circuit, the capacitive high-pass filter's cut-off frequency can be found with the same formula.

For better performance yet, we might like to have some kind of filter circuit capable of passing frequencies that are between low and high frequencies so that none of the low- or high-frequency signal power is wasted. What we would be looking for is called a *Band-Pass* filter, which brings us to the next section.

Band-Pass Filters

There are applications where a particular band, or spread, or frequencies need to be filtered from a wider range of mixed signals. Filter circuits can be designed to accomplish this task by combining the properties of low-pass and high-pass into a single filter. The result is called a *Band-Pass* filter.

Creating a band-pass filter from a low-pass and high-pass filter can be illustrated using a block diagram as in figure 16.8.

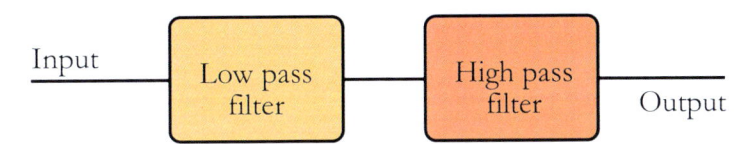

Figure 16.8 – Block Diagram of Simple Band Pass Filter

What emerges from the series combination of these two filter circuits is a circuit that will only allow passage of those frequencies that are neither too high nor too low. Using real components, figure 16.9 illustrates what a typical schematic might look like.

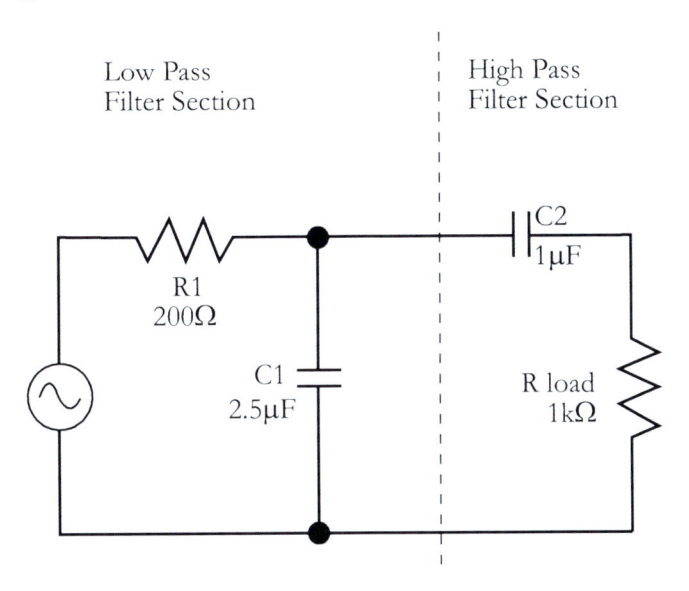

Figure 16.9 – Simple Capacitive Band Pass Filter

CHAPTER SIXTEEN
FILTERS

Band-pass filters can also be constructed using inductors, but as mentioned before, the reactive *purity* of capacitors gives them a design advantage and inductive band pass filters are rare.

The fact that the low-pass section comes first in this design instead of the high-pass section makes no difference in its overall operation. It will still filter out all frequencies too high or too low, as required.

Once again a combination of LC circuits can be used in band pass filters. Hopefully you will remember from chapter 14, that different combinations of capacitors and inductors produce circuits known as *acceptors* and *rejecters*. Figure 16.10.

The Rejecter circuit is the parallel combination L1 and C1 which is connected across the output. The acceptor circuit is the series combination L2 and C2 which is connected in series with the output. Both circuits are tuned to the same frequency, the centre frequency of the required band.

Importantly no mutual coupling exists between L1 and L2.

As we know the acceptor circuit offers low impedance to the band of resonant frequencies and passes them onto the output terminals. To all the other input frequencies away from resonance the acceptor, offers a high impedance and blocks or attenuates the signals from reaching the output.

Conversely, the rejecter circuit offers low impedance to the unwanted frequencies above and below the band and so they are filtered off through it. At resonance the rejecter circuit has a high impedance and the signal is passed to the output via the low impedance acceptor.

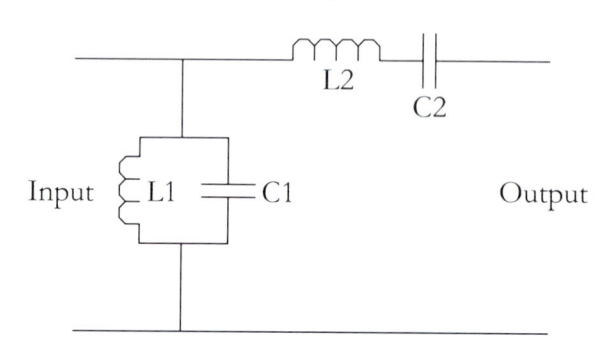

Figure 16.10 – LC Band Pass Filter.

Band-Stop Filters

Band Stop filters are also called ***Band-Elimination***, ***Band-Reject***, or ***Notch*** filters, and this kind of filter passes all frequencies *above* and *below* a particular range, set by the component values. Not surprisingly, it can be made out of a low-pass and a high-pass filter, just like the band-pass design, except that this time the two filter sections are connected in parallel with each other instead of in series. Constructed using two capacitive filter sections, a typical filter looks like that in figure 16.11.

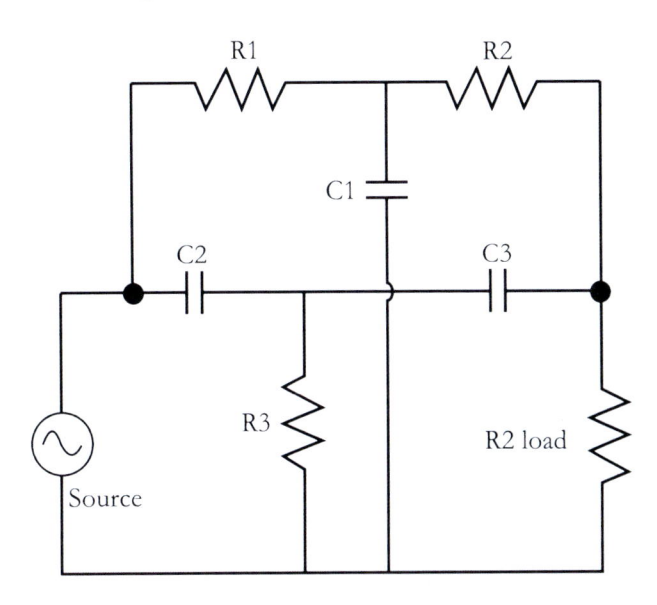

Figure 16.11 – Twin 'T' Band-Stop Filter

The low-pass filter section is comprised of R1, R2, and C1 in a 'T' configuration while the high-pass filter section is comprised of C2, C3, and R3 in a 'T' configuration as well. Together, this arrangement is commonly known as a ***'Twin-T' filter*** giving a sharp response when the component values are chosen in the ratios C1 = C2 = C3 = ½ C1.

Once again a combination of LC circuits can be used in band stop filters. Figure 16.12.

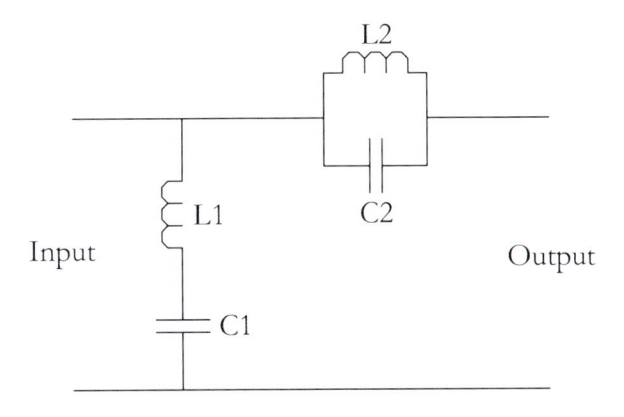

Figure 16.12 – LC Band Stop Filter

As you can see, the acceptor and rejecter circuits have swapped positions. The rejector combination is now in series with the output, whilst the acceptor is connected across (in parallel) the output. This time rejecter circuit offers low impedance to all the required frequencies (away from resonance) and so passes them onto the output terminals, but offers high impedance to the unwanted band of frequencies (at resonance).

The acceptor circuit L1 and C1 offers low impedance to the unwanted band of frequencies and so they are filtered off through it; it offers high impedance to the wanted frequencies and so, does not attenuate them appreciably.

The circuit symbol and frequency response of all these four filter types is shown in figure 16.13.

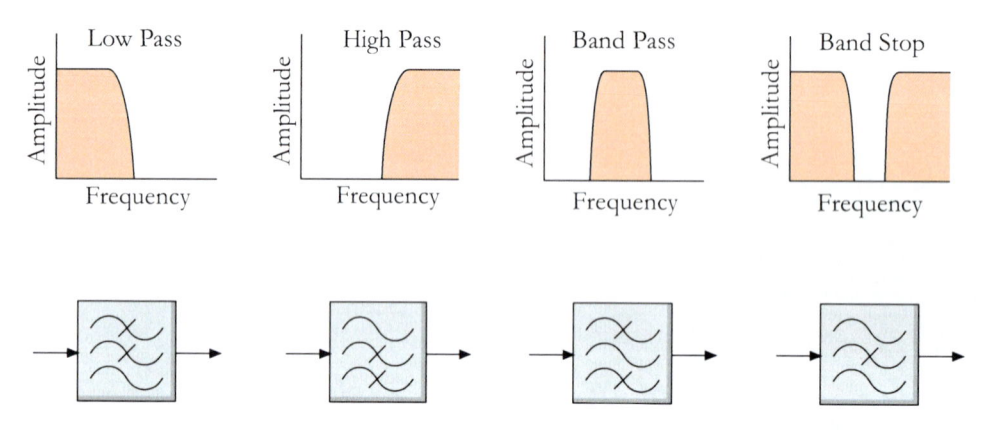

Figure 16.13 – Filter Frequency Responses

Filter Applications

In the world of electrics and electronics, especially in an aircraft with confined space, there are many signals transmitted around the aircraft they may need separating. By the appropriate use of filters, unwanted signals can be eliminated, or at least attenuated to an insignificant level, so that only the correct ones are amplified and processed.

Filters are also used in rectification circuits to remove any unwanted ripple in order to obtain a smooth DC output. The simplest filter for this purpose is a capacitor across the supply output terminals, as shown in Figure 16.14.

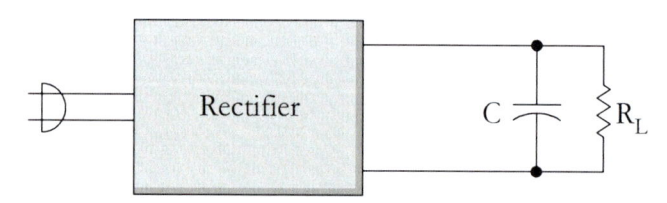

Figure 16.14 – Capacitor Smoothing Filter

Here the capacitor appears like an open circuit to DC, while it looks almost like a short circuit to AC. In this way any AC ripple flows through the capacitor and is removed from the output. This simple capacitor filter is most effective for small load currents.

When high load currents are involved an LC circuit is typically used, Figure 16.15.

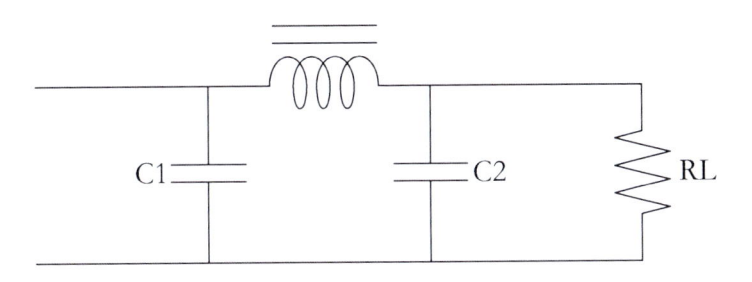

Figure 16.15 – Typical LC Smoothing Circuit

Here a filter or reservoir capacitor C1, is connected across the rectifier output, an inductor L, is in series with the output and another filter or smoothing capacitor, C2, is connected across the load, RL. Capacitor C1 offers low reactance to the AC component of the rectifier output while it offers infinite resistance to the DC component. As a result the capacitor filters off or shunts an appreciable amount of the AC component while the DC component carries on to the inductor L.

The inductor L offers high reactance to any AC component but almost zero resistance to the DC component. As such the DC component flows through the inductor while the AC component is blocked.

Capacitor C2 filters any AC component which the inductor had failed to block, ensuring only the DC component appears across the load RL.

Revision

Filters

Questions

1. A typical electronic filter:

    a) Mixes several different frequencies

    b) Passes or stops selective frequencies

    c) Stops all frequencies

2. With an inductive low-pass filter the inductor's impedance:

    a) Increases with increasing frequency

    b) Decreases with increasing frequency

    c) Is unaffected by frequency

3. With a capacitive high-pass filter:

    a) The capacitor's impedance increases with decreasing frequency

    b) The capacitor's impedance decreases with decreasing frequency

    c) Is unaffected by frequency

4. The cut-off frequency of a filter circuits is at:

    a) ±0.707 of the signal

    b) ±0.637 of the signal

    c) ±0.5 of the signal

5. The graph in the figure below shows the response of a:

   a) Band Pass filter

   b) Band Stop filter

   c) High Pass filter

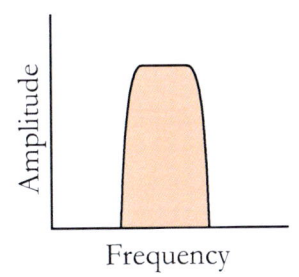

CHAPTER SIXTEEN
FILTERS

Revision

Filters

Answers

1. **B**
2. **A**
3. **A**
4. **A**
5. **A**

AC Generators

Introduction

Most of the electrical power used in modern, large transport aircraft, as well as in domestic applications, is *Alternating Current (AC or ac)*. As a result, the *AC Generator* is the most important means of producing electrical power. In General Aviation they are sometimes called *Alternators* and vary in size depending upon the power load requirement. The typical aircraft ac system generates a sine wave of a given voltage, typically 115 V and 26 V, and in most case, of a constant frequency. The majority of aircraft that use ac as the primary power source use a 3-phase system, i.e. the generator produces three (3) sine waves that are at 120° with respect to each other. Many of the terms and principles covered in this chapter should be familiar to you from earlier chapters as they are the same as those covered in the chapter on dc generators. However, before we go on any further, we need to revisit 3-phase ac to look at how practical systems are connected.

3-Phase Revisited

In an earlier chapter we touched on the concept of 3-phase ac by looking at the difference between single ac and 3-phase ac voltages. This figure is reproduced here as figure 17.1 for convenience's sake.

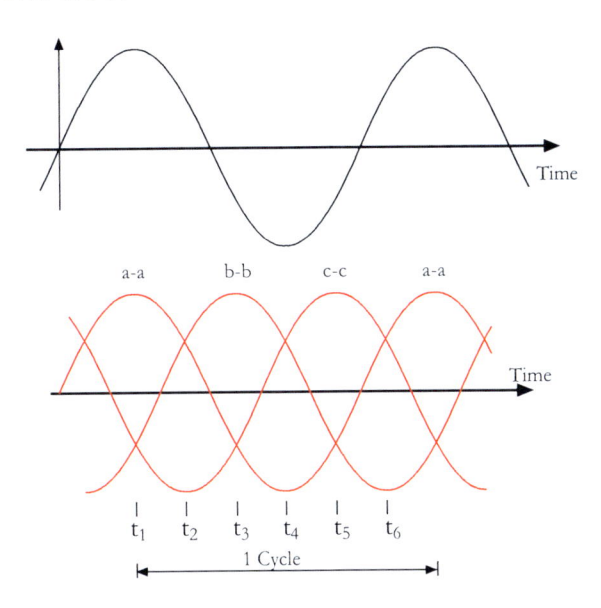

Figure 17.1 – Single Versus 3-Phase AC

As shown in figure 17.1, the three (3) waveforms are identical in shape and can be used to provide three separate outputs.

There are two (2) ways of connecting the three output windings of a 3-phase generator.

- Star or 'Y' wound
- Delta wound

Star or 'Y' Wound 3-Phase AC Generator

With this type of generator, each of the three windings are connected together at a common or neutral point as illustrated in figure 17.2 below:

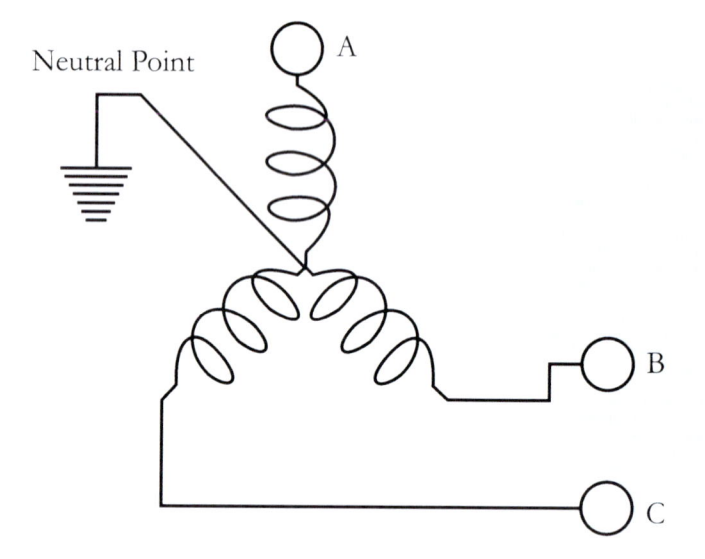

Figure 17.2 – Star or 'Y' Wound Generator

As shown in figure 17.2, the other three ends of the windings are brought out as its output leads, each of which is now across two of the windings in series. However, the output voltage will **never** be twice that of one windings as the voltages are at 120° with respect to each other. It will instead be 1.73 times that of the single winding.

Aircraft generators produce **115 V ac** from each winding at a frequency of **400Hz**. Therefore, the output across outputs A to B or B to C is approximately 200 V ac.

Since the windings are in series between two of the output leads, the output current is the same as the phase current.

Note: For the remainder of these notes, I will refer to the above type of generator as a '*Star*' wound generator.

Delta Wound Generator

With the Delta wound generator, both ends of each winding can be connected to the ends of the other windings to form a *Delta* connection. This name is derived from the resemblance of the coil assembly to the Greek letter Δ, as shown in figure 17.3 below.

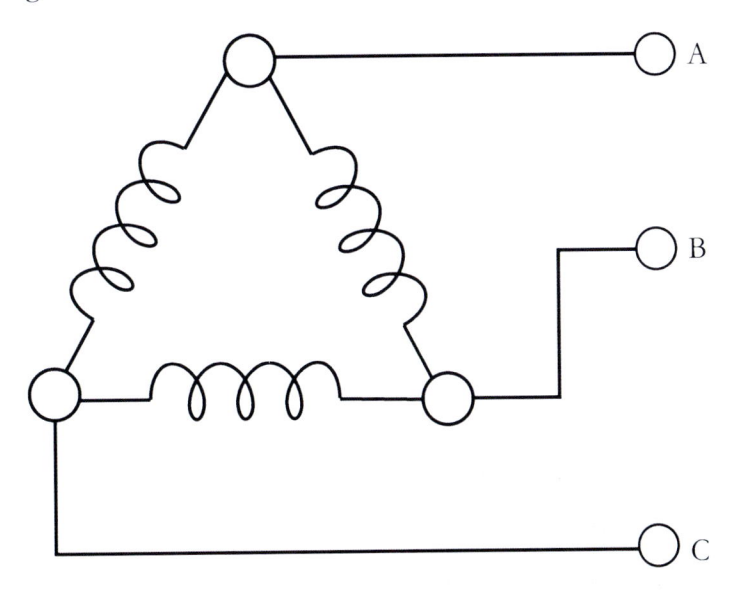

Figure 17.3 – Delta Connected 3-Phase Generator

With this construction, an output lead is brought from each junction so that the output voltage will always be the same as the phase voltage. As shown in figure 17.3, there are two coils in series across or in parallel with each of the phase windings. Since the current in each of the windings is 120° out of phase with that in the other windings, the output current is also 1.73 times that of the current in the phase winding.

AC Versus DC Advantages

The main advantage of ac over dc is that it operates at a higher voltage, typically 115 V ac versus 28 V dc. The use of a higher voltage is not necessarily an advantage in itself as higher voltage, by its very nature, requires better standards of insulation. However, it has a distinct advantage when looking at large *amounts* of power as higher voltage does mean lower currents (P=VI). The lower the current, the lower the I^2R losses and the lower voltage drop. In addition, as conductor size is directly proportional to current flow, the reduction in current of an AC system means a significant weight saving over a corresponding DC system.

Basic AC Generators

Regardless of size, all electrical generators, whether dc or ac, depend upon the principle of magnetic induction. An emf is induced in a coil as a result of:

- A coil cutting through a magnetic field
- A magnetic field cutting through a coil

As long as there is relative motion between a conductor and a magnetic field, a voltage will be induced in the conductor. That part of a generator that produces the magnetic field is called the *field* and that part in which the voltage is induced is called the *armature*. For relative motion to take place between the conductor and the magnetic field, all generators must have two mechanical parts, a rotor and a stator. The rotor is the part that rotates; the stator is the part that remains stationary. In a dc generator, the armature is always the rotor, but with ac generators, the armature may be either the rotor or stator.

Rotating-Armature Generators

The rotating-armature generator is similar in construction to the dc generator in that *the armature rotates* in a stationary magnetic field as shown in figure 17.4.

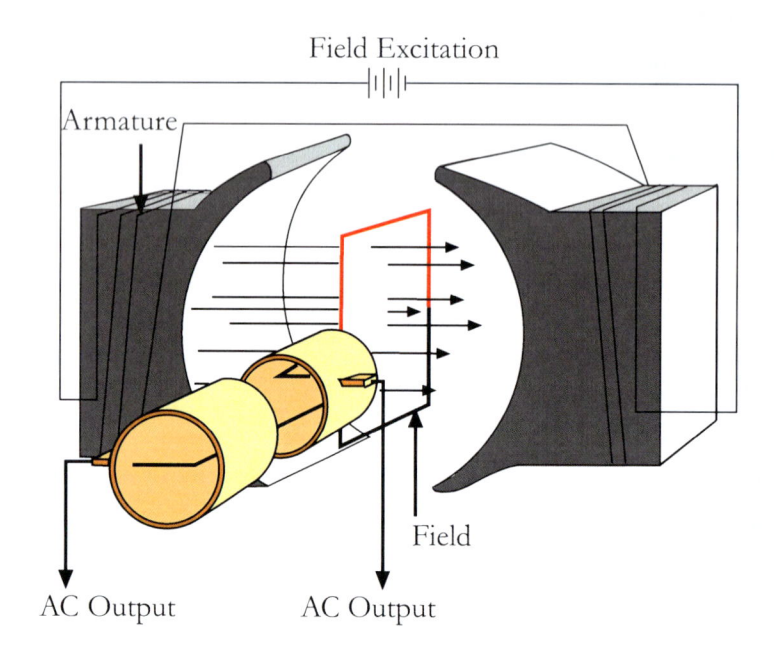

Figure 17.4 – Simple Rotating Armature Generator

In the DC generator, the emf generated in the armature windings is converted from ac to dc by means of the commutator. In the generator, the generated ac is brought to the load unchanged by using *slip rings*.

The rotating armature is only found in generators of low power rating and generally is not used to supply electric power in large quantities.

A major disadvantage of a rotating armature it that it requires slip rings and brushes to conduct the current from the armature to the load. The armature, brushes, and slip rings are difficult to insulate, and *arc-overs* and *short circuits* can result at high voltages. For this reason, high-voltage generators are usually of the rotating-field type.

Rotating-Field Generators

The rotating-field generator has a *stationary armature winding* and a rotating-field winding as shown in figure 17.5.

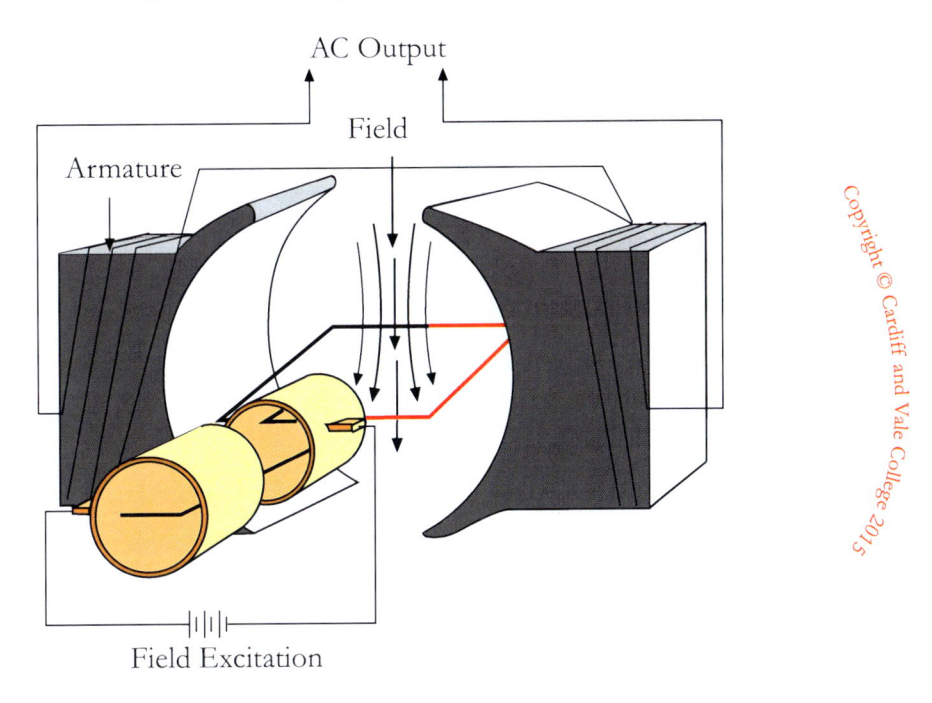

Figure 17.5 – Simple Rotating Field Generator

The advantage of having a stationary armature winding is that the generated voltage can be connected directly to the load and since the voltage applied to the rotating field is low voltage dc, the problem of high voltage arc-over at the slip rings does not exist.

The stationary armature, or *stator*, of this generator type holds the windings that are cut by the rotating magnetic field. The voltage generated in the armature as a result of this cutting action is the ac power that will be applied to the load. The stators of all rotating-field generators are basically the same.

The stator consists of a *laminated iron core* with the armature windings embedded in this core, as shown in figure 17.6; which is secured to the stator frame.

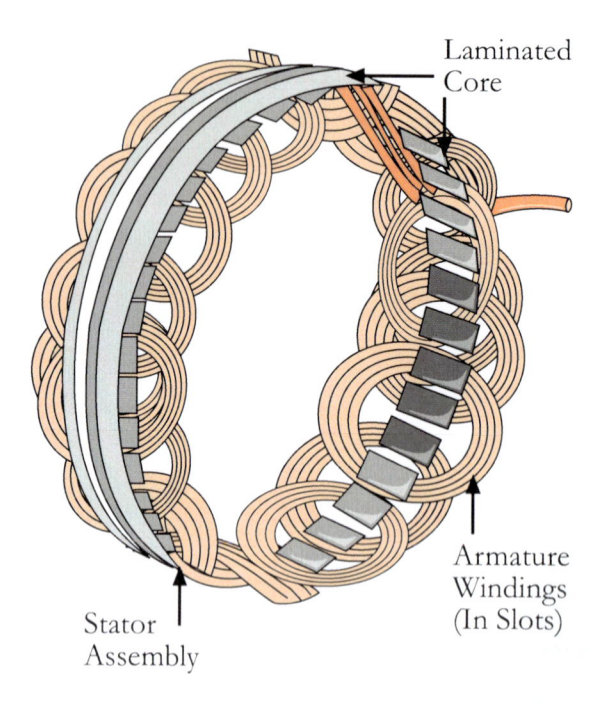

Figure 17.6 - Stationary Armature Windings

Practical Generators

The generators described so far in this chapter are *elementary* or simple in nature; they are seldom used except as examples to aid in understanding practical generators. The remainder of this chapter will relate the principles of the elementary generator to the generators actually in use in the civilian aircraft industry. The following paragraphs in this chapter will introduce such concepts as prime movers, field excitation, armature characteristics and limitations, single-phase and polyphase generators, controls, regulation, and parallel operation.

Functions of Generator Components

A typical rotating-field ac generator consists of an ac generator and a smaller dc generator built into a single unit. The output of the generator section supplies alternating voltage to the load. The only purpose for the dc generator is to supply the direct current required to maintain the generator field. This dc generator is referred to as the *Exciter*. The construction of a typical ac generator is shown in figure 17.7.

CHAPTER SEVENTEEN
AC GENERATORS

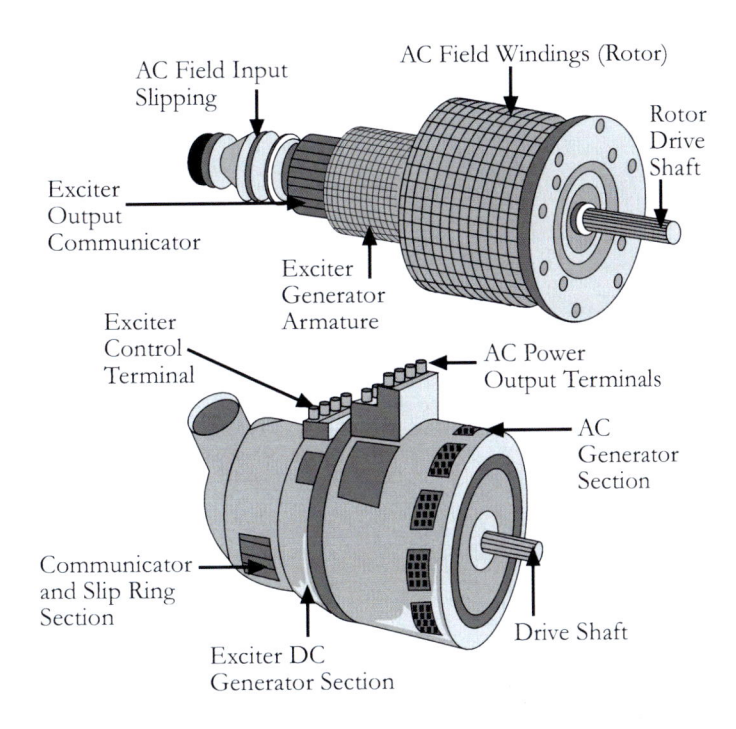

Figure 17.7 – Typical AC Generator

In order to make things easier to follow in the ensuing explanation, I have included a schematic of the generator in figure 17.8.

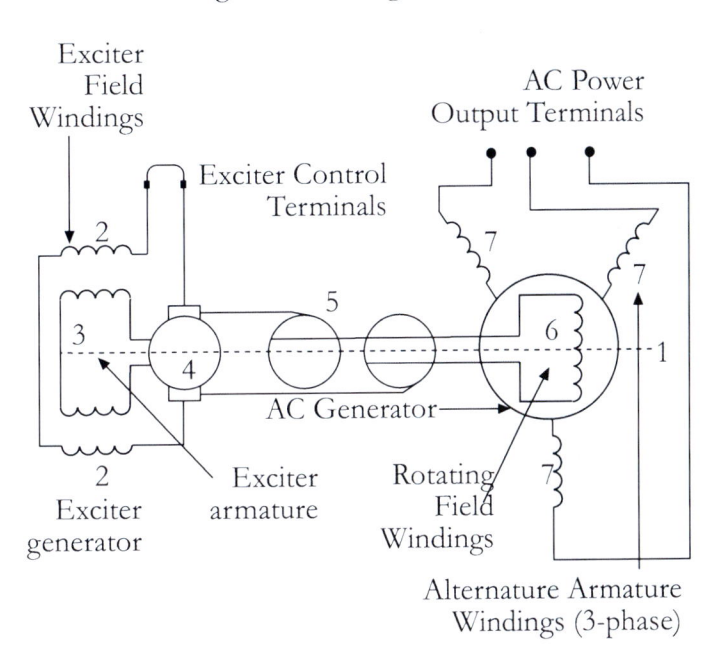

Figure 17.8 – Generator Schematic

The exciter is a dc, shunt-wound, self-excited generator and the exciter shunt field (2) creates an area of intense magnetic flux between its poles. When the exciter armature (3) is rotated in the exciter-field flux, voltage is induced in the exciter armature windings. The output from the exciter commutator (4) is

CHAPTER SEVENTEEN
AC GENERATORS

connected through brushes and slip rings (5) to the ac generator field. Since this is direct current already converted by the exciter commutator, the current always flows in one direction through the generator field (6). Therefore, a fixed-polarity magnetic field is maintained at all times in the generator field windings. When the generator field is rotated, its magnetic flux is passed through and across the generator armature windings (7).

In some generators, the exciter is supplied directly off the aircraft's batteries but as this can drag the battery voltage down considerably, is only used for smaller generators.

The armature is wound for a three-phase output, which will be covered later in this chapter. However, you should remember as I have pointed out several times before, *a voltage is induced in a conductor if it is stationary and a magnetic field is passed across the conductor, the same as if the field is stationary and the conductor is moved*. The alternating voltage in the ac generator armature windings is connected through fixed terminals to the ac load.

Prime Movers

All generators, large and small, ac and dc, require a source of *mechanical* power to turn their rotors; this source of mechanical energy is called a *Prime Mover*. In an aircraft environment, this is obviously the aircraft's engines or its APU. The prime mover plays an important part in the design of generators since the speed at which the rotor is turned determines certain characteristics of generator construction and operation.

Generator Rotors

There are two types of rotors used in rotating-field generators.

- The turbine-driven Rotor
- Salient-pole rotors

Figure 17.9 shows the construction of these two rotors.

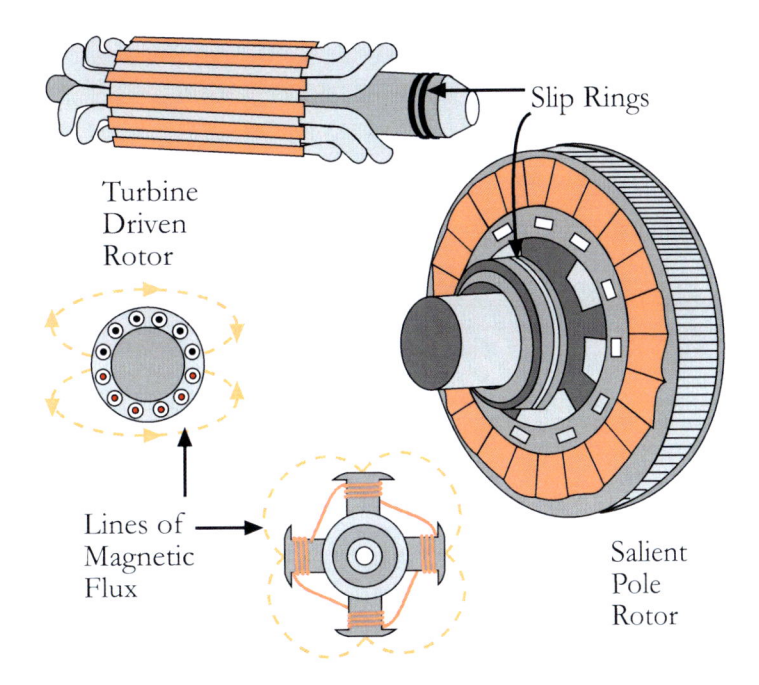

Figure 17.9 – Typical Rotor Construction

As you may have guessed, the turbine-driven rotor shown in figure 17.9, is used when the prime mover is a high-speed turbine. The windings in the turbine-driven rotor are arranged to form two or four distinct poles. The windings are firmly embedded in slots to withstand the tremendous centrifugal forces encountered at high speeds.

The salient-pole rotor shown in figure 17.9, is used in low-speed generators and often consists of several separately wound pole pieces, bolted to the frame of the rotor.

If we could compare the physical size of the two types of rotors with the same electrical characteristics, we would see that the salient-pole rotor would have a greater diameter. At the same number of revolutions per minute, it has a greater centrifugal force than the turbine-driven rotor. To reduce this force to a safe level so that the windings will not be thrown out of the machine, the salient pole is used in low-speed designs, usually with a **Constant Speed Drive (CSD)** unit to keep the speed constant.

Generator Characteristics and Limitations

Generators are rated according to the voltage they are designed to produce and the maximum current they are capable of providing. The maximum current the generator can supply depends upon the maximum heating loss that can be sustained in the armature. This heating loss, (I^2R power loss), acts to heat the conductors, and if excessive, destroys the insulation. Therefore, generators are rated in terms of this current and the voltage output, (the generator) is rated in **Volt-Amperes (VA)**, or more typically for large aircraft **kilo Volt-Amperes (kVA)**.

CHAPTER SEVENTEEN
AC GENERATORS

When a generator is fitted to an aircraft, it is already destined to do a specific task. The speed at which it is designed to rotate, the voltage it will produce, the current limits, and other operating characteristics are already built in. This information is usually stamped on a nameplate on the case so that the end-user knows its limitations.

Single-Phase Generators

A generator that produces a single, continuously alternating voltage is known as a *Single-Phase* generator. All of the generators that have been discussed so far fit this definition. The stator, (armature, windings) are connected in series.

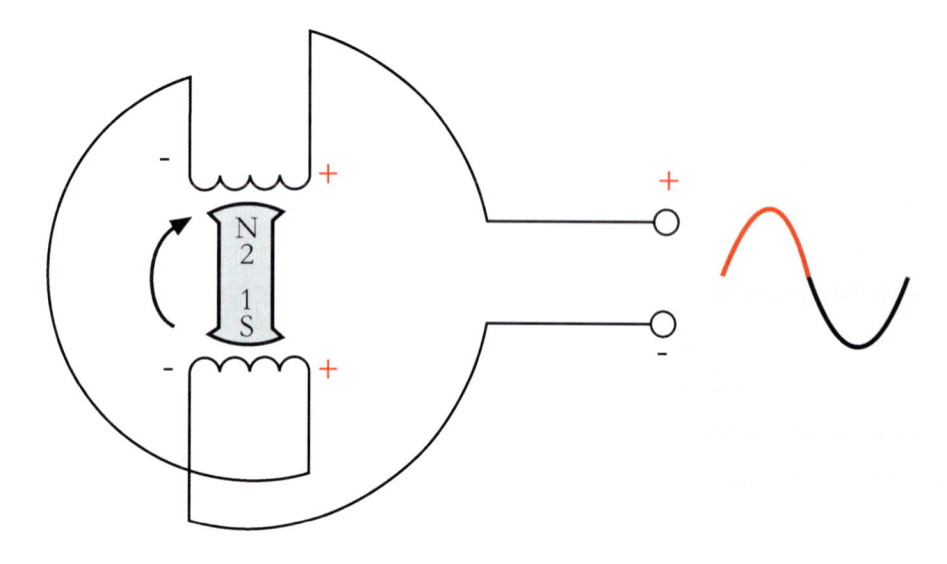

Figure 17.10 - Single-Phase Generator

The individual voltages, therefore, add to produce a single-phase ac voltage. Figure 17.10 shows a basic generator with its single-phase output voltage.

The definition of phase in this context is not the same as discussing different values being in-phase or out-of-phase with each other. In this context, the word *phase* means voltage as in single voltage.

Single-phase generators are found in smaller ac aircraft and they are most often used when the loads being driven are relatively light. The reason for this will become apparent as we get into *multiphase generators*, also called *polyphase generators*.

Power that is used in some aircraft, the home, shops and to operate portable tools and small appliances is single-phase power. Single-phase power generators always generate single-phase power. However, all single-phase power does not necessarily come from single-phase generators. This will sound more reasonable as we get into the next subjects.

Two-Phase Generators

Two phase implies two voltages if we now apply our new definition of phase and yes, it is that simple. A two-phase generator is designed to produce two completely separate voltages. Each voltage, by itself, may be considered as a single-phase voltage and each is generated completely independent of the other. Figure 17.11 shows a simplified two-pole, two-phase generator.

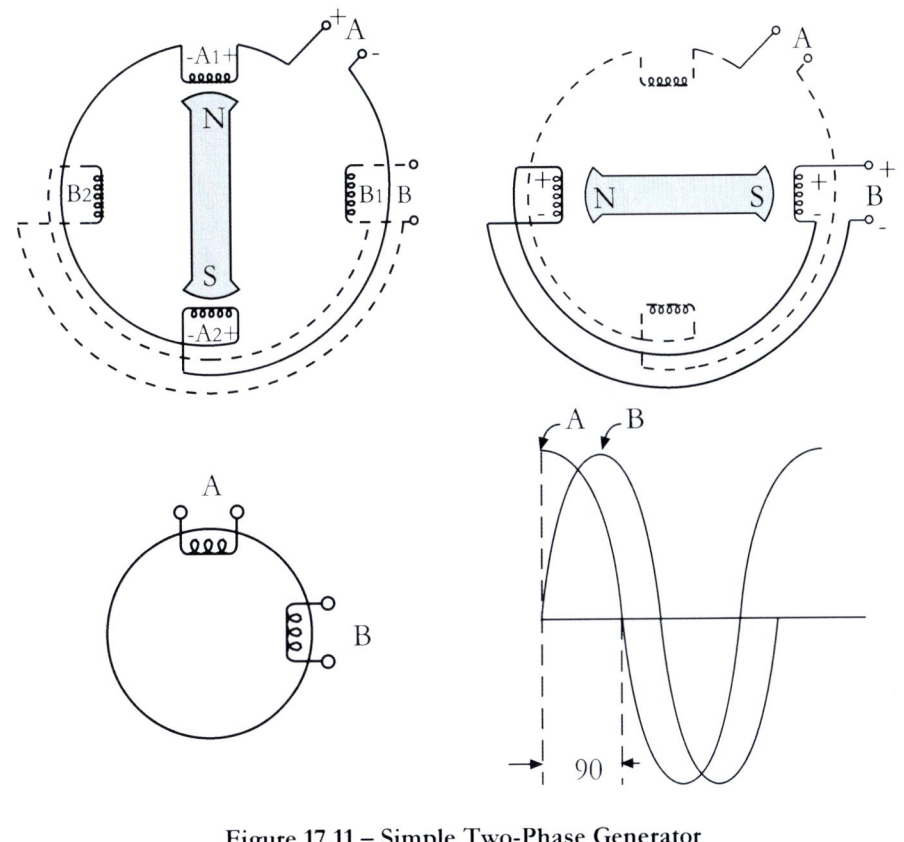

Figure 17.11 – Simple Two-Phase Generator

Note that the windings of the two phases are physically at right angles, i.e. 90°, to each other and so perhaps as you would expect, the outputs of each phase are 90° apart. The graph in figure 17.11 shows the two phases to be 90° apart, with A leading B. Note that by using the *original* definition of phase, we can say that A and B are 90° out of phase and by design, this will always be the case between the phases of a two-phase generator.

The stator in figure 17.11 consists of two single-phase windings completely separated from each other. Each winding is made up of two windings that are connected in series so that their voltages add. The rotor is identical to that used in the single-phase generator.

In the top left-hand schematic of figure 17.11, the rotor poles are opposite *all* the windings of phase A. Therefore, the voltage induced in phase A is maximum, and the voltage induced in phase B is at zero. As the rotor continues rotating counter-clockwise, it moves away from the A windings and approaches the B windings. As a result, the voltage induced in phase A

CHAPTER SEVENTEEN
AC GENERATORS

decreases from its maximum value, and the voltage induced in phase B increases from zero.

In the top right-hand schematic, the rotor poles are opposite the windings of phase B and now the voltage induced in phase B is maximum, whereas the voltage induced in phase A has dropped to zero.

Notice that a 90° rotation of the rotor corresponds to one-quarter of a cycle, or 90° *electrically*. The waveform picture of figure 17.11 shows the voltages induced in phase A and B for one cycle, which is 90° out of phase, and are independent of each other. **Each output is a single-phase voltage, just as if the other did not exist.**

The obvious advantage, so far, is that there are two separate output voltages and there is some saving in having one set of bearings, one rotor, one housing, and so on, to do the work of two. However, there is the disadvantage of having twice as many stator coils, which require a larger and more complex stator. The schematic in figure 17.12 shows four separate wires brought out from the A and B stator windings; the same as in figure 17.11.

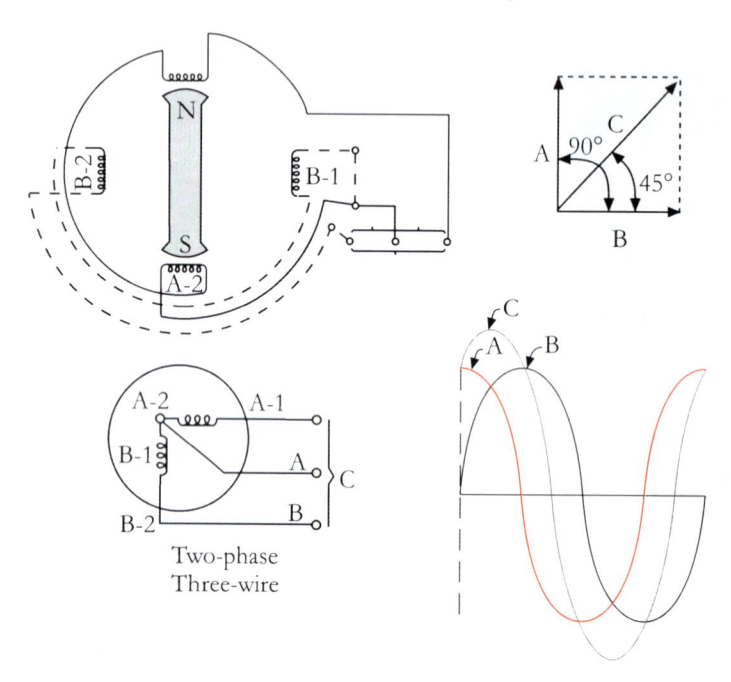

Figure 17.12 - Connections of a 2-Phase, 3-Wire Generator

Notice, however, that the dotted wire now connects one end of B1 to one end of A2. The effect of making this connection is to provide a new output voltage, which is a sine-wave voltage, shown as C in figure 17.12. It is larger than either A or B and is the result of adding the instantaneous values of phase A and phase B. For this reason it appears exactly half way between A and B and so must lag A by 45° and lead B by 45°, as shown in the small vector diagram.

Now if we look at the smaller schematic diagram in figure 17.12, only three connections have been brought out from the stator. Electrically, this is the same as the large diagram above it. Now instead of being connected at the output terminals, the B1-A2 connection is made internally during manufacture. A two-phase generator connected in this manner is called a *two-phase, three-wire generator.*

This three-wire connection makes it possible to have three different load connections, A and B across each phase, and C across both phases. The output at C is always 1.414 times, i.e. $\sqrt{2}$ the voltage of either phase. These multiple outputs are additional advantages of the two-phase generator over the single-phase.

Hopefully, now you can see why single-phase power does not *always* come from single-phase generators.

The two-phase generator discussed above is seldom seen in practical aircraft use, but the operation of polyphase generators is more easily explained using two phases than three phases. The three-phase generator, which will be covered next, is by far the most common of all generators in use on today's modern aircraft.

The Three-Phase Generator

The three-phase generator, as the name implies, has three single-phase windings spaced such that the voltage induced in any one phase is displaced by 120° from the other two. A schematic diagram of a three-phase stator showing all the coils becomes complex and it is difficult to see what is actually happening. The simplified schematic of figure 17.13, view A, shows all the windings of each phase lumped together as one winding with the rotor omitted for simplicity.

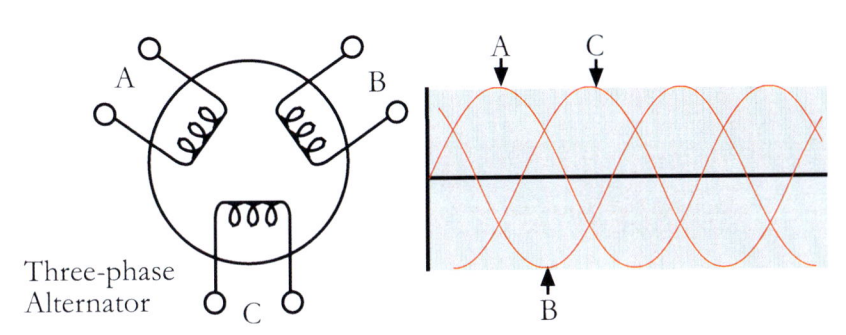

Figure 17.13 - Three-Phase Generator Connections

The voltage waveforms generated across each phase are shown on the graph in figure 17.13, phase-displaced 120° from each other. The three-phase generator as shown in this schematic is made up of three single-phase generators whose generated voltages are out of phase by 120°. The three phases are independent of each other.

Rather than having six leads coming out of the three-phase generator, the same leads from each phase may be connected together to form a Star, i.e. Y, connection, as shown in figure 17.14.

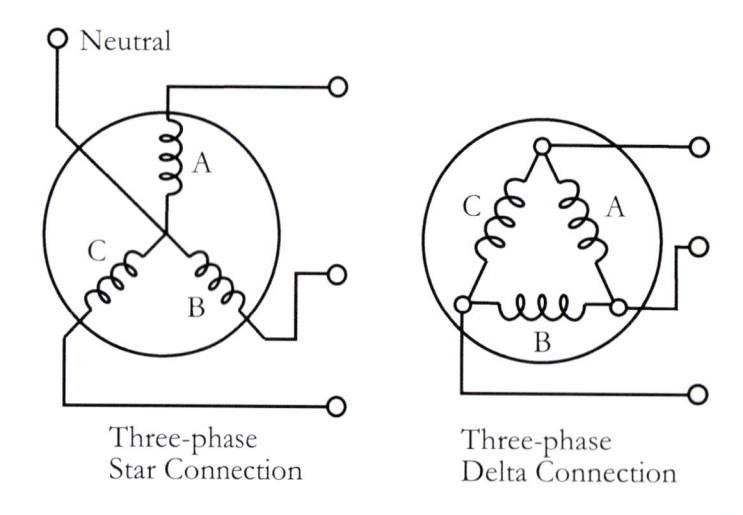

Figure 17.14 – Star & Delta Wound Generator Schematics

As already discussed earlier, with this arrangement, the neutral connection is brought out to a terminal when a single-phase load is supplied and single-phase voltage is available from neutral to A, neutral to B, and neutral to C.

In a three-phase, Y-connected generator, the total voltage, or line voltage, across any two of the three line leads is the vector sum of the individual phase voltages. Each line voltage is 1.73 times one of the phase voltages. Because the windings form only one path for current flow between phases, the line and phase currents are the same. A three-phase stator can also be connected so that the phases are connected end-to-end; it is now delta connected, as also shown in figure 17.14. Remember in the delta connection, line voltages are equal to phase voltages, but each line current is equal to 1.73 times the phase current.

The output of an aircraft AC generator is normally Star wound, producing a single phase output of 115 Volts at 400Hz and a three phase output of 200 Volts at 400Hz

Frequency of Operation

The output frequency of generator voltage depends upon the speed of rotation of the rotor and the number of poles. The faster the speed, the higher the frequency and the more poles there are on the rotor, also the higher the frequency is for a given speed. When a rotor has rotated through an angle such that two adjacent rotor poles, i.e. a north and a south pole, have passed one winding, the voltage induced in that winding will have varied through one complete cycle. For a given frequency, typically 400 Hz in an aircraft, the more pairs of poles there are, the lower the speed of rotation. This principle is illustrated in figure 17.15.

Both alternators are rotating at 120 RPM

0°   180°   360°   0°   180°   360°
8-pole low speed   2-pole low speed

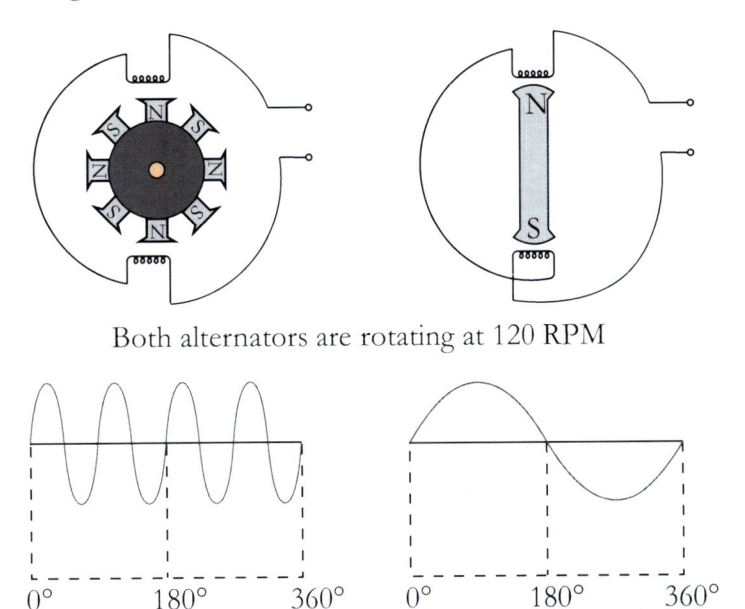

Figure 17.15 – Frequency Regulation

As shown in figure 17.15, a two-pole generator must rotate at four times the speed of an eight-pole generator to produce the same frequency of generated voltage. The frequency of any ac generator in hertz (Hz) is related to the number of poles and the speed of rotation, as expressed by the equation:

$$\text{Frequency} = \frac{NP}{60}$$

Where: P is the number of **pairs** of poles

N is the speed of rotation in rpm

As N is the rotational speed per minutes, we must divide by 60 to calculate the frequency ie cycles per second.

If 'P' is used to denote the total number of poles the formula becomes

$$\text{Frequency} = \frac{NP}{120}$$

120 is a constant to allow for the conversion of minutes to seconds and from poles to pairs of poles

However, some aircraft generators do not require a fixed frequency, eg those used for windscreen heating, and so where this is provided it is the aircraft engine's speed that influences the frequency.

Generator Loads

An AC generator is subjected to 3 types of electrical loads:

- Resistive Loads
- Inductive Loads
- Capacitive Loads

Inductive and Capacitive loads are normally classified under one heading and known as "Reactive Loads"

No Load Condition

Under no load conditions, no load current is being drawn therefore there is no distortion of the magnetic field as shown in 17.16 new.

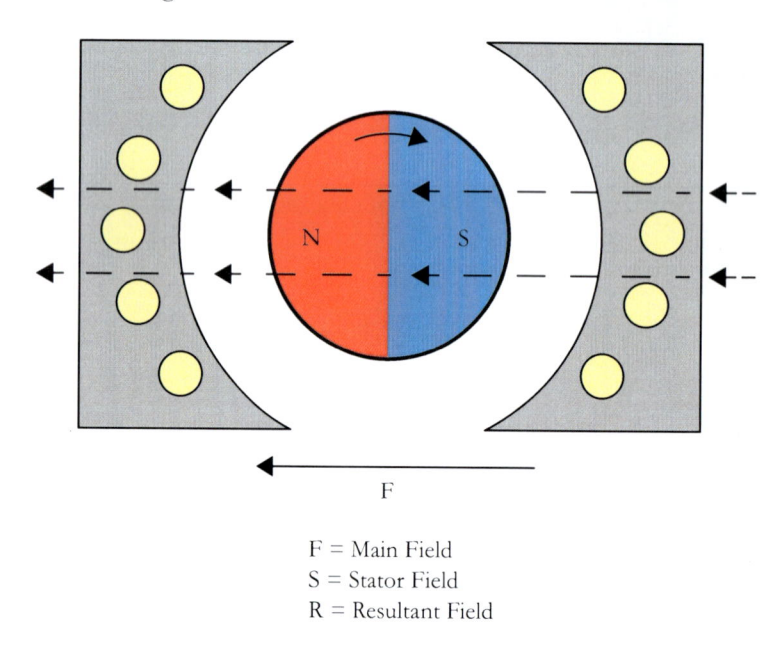

F = Main Field
S = Stator Field
R = Resultant Field

Figure 17.16 – AC Generator No Load Condition

Resistive Loads

In an AC circuit containing **pure resistance** only, such as lighting or de-icing systems, the current and voltage rise and fall together. A resistive load causes no phase shift between voltage and current. Power that results from a resistive load is the power that performs work in a circuit and, as such, is known as "true" power.

Resistive loads tend to slow the generator down, reducing both the output frequency and voltage causing the rotating field to twist against the direction of rotation as shown in Figure 17.17.

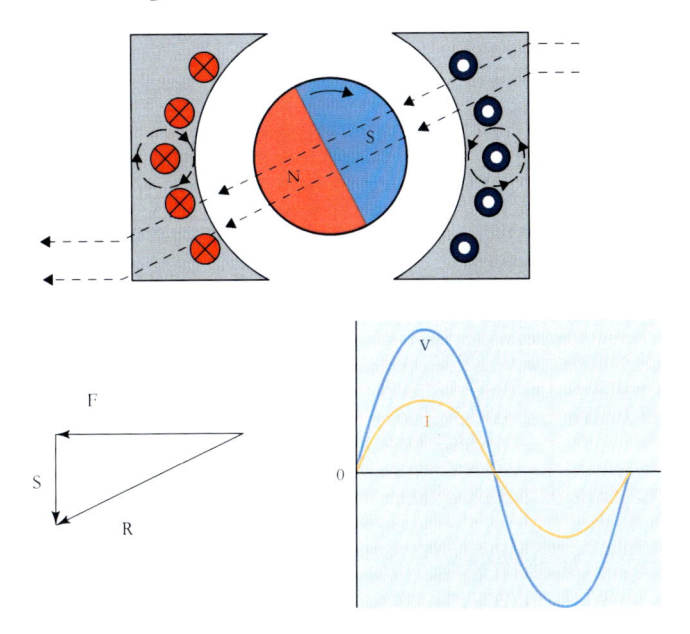

Figure 17.17 – AC Generator Resistive Load

Resistive loads are sometimes referred to as "Real Loads". An increase in the prime mover torque will regain the original speed of rotation and therefore the original frequency and voltage.

Reactive Loads

When a **purely inductive** load is placed on a generator the current in the stator lags the voltage by 90°. This causes the stator field to move around 90°. The stator field opposes the main field, producing a weaker field and a **reduction** in output voltage, Figure 17.18.

CHAPTER SEVENTEEN
AC GENERATORS

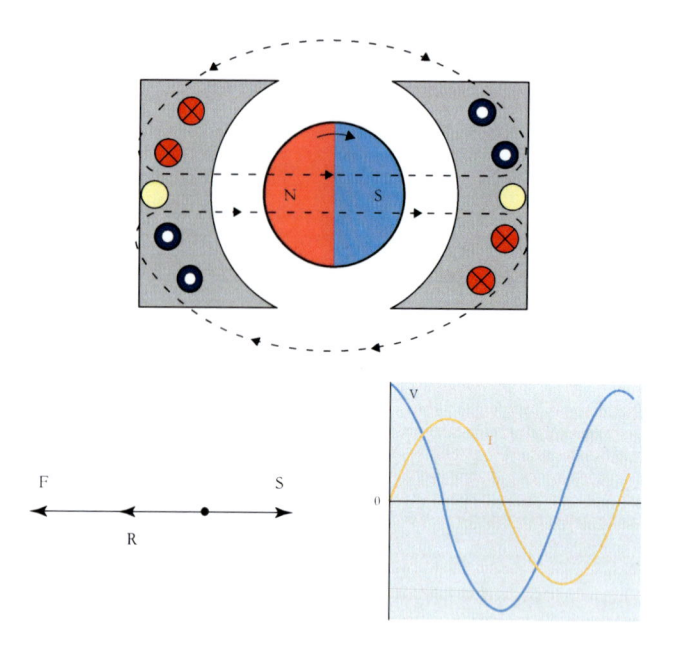

Figure 17.18 – AC Generator Inductive Load

If a capacitive load is experienced, the stator field is advanced by 90°. This assists the main field, and increases the overall field strength increasing the generator output voltage, Figure 17.19.

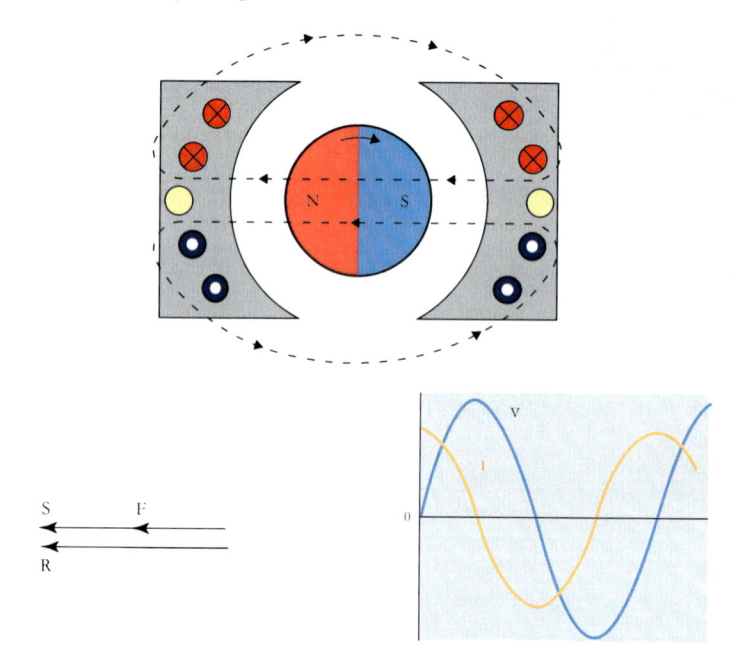

Figure 17.19 – AC Generator Capacitive Load

It is also worth noting that any reactive load, either capacitive or inductive, will not cause the generator to slow down, therefore only the output voltage is affected and the frequency stays the same. As such, we cannot use the prime mover to correct for the change in generator output voltage due to a reactive load. This is achieved by controlling the generators field current using a Voltage Regulator.

Voltage Regulation

As we have seen before, when the load on a generator is changed, the terminal voltage varies with the amount of variation dependent on the design of the generator.

The voltage regulation of a generator is the change of voltage from full load to no load, expressed as a percentage of full-load volts, when the speed and dc field current are held constant; given by the following formula:

$$\% \text{ of Regulation} = \frac{E_{nl} - E_{fl}}{E_{fl}} \times 100$$

Remember, the lower the percentage of regulation, the better it is in most applications.

Principles of AC Voltage Control

In a generator, an alternating voltage is induced in the armature windings when magnetic fields of alternating polarity are passed across these windings. The amount of voltage induced in the windings depends mainly on three things:

- The number of conductors in series per winding
- The speed at which the magnetic field cuts the winding
- The strength of the magnetic field

Any of these three factors could be used to control the amount of voltage induced in the generator windings.

The number of windings, of course, is fixed when the generator is manufactured and in addition, the output frequency is required to be a constant value of 400 Hz and so the speed of the rotating field must be held constant. **This prevents the use of the generator rpm as a means of controlling the voltage output.**

Therefore, the only practical method of voltage control is to manage the strength of the rotating magnetic field. This is achieved by changing the amount of current flowing through the field coil, accomplished by varying the amount of voltage applied across the field coil.

Permanent Magnetic Generator (PMG)

With larger types of rated output generators, a separate generator, know as a **Permanent Magnet Generator (PMG)** is mounted on the common shaft. The advantage of using a PMG is that its output is used for the excitation current and also the system's protection circuits. Figure 17.20 shows a schematic diagram of a brushless AC generator fitted with a permanent magnet generator.

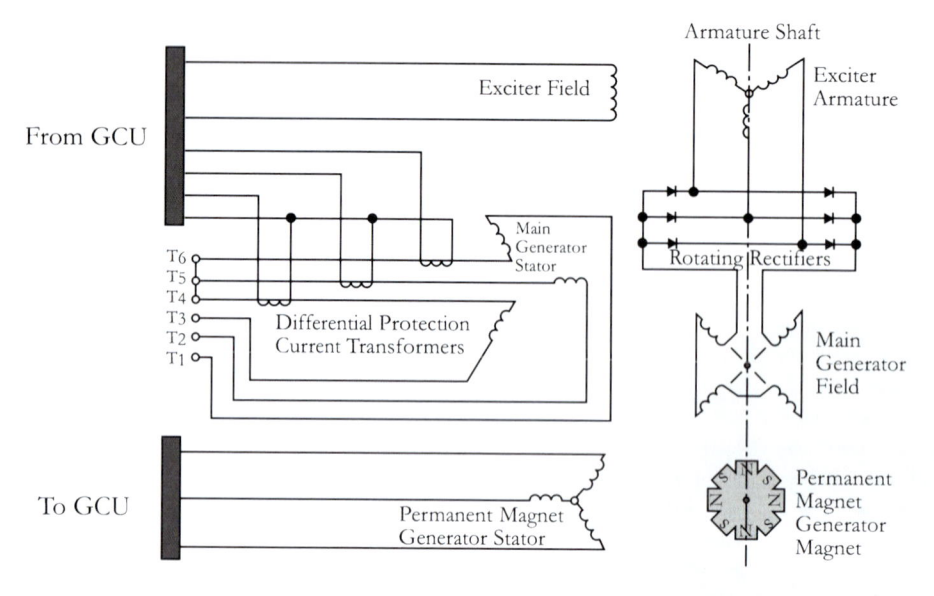

Figure 17.20 - Brushless AC Generator with PMG

The permanent magnet is mounted on the same shaft as the exciter armature, diodes and main field winding, and is driven by the prime mover. The PMG consists of a multi salient pole rotor with its own three phase output windings on the stator. The output frequency is higher than the generator output frequency, typically double, because of the number of salient poles.

In effect we now have three generators combined to make up a single one – a permanent magnet generator (shown at the bottom), an exciter generator (shown at the top) and the main generator (centre).

The output of the PMG is taken from the stator winding (bottom left) and fed to the exciter field (top left) via a Generator Control Unit (GCU) which rectifies and regulates the output (not shown).

The dc exciter field induces an alternating current into the rotating exciter armature. The exciter armature is connected to the rotating rectifier, which changes the alternating current to direct current and sends a current to the main generator field. The main field induces an AC voltage into the main generator armature.

The GCU monitors the main generator output and in turn regulates the exciter field current as required. If a higher generator output is needed, the GCU increases the exciter field current; which, in turn, increases the exciter armature output and the main field current. This results in a stronger main field which increases the main armature's output.

If less generator output is required, the GCU will weaken the exciter field current, and the generator output will decrease.

The advantage of using a permanent magnet generator is that its output is used for the *excitation current* and also the system protection circuits.

Differential protection current transformers are built into the generator, which monitor the current flowing through the output windings as well as the load current.

Faults

If any of the rotating diodes were to become open-circuited, the main field excitation will reduce, but this drop will be compensated for by the voltage regulator. If however a diode became short-circuited, this will cause some AC to appear in the rotating field and result in a harmonic frequency to appear in the AC output. This would cause the generator to trip off line.

Revision

AC Generators

Questions

1. The typical voltage outputs from an aircraft's AC generator is:

    a) 115 V ac 400 Hz

    b) 115 V ac 60 Hz

    c) 26 V ac 400 Hz

2. The three phases in a 3-phase generator are at:

    a) 90° to each other

    b) 180° to each other

    c) 120° to each other

3. In an ac generator, connected as a Star, each of the three windings are connected:

    a) Together at a common or neutral point

    b) In parallel with each other

    c) Insulated from each other

4. If the phase voltage of a star wound generator is 100 V ac, its line voltage is:

    a) 100 V

    b) 141.4 V

    c) 173 V

5. If the phase current of a delta wound generator is 10 A, its line current is:

   a) 10 A

   b) 17.3 A

   c) 14.14 A

6. When comparing the conductor size between a dc generator and ac generator that have similar power outputs, the cross-sectional-area of the conductor:

   a) Is larger in the ac generator

   b) Is smaller in the ac generator

   c) Is the same size in both generators

7. With a typical AC generator, the generated ac is brought to the load by means of a:

   a) Commutator

   b) Slip rings

   c) Stator

8. In a practical generator, the Armature is usually:

   a) Compensation windings

   b) The rotor

   c) The stator

9. In the armature core, eddy current losses are improved by:

   a) Increasing its cross-sectional area

   b) Reducing the number of laminations

   c) Increasing the number of laminations

CHAPTER SEVENTEEN
AC GENERATORS

10. An AC generator's exciter is usually powered from:

    a) The aircraft's batteries

    b) The APU

    c) A dc generator

11. One example of an generator's prime mover is:

    a) A dc generator

    b) APU

    c) Aircraft batteries

12. The output of an generator in a large aircraft is typically measured in:

    a) kW

    b) kVA

    c) kV

13. With a two-phase generator, the windings are physically mounted at:

    a) 90°

    b) 180°

    c) 120°

14. With a two-phase, three-wire generator, if the output is taken at the 'C' phase, its voltage is:

    a) 1.73 × Phase voltage

    b) 1.414 × Phase voltage

    c) Phase voltage

15. If the no-load voltage of a generator is 110 volts and its full-load voltage of is 100 volts, its % of regulation is:

   a)   20%

   b)   10%

   c)   5%

16. What is the frequency output of an ac generator is it is spinning at 360 RPM and has 10 poles:

   a)   300 Hz

   b)   30 Hz

   c)   150 Hz

17. The only practical method of voltage control in a generator is:

   a)   To manage the number of poles in the rotating magnetic field

   b)   To manage the speed of the rotating magnetic field

   c)   To manage the strength of the rotating magnetic field

18. A practical method of Frequency control in a generator is:

   a)   To manage the number of poles in the rotating magnetic field

   b)   To manage the speed of the rotating magnetic field

   c)   To manage the strength of the rotating magnetic field

CHAPTER SEVENTEEN
AC GENERATORS

Revision

AC Generators

Answers

1. **A**
2. **C**
3. **A**
4. **C**
5. **B**
6. **B**
7. **C**
8. **C**
9. **C**
10. **C**
11. **B**
12. **B**
13. **A**
14. **B**
15. **B**
16. **B**
17. **C**
18. **B**

AC Motors

Introduction

Most of the power-generating systems on modern aircraft produce AC; indeed even primary DC aircraft usually require an element of AC for reference and other values. For this reason a majority of the motors used throughout the aircraft are designed to operate on AC. However, there are other advantages in the use of AC motors besides the wide availability of AC power. In general, AC motors cost less than DC motors and some types of AC motors do not use brushes and commutators. This eliminates many problems of maintenance, wear, etc and also eliminates the problem of dangerous sparking.

An AC motor is particularly well suited for constant-speed applications as its speed is determined by the frequency of the AC voltage applied to the motor terminals. Obviously then, the DC motor is better suited for uses that require variable-speeds. However, an AC motor can also be made with variable speed characteristics but only within certain limits.

Industry builds AC motors in different sizes, shapes, and ratings for many different applications and these are designed for use with either polyphase or single-phase power systems. It is not possible here to cover all aspects of AC motors and so we will only look at those that apply to aviation, which cover the principles of the most commonly used types.

In this chapter, AC motors are divided into *Series*, *Synchronous* and *Induction* motors, we will also discuss single-phase, and polyphase motors.

Synchronous motors, for purposes of this chapter, may be considered as polyphase motors, of constant speed, whose rotors are energised with DC voltage.

Induction motors, single-phase or polyphase, whose rotors are energised by induction, are the most commonly used AC motor. The series AC motor, in a sense, is a familiar type of motor as it is similar to the DC motor covered in a previous chapter.

CHAPTER EIGHTEEN
AC MOTORS

Series AC Motor

A series AC motor is the same *electrically* as a DC series motor, figure 18.1.

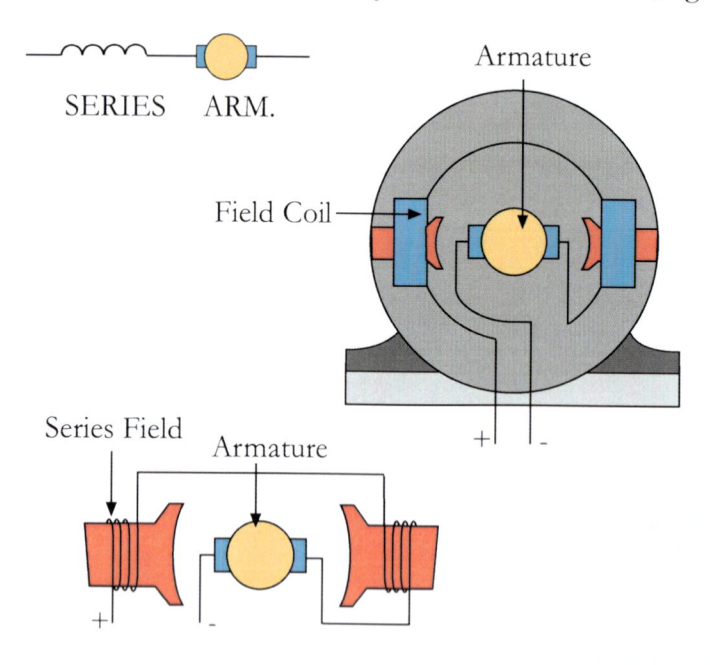

Figure 18.1 - Series AC Motor

Looking at figure 18.1, if we use the *right-hand gripping rule* for the polarity of coils we can see that the instantaneous magnetic polarities of the armature and field oppose each other, and motor action results.

As this is a series arrangement, if the current is reversed by swapping the polarity of the input, the field magnetic polarity still opposes the armature magnetic polarity because the reversal affects both the armature and the field. Obviously, with an AC input it causes these reversals to take place continuously and the motor moves.

The construction of the AC series motor differs slightly from the dc series motor as special metals, laminations, and windings are used to reduce losses caused by eddy currents, hysteresis, and high reactance. DC power can be used to drive an AC series motor efficiently, but the opposite is not true.

The characteristics of a series AC motor are similar to those of a series dc motor and it is a varying-speed machine. It has low speeds for large loads and high speeds for light loads and the starting torque is very high. Series motors are used for driving fans, electric drills, and other small appliances.

Rotating Magnetic Fields

The principle of rotating magnetic fields is the key to the operation of most AC motors as synchronous and induction types rely on rotating magnetic fields in their stators to cause their rotors to turn. The idea is simple. A magnetic field in a stator can be made to rotate electrically, around and around. Another magnetic field in the rotor can be made to chase it by being attracted and repelled by the stator field. Because the rotor is free to turn, it follows the rotating magnetic field in the stator.

Single Phase Rotating Magnetic Field

To produce a rotating field from a single phase AC supply requires a minimum of two pairs of field windings and a four pole stator.

In order to create the rotating field, we must ensure that the current in one pair of field windings is in quadrature with the current in the other pair. To achieve this an inductor or capacitor is connected in series with one pair of field windings. A capacitor is generally used because it is more efficient, Figure 18.2.

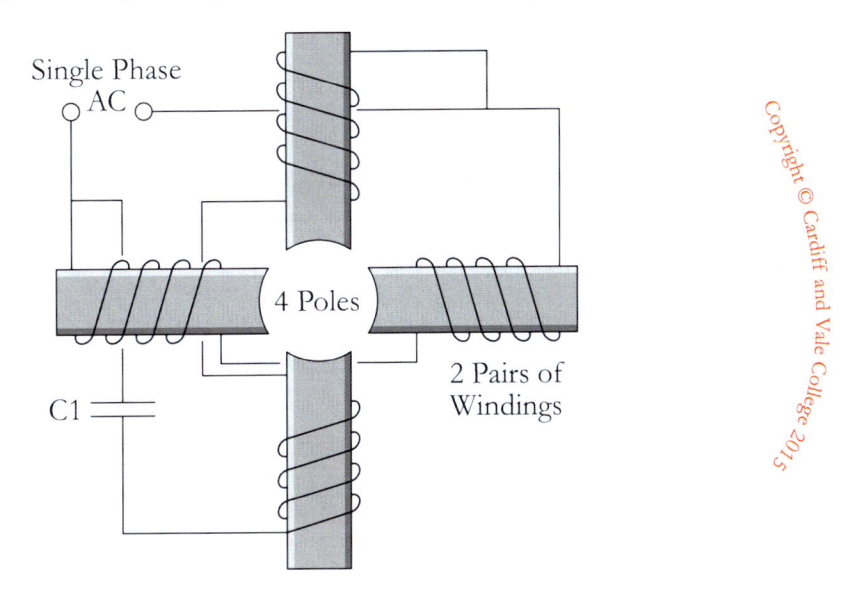

Figure 18.2 – Single Phase Rotating Field

The direction of rotation of the magnetic field depends on the order in which the poles become magnetised.

Changing the direction of rotation of the field can be achieved by swapping the supply to one pair of field windings, or by switching the capacitor from one field winding to the other. If the supply to both field windings is reversed, the motor will run in the same direction.

Rotating magnetic fields may be easily set up in two-phase or three-phase machines. To establish a rotating magnetic field in a motor stator, the number of pole pairs must be the same as, or a multiple of, the number of phases in the applied voltage. The poles must then be displaced from each other by an angle equal to the phase angle between the individual phases of the applied voltage.

Two-Phase Rotating Magnetic Field

A rotating magnetic field is probably most easily seen in a two-phase stator and the stator of a two-phase induction motor is made up of two windings, or a multiple of two. They are placed at right angles to each other around the stator as illustrated in the simplified drawing in figure 18.3, which shows a two-phase stator.

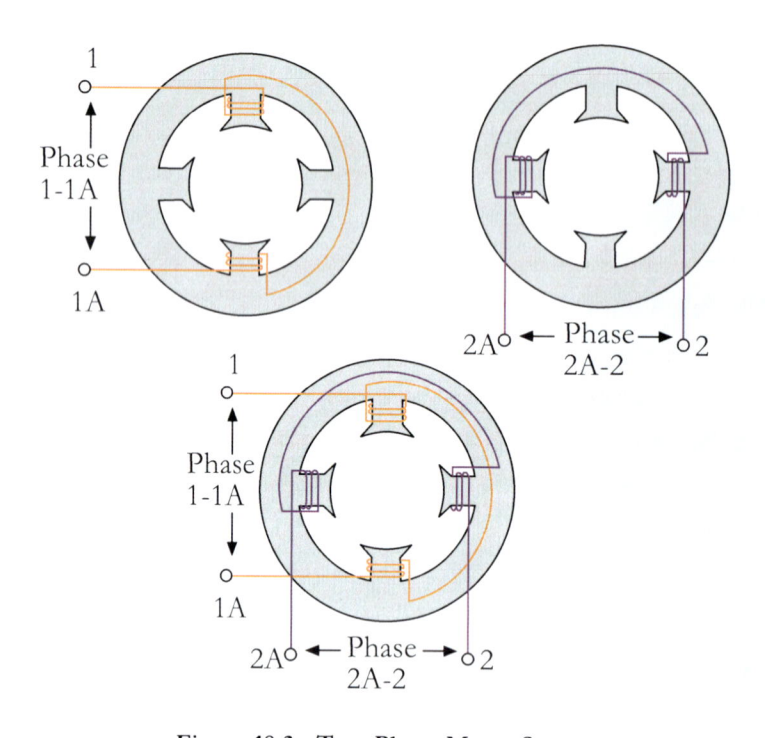

Figure 18.3 - Two-Phase Motor Stator

If the voltages applied to phases 1 to 1A and 2 to 2A are 90° out of phase, the currents that flow in the phases are displaced from each other by 90°. Since the magnetic fields generated in the coils are in phase with their respective currents, the magnetic fields are also 90° out of phase with each other. These two out-of-phase magnetic fields, whose coil axes are at right angles to each other, add together at every instant during their cycle, ie they produce a resultant field that rotates one revolution for each cycle of AC. To analyse the rotating magnetic field in a two-phase stator, figure 18.4, we need to look at a graph of its input voltage against time.

CHAPTER EIGHTEEN
AC MOTORS

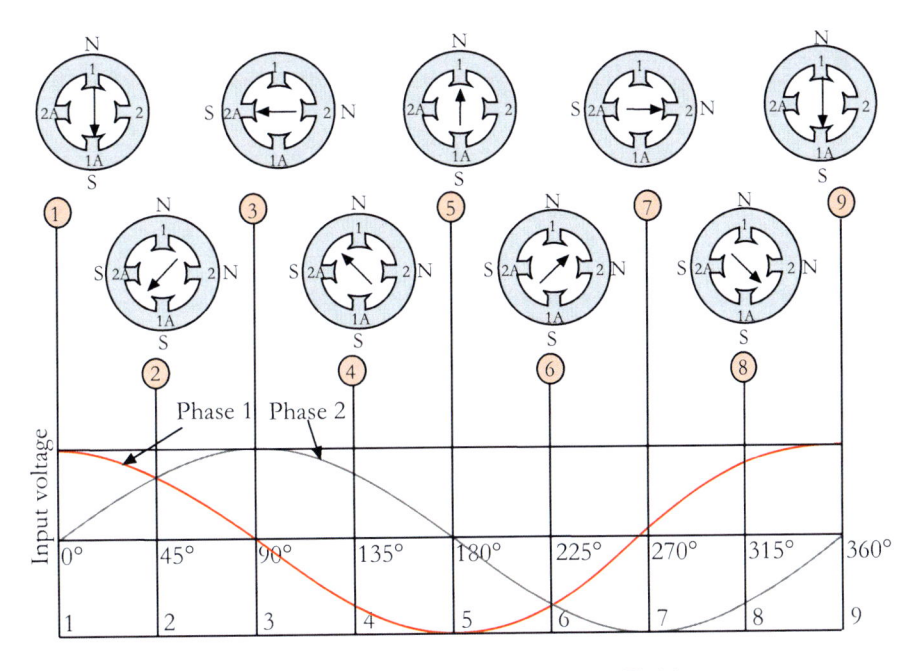

Figure 18.4 - Two-Phase Rotating Field

The rotating arrow in the figure above represents the rotor. For each point set up on the voltage chart, consider that current flows in a direction that will cause the magnetic polarity indicated at each pole piece. Note that from one point to the next, the polarities are rotating from one pole to the next in a clockwise way. One complete cycle of input voltage produces a 360° rotation of the pole polarities.

The waveforms in figure 18.4 are of the two input phases, displaced 90°, because of the way they were generated in a two-phase alternator. The waveforms are numbered to match their associated phase. Although not shown in this figure, the windings for the poles 1 to 1A and 2 to 2A would be as shown in the previous figure. At position 1, the current flow and magnetic field in winding 1 to 1A is at maximum because the phase voltage is maximum, while the current flow and magnetic field in winding 2 to 2A is zero because the phase voltage is zero. The resultant magnetic field is therefore in the direction of the 1 to 1A axis.

At the 45° point, i.e. position 2, the resultant magnetic field lies midway between windings 1 to 1A and 2 to 2A and the coil currents and magnetic fields are equal in strength. At 90°, i.e. position 3, the magnetic field in winding 1 to 1A is zero while the magnetic field in winding 2 to 2A is at maximum.

Now the resultant magnetic field lies along the axis of the 2 to 2A winding as shown and has rotated clockwise through 90° to get from position 1 to position 3. When the two-phase voltages have completed one full cycle, i.e. position 9, the resultant magnetic field has rotated through 360°. Therefore, by placing two windings at right angles to each other and exciting these windings with voltages 90° out of phase, a rotating magnetic field results.

423

CHAPTER EIGHTEEN
AC MOTORS

The only way to reverse the direction of rotation of a two phase motor is to swap the supply connections to one pair of field windings.

Two-phase motors are rarely used except in special-purpose equipment, but they are discussed here to aid in understanding rotating fields. You will, however, encounter many single-phase and three-phase motors.

Three-Phase Rotating Fields

The three-phase induction motor also operates on the principle of a rotating magnetic field and figure 18.5 below shows the individual windings for each phase and how the three phases are tied together in a 'Y' or Star wound stator.

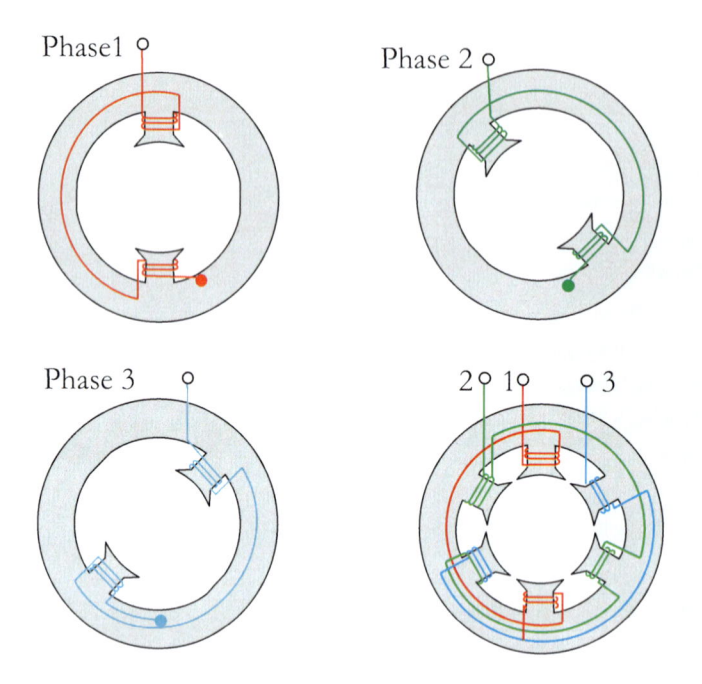

Figure 18.5 - Three-phase, Y or Star Wound Stator

The dot in each diagram indicates the common point of the Star-connection.

As shown in this figure, the individual phase windings are equally spaced around the stator, i.e. at 120° apart.

The three-phase input voltage to the stator of figure 18.5 is shown in the graph in figure 18.6.

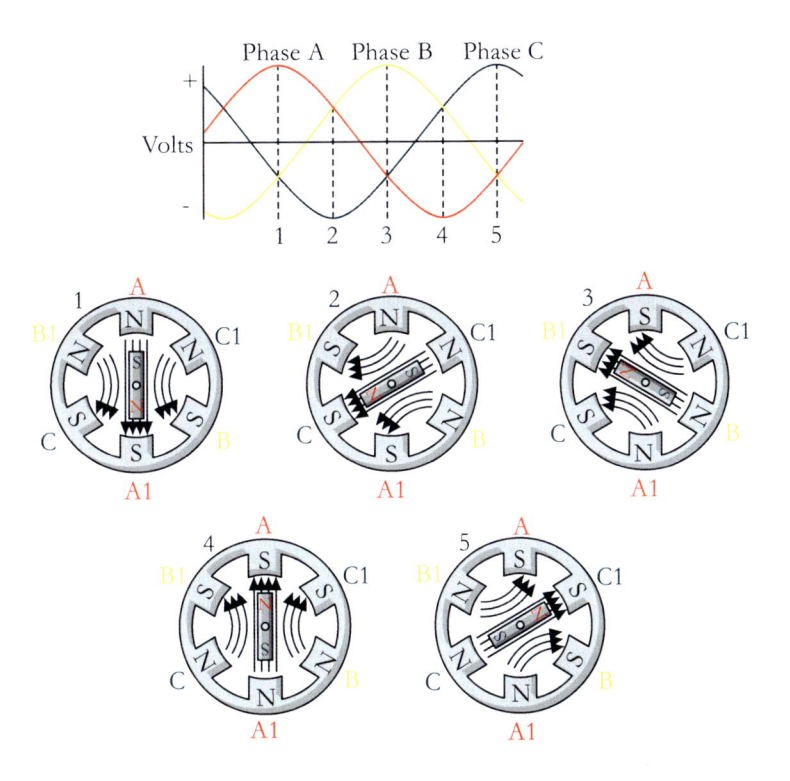

Figure 18.6 - 3-phase Rotating-Field Polarities & Input Voltages

At the instant of time shown as 1 in Figure 18.6, the resultant magnetic field produced by the application of the three voltages has its greatest intensity in a direction extending from pole A to pole A1. Under this condition, pole A can be considered as a North pole and pole A1 as a South pole.

At the instant of time shown as 2, the resultant magnetic field will have its greatest intensity in the direction extending from pole C1 to pole C; in this case, pole C1 can be considered as a North Pole and pole C as a south pole. We can see from this that between instant 1 and instant 2, the magnetic field has rotated clockwise.

At instant 3, the resultant magnetic field has its greatest intensity in the direction from pole B to pole B1, and the resultant magnetic field has continued to rotate clockwise.

At instant 4, poles A1 and A can be considered as north and south poles, respectively, and the field has rotated still farther.

At later instants of time, the resultant magnetic field rotates to other positions while travelling in a clockwise direction, a single revolution of the field occurring in one cycle. If the exciting voltages have a frequency of 100 Hz the magnetic field makes 100 revolutions per second, or 6,000 rpm. This speed is known as the *synchronous speed* of the rotating field.

When selecting a three phase motor, we must also consider the number of poles required to achieve the desired speed of rotation. Now you may think that increasing the number of pairs of poles per phase would also increase the rotation speed, however you would be wrong!

Transposition of the formula used to calculate the frequency in an AC Generator shows that the speed of rotation can be determined by:

$$NS = 60 \times (f/P)$$

Where NS = synchronous speed; f = supply frequency & P = pairs of poles per phase.

For example a 50Hz supply to a machine using one pair of poles per phase would produce a synchronous speed of:

$$NS = 60 \times (50/1)$$
$$\text{Therefore } NS = 3000 \text{rpm}$$

What will happen if we double the number of pole pairs per phase?

$$NS = 60 \times (50/2)$$
$$\text{Therefore } NS = 1500 \text{rpm}$$

As you can see, the rotational speed is halved when the number of pole pairs per phase is doubled. If you consider figure 18.6, then you can see that it takes each phase the full 360° of mechanical rotation of the machine to produce one cycle. If however we now double the number of pole pairs per phase, then each phase cycle is produced in 180° of mechanical rotation of the machine. As a result increasing the number of pole pairs reduces the speed of the motor. To increase the speed, we must increase the supply frequency.

The direction of rotation of the field depends on the order in which the windings are energised. To reverse the direction of rotation, it is only necessary to swap the connection to any two of the field windings, Figure 18.7.

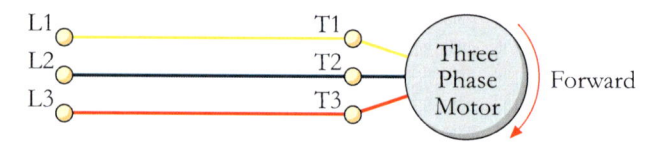

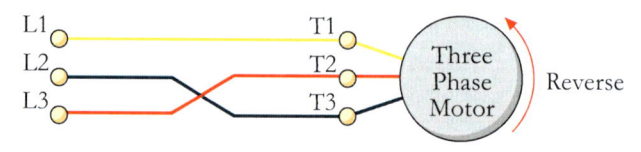

Figure 18.7 – Reversing the Direction of a Three Phase Motor

Rotor Behaviour in a Rotating Field

For purposes of explaining rotor movement, let us assume that we can place a bar magnet in the centre of the stator diagrams of figure 18.6, mounted so that it is free to rotate in this area. Let us also assume that the bar magnet is aligned so that at point 1 its south pole is opposite the large N of the stator field.

Now as we already know, unlike poles attract, and the two fields are aligned so that they are attracting each other. Now, going from point 1 through point 5, as before, the stator field rotates clockwise. The bar magnet, free to move, will follow the stator field, because the attraction between the two fields continues to exist and a shaft running through the pivot point of the bar magnet would rotate at the same speed as the rotating field. This speed is known as *synchronous speed*, the shaft representing the shaft of an operating motor to which the load is attached.

This explanation is an oversimplification and is meant to show how a rotating field can cause mechanical rotation of a shaft. Such an arrangement would work, but it is not used, as there are limitations to a permanent magnet rotor. Practical motors use other methods, as we shall see in the next paragraphs.

Synchronous Motors

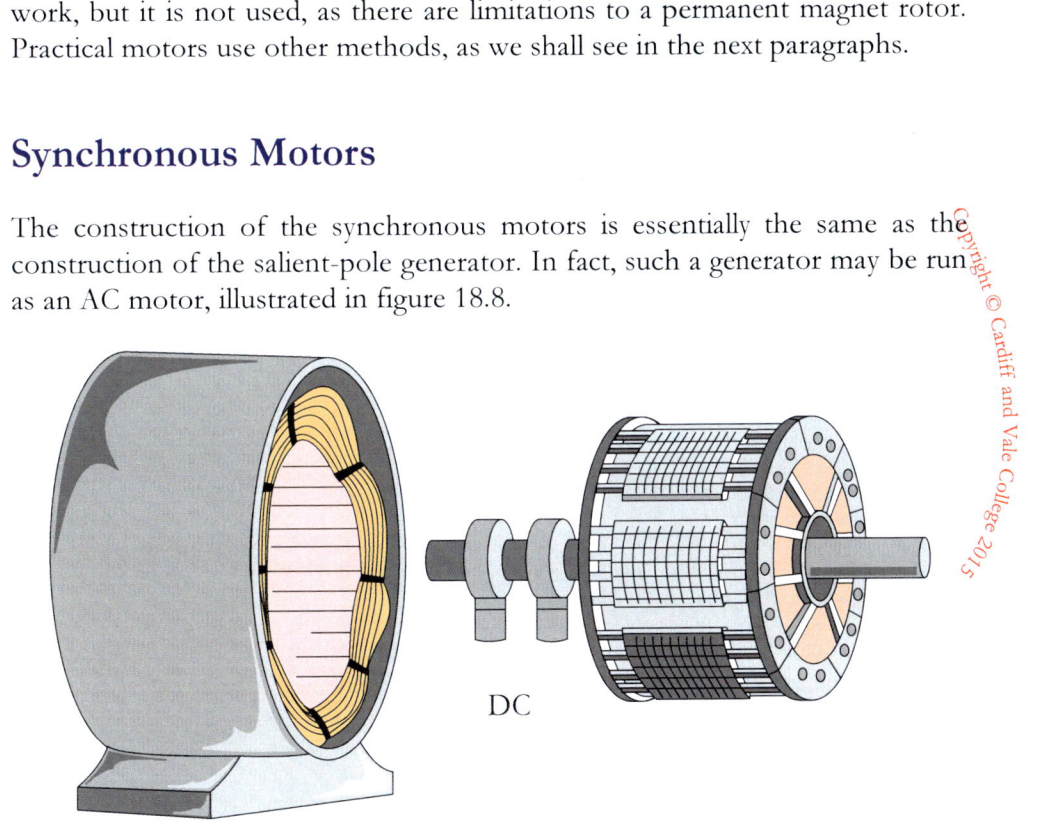

The construction of the synchronous motors is essentially the same as the construction of the salient-pole generator. In fact, such a generator may be run as an AC motor, illustrated in figure 18.8.

Figure 18.8 - Revolving-Field Synchronous Motor

Synchronous motors have the characteristic of constant speed between no load and full load conditions and they are capable of correcting the low power factor of an inductive load when they are operated under certain conditions. They are often used to drive DC generators.

CHAPTER EIGHTEEN
AC MOTORS

Synchronous motors are designed in sizes up to thousands of horsepower. They may be designed as either single-phase or multiphase machines. The following discussion is based on a three-phase design.

To understand how the synchronous motor works, assume that the application of three-phase AC power to the stator causes a rotating magnetic field to be set up around the rotor. The rotor is energised with dc, i.e. it acts like a bar magnet, and the strong rotating magnetic field attracts the strong rotor field activated by the dc. This results in a strong turning force, i.e. torque, on the rotor shaft and it is therefore able to turn a load as it rotates in step with the rotating magnetic field. Anyway, it works this way once it's started.

However, one of the disadvantages of a synchronous motor is that it cannot be started from a standstill by applying three-phase AC power to the stator. When AC is applied to the stator, a high-speed rotating magnetic field appears immediately. This rotating field rushes past the rotor poles so quickly that the rotor does not have a chance to get started and in effect, the rotor is repelled first in one direction and then the other. A synchronous motor in its purest form has no starting torque and only has any torque when it is running at synchronous speed.

A *squirrel-cage* type of winding is added to the rotor of a synchronous motor to cause it to start. The squirrel cage is shown as the outer part of the rotor in figure 18.9 below.

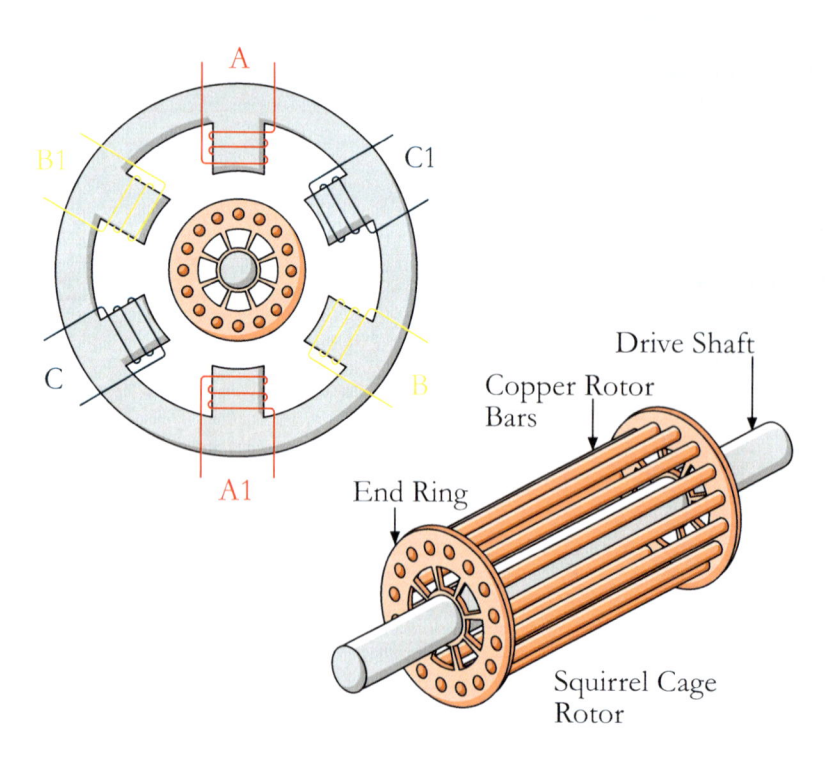

Figure 18.9 - Self-Starting Synchronous AC Motor

It is so named because it is shaped and looks like a turning squirrel or hamster cage. Simply, the windings are heavy copper bars shorted together by copper rings and a low voltage is induced in these shorted windings by the rotating three-phase stator field. Because of the short circuit, a relatively large current flows in the squirrel cage.

To start a practical synchronous motor, the stator is energised, but the dc supply to the rotor field is not energised. The squirrel-cage windings bring the rotor to near synchronous speed and at that point, the dc field is energised. This locks the rotor in step with the rotating stator field, full torque is developed, and the load is driven. A mechanical switching device that operates on centrifugal force is often used to apply dc to the rotor as synchronous speed is reached.

Without the dc exciter voltage the motor would be unable to run at synchronous speed. If the rotor was operating at exactly the same rate as the rotating field, the no current would flow in the squirrel cage as no changing lines of flux are acting on the rotor.

Therefore a practical synchronous motor has the disadvantage of requiring a dc exciter voltage for the rotor. This voltage may be obtained either externally or internally, depending on the design of the motor.

Induction Motors

The induction motor is the most commonly used type of AC motor. Its simple, rugged construction costs relatively little to manufacture. The induction motor has a rotor that is not connected to an external source of voltage. The induction motor derives its name from the fact that AC voltages are induced in the rotor circuit by the rotating magnetic field of the stator. In many ways, induction in this motor is similar to the induction between the primary and secondary windings of a transformer that we have already discussed.

Large motors and permanently mounted motors that drive loads at fairly constant speed are often induction motors. The stator construction of the three-phase induction motor and the three-phase synchronous motor are almost identical. However, their rotors are completely different, figure 18.10.

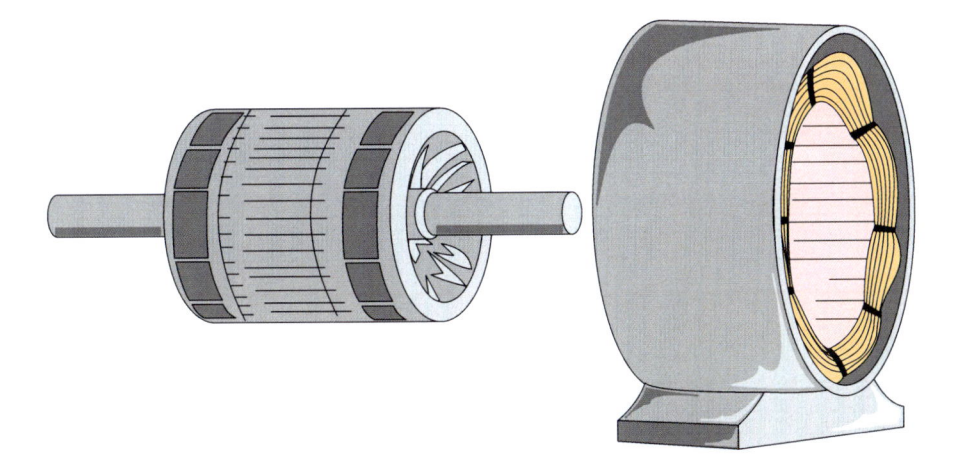

Figure 18.10 – Typical Induction Motor

CHAPTER EIGHTEEN
AC MOTORS

The induction rotor is made of a laminated cylinder with slots in its surface. The windings in these slots are one of two types, figure 18.11.

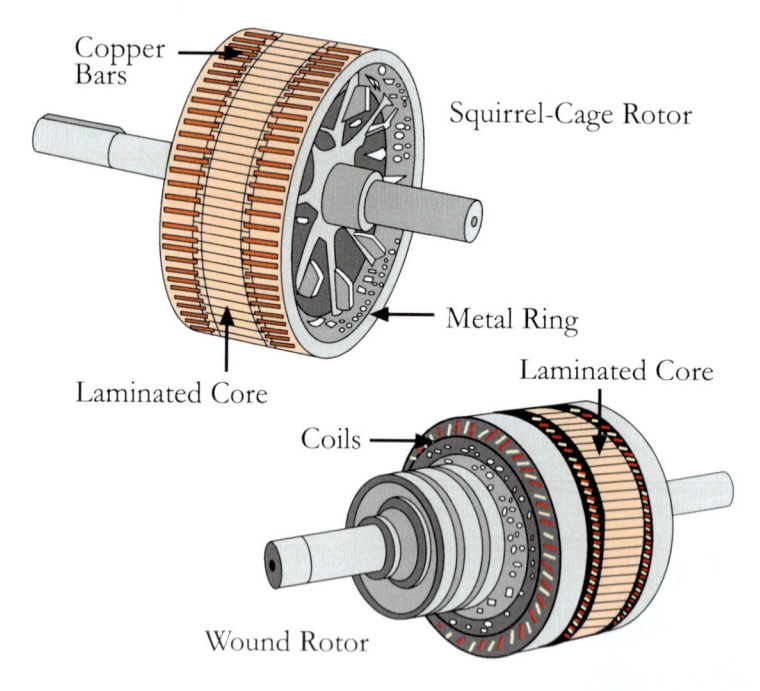

Figure 18.11 - Types of AC Induction Motor Rotors

The most common is the squirrel-cage winding, which is entirely made up of heavy copper bars connected together at each end by a metal ring made of copper or brass. No insulation is required between the core and the bars because of the very low voltages generated in the rotor bars. The other type of winding contains actual coils placed in the rotor slots and so is then called a wound rotor. Regardless of what type of rotor is used, the basic principle is the same.

The rotating magnetic field generated in the stator induces a magnetic field in the rotor. The two fields interact and cause the rotor to turn. To obtain maximum interaction between the fields, the air gap between the rotor and stator is very small.

As you know from **Lenz's law**, any induced emf tries to oppose the changing field that induces it. In the case of an induction motor, the changing field is the motion of the resultant stator field. A force is exerted on the rotor by the induced emf and the resultant magnetic field and this force tends to cancel the relative motion between the rotor and the stator field. The rotor, as a result, moves in the same direction as the rotating stator field.

It is impossible for the rotor of an induction motor to turn at the same speed as the rotating magnetic field. If the speeds were the same, there would be no **relative** motion between the stator and rotor fields; without relative motion there would be no induced voltage in the rotor. In order for relative motion to exist between the two, the rotor must rotate at a speed slower than that of the rotating magnetic field. The **difference** between the speed of the rotating stator field and the rotor speed is called **slip**. The **smaller** the slip, the closer the rotor speed approaches the stator field speed.

The speed of the rotor depends upon the torque requirements of the load. The bigger the load, the stronger the turning force needed to rotate the rotor.

The turning force can increase *only* if the rotor-induced emf increases. This emf can increase *only* if the magnetic field cuts through the rotor at a faster rate. To increase the *relative* speed between the field and rotor, the rotor must slow down. Therefore, *for heavier loads the induction motor turns slower than for lighter loads*.

We have already discussed that slip is directly proportional to the load on the motor. Actually only a slight change in speed is necessary to produce the usual current changes required for normal changes in load. This is because the rotor windings have such a low resistance and as a result, induction motors are called *constant-speed motors*.

Single-Phase Induction Motors

There are probably more single-phase AC induction motors in use today than the total of all the other types put together. Obviously, it is logical that the least expensive, lowest maintenance type of AC motor should be used most often; the single-phase AC induction motor fits that description.

Unlike polyphase induction motors, the stator field in the single-phase motor does not rotate. Instead it simply alternates polarity between poles as the AC voltage changes polarity.

Voltage is induced in the rotor as a result of magnetic induction, and a magnetic field is produced around the rotor. This field will always be in opposition to the stator field, as Lenz's law applies, but the interaction between the rotor and stator fields will not produce rotation. The double-ended arrow in figure 18.12, view A shows the interaction.

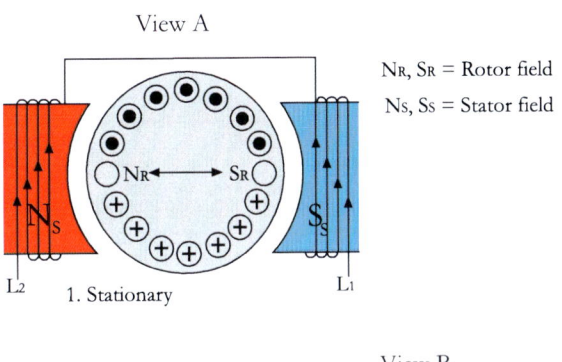

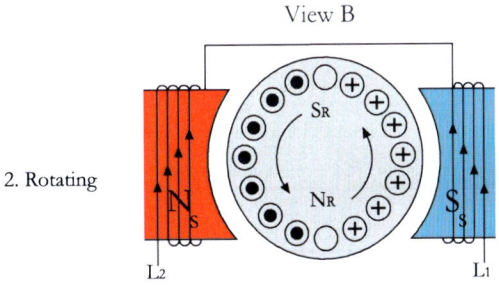

Figure 18.12 - Rotor Currents in a Single-Phase AC Induction Motor

Because this force is across the rotor and through the pole pieces, there is no rotary motion, just a push and/or pull along this line.

Now, if some outside force rotates the rotor, the push-pull along the line in figure 18.12, view A, is disturbed. If we now look at the rotating fields as shown in figure 18.12, view B, at this instant, the South Pole on the rotor is being attracted by the left-hand pole and the north rotor pole is being attracted to the right-hand pole. All of this is a result of the rotor being rotated 90° by an outside force. The pull that now exists between the two fields becomes a rotary force, turning the rotor and because the two fields continuously alternate, they will never actually line up, and the rotor will continue to turn, once started.

There are several types of single-phase induction motors in use today. They are basically identical except for their means of starting. In this chapter, we will look at the split-phase and shaded-pole motors; named because of the methods used to get them started. Once they are up to operating speed, all single-phase induction motors operate the same.

Split-Phase Induction Motors

Split-Phase Motors are designed to use inductance, capacitance, or resistance to develop a starting torque; all principles we have already learned in earlier chapters.

Capacitor-Start

The first type of split-phase induction motor that we will cover is the *capacitor-start type*. Figure 18.13 shows a simplified schematic of a typical capacitor-start motor.

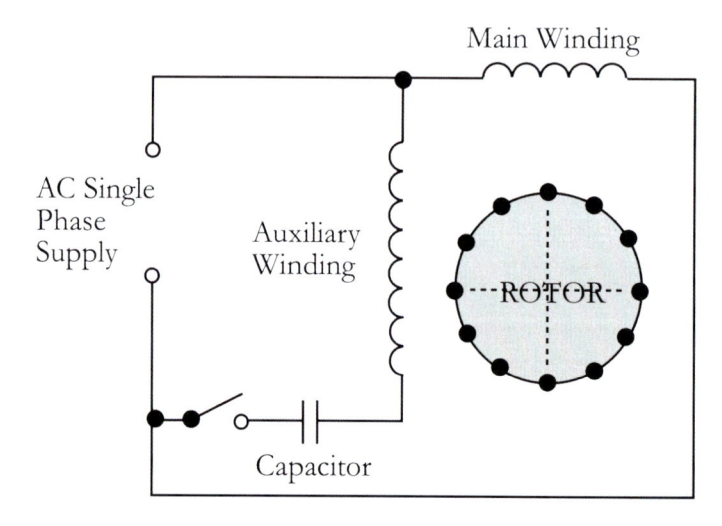

Figure 18.13 - Capacitor-Start, AC Induction Motor

The stator consists of the main winding and a starting winding, sometimes called an *auxiliary winding*. The auxiliary winding is connected in parallel with the main winding and is placed physically at right angles to it. A 90° electrical phase difference between the two windings is achieved by connecting the auxiliary winding in series with a capacitor and starting switch. When the motor is first energised, the starting switch is closed. This places the capacitor in series with the auxiliary winding. The capacitor's value is such that the auxiliary circuit is effectively a resistive-capacitive circuit that has capacitive reactance, expressed as X_C.

From our earlier work, you know that in this circuit the current **leads** the line voltage by about 45°, because X_C is made to approximately equal R. The main winding has enough resistance-inductance, i.e. inductive reactance expressed as X_L, to cause the current to lag the line voltage by about 45°, again because X_L approximately equals R. The currents in each winding are therefore 90° out of phase and so are the generated magnetic fields. The effect is that the two windings act like a two-phase stator and produce the rotating field required to start the motor. When nearly full speed is obtained, a centrifugal device, e.g. the starting switch, cuts out the starting winding. The motor then runs as a plain single-phase induction motor. Since the auxiliary winding is only a light winding, the motor does not develop sufficient torque to start heavy loads and so split-phase motors only come in small sizes, e.g. avionic cooling blower motors.

Resistance-Start

Another type of split-phase induction motor is the **resistance-start motor**. This motor also has a starting winding, illustrated in figure 18.14, in addition to the main winding.

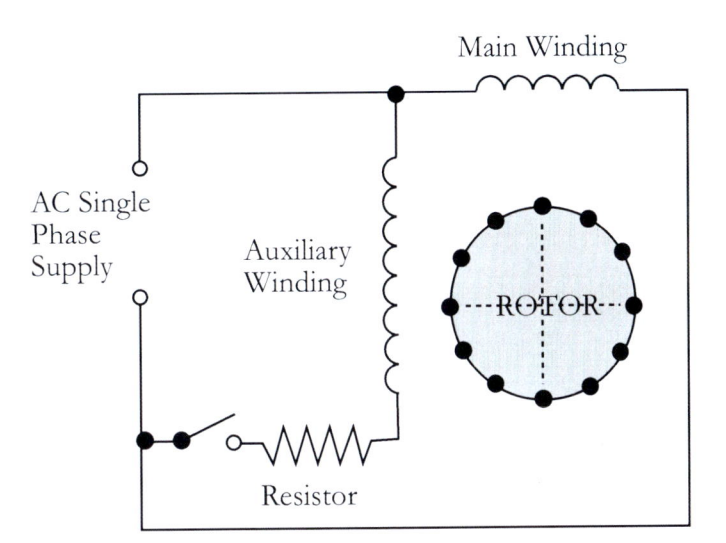

Figure 18.14 - Resistance-Start AC Induction Motor

The resistor is switched in and out of the circuit just as it is in the capacitor-start motor. The starting winding is positioned at right angles to the main winding and the electrical phase shift between the currents in the two windings is achieved by making the impedance of the windings unequal.

The main winding has a high inductance and a low resistance. The current, therefore, lags the voltage by a large angle. The starting winding is designed to have a fairly low inductance and high resistance and so here the current lags the voltage by a smaller angle. For example, suppose the current in the main winding lags the voltage by 70°; the current in the auxiliary winding lags the voltage by 40°. The currents are, therefore, out of phase by 30° and the magnetic fields are out of phase by the same amount. Although the ideal angular phase difference is 90° for maximum starting torque, the 30°-phase difference still generates a rotating field. This supplies enough torque to start the motor.

When the motor comes up to speed, a speed-controlled switch disconnects the starting winding from the line, and the motor continues to run as an induction motor. However, with this arrangement, the starting torque is not as great as it is in the capacitor-start.

Shaded-Pole Induction Motors

The *shaded-pole induction motor* uses a unique method to start the rotor turning. Constructing the stator in a special way produces the effect of a moving magnetic field. This motor has projecting pole pieces just like some dc motors. In addition, a copper strap called a shading coil surrounds portions of the pole piece surfaces. This is illustrated in figure 18.15.

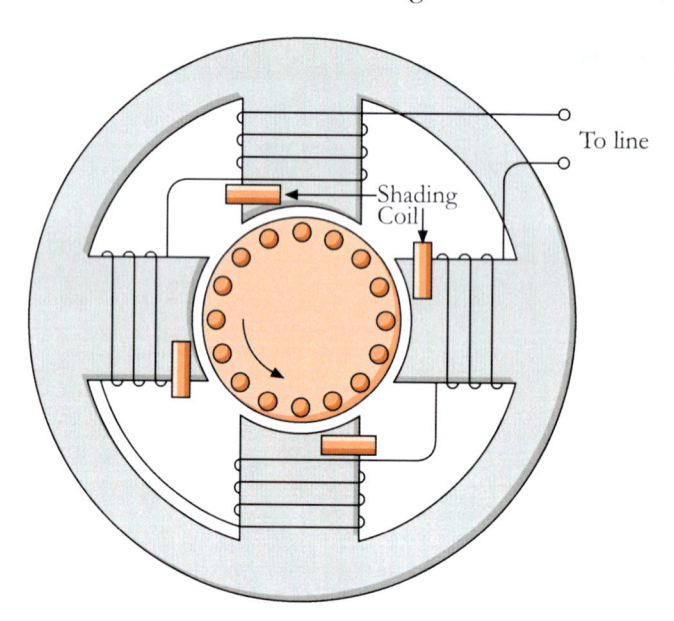

Figure 18.15 - Shaded Pole Induction Motor

The continuous copper loop around a small portion of each of the poles "shades" that portion of the pole, causing the magnetic field in the shaded area to lag behind the field in the unshaded area. And the reaction of the two fields gets the shaft rotating, Figure 18.16.

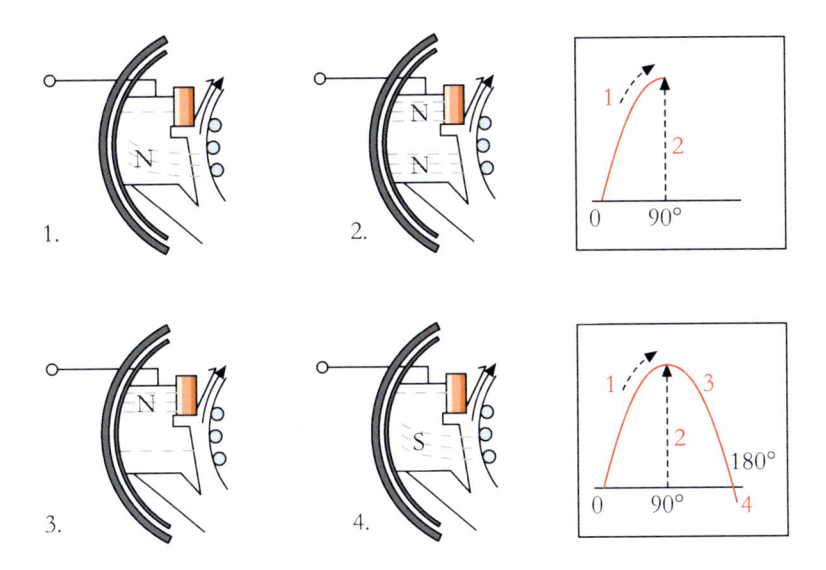

Figure 18.16 - Shaded Pole AC Induction Motor

The strap causes the field to move back and forth across the face of the pole piece. Note the numbered sequence and points on the magnetisation curve in the figure. As the alternating stator field starts increasing from zero (1), the lines of force expand across the face of the pole piece, cut through the strap and so a voltage is induced in it. The current that results generates a field that opposes the cutting action, and decreases the strength, of the main field. This produces the following actions:

As the field increases from zero to a maximum at 90°, a large portion of the magnetic lines of force are concentrated in the unshaded portion of the pole (1). At 90° the field reaches its maximum value. Since the lines of force have stopped expanding, no emf is induced in the strap, and no opposing magnetic field is generated. As a result, the main field is uniformly distributed across the pole (2).

From 90° to 180°, the main field starts decreasing or collapsing inward and the field generated in the strap opposes the collapsing field. The effect is to concentrate the lines of force in the shaded portion of the pole face (3).

You can see that from 0° to 180°, the main field has shifted across the pole face from the unshaded to the shaded portion. From 180° to 360°, the main field goes through the same change as it did from 0° to 180°; however, it is now in the opposite direction (4). The direction of the field does not affect the way the shaded pole works. The motion of the field is the same during the second half-cycle as it was during the first half of the cycle.

The motion of the field backwards and forwards between the shaded and unshaded portions produces a weak torque to start the motor. Because of the weak starting torque, shaded-pole motors are only built in small sizes and in aviation, are only used to drive devices such as fans, clocks, blowers, etc.

Hysteresis Motor

A hysteresis motor is a synchronous motor which does not require dc excitation to the rotor. The motor starts by virtue of the hysteresis losses induced in its hardened steel rotor by the rotating field. The can operates at synchronous speed because of the high retentivity of the rotor. Put simply it is an AC motor whose operation depends upon the hysteresis effect i.e., magnetization produced in a ferromagnetic material lags behind the magnetizing force.

These motors are typically found in servomechanisms, which we shall cover in Module 4, and may be described as 2 or 3 phase self-starting synchronous machines.

In the example detailed shown in Figure 18.17, a two-phase supply is required, the reference phase voltage being applied to phase windings A and A1, and the error-controlled voltage to B and B1. The error voltage is used to drive the motor. At one instant, A will act as a North pole and A1 as a South pole; B and B1 will be neutral.

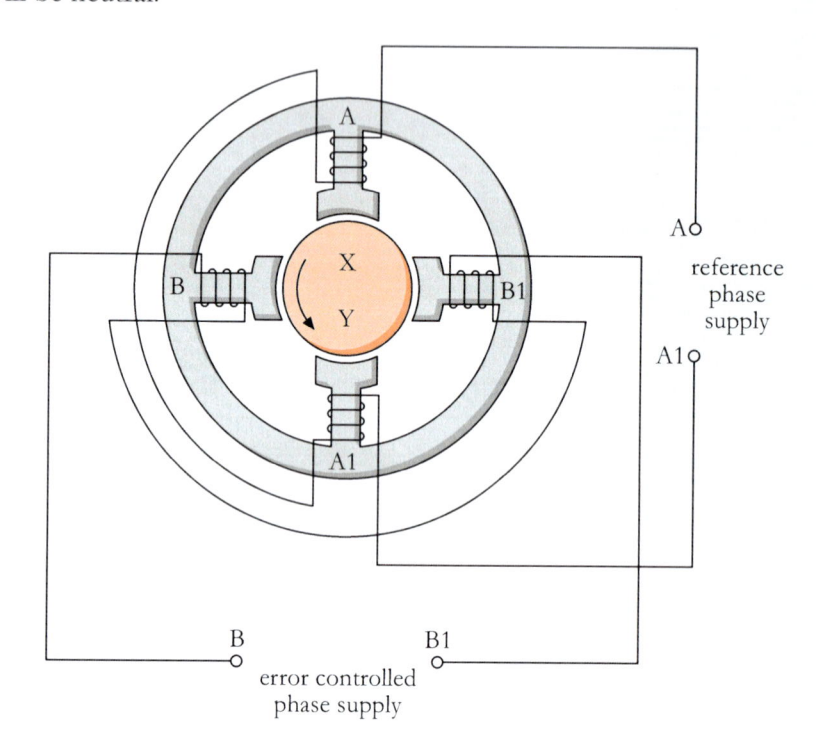

Figure 18.17 – Hysteresis Motor

The rotor, typically a cobalt steel ring, will therefore have a South Pole induced at X and a North Pole at Y. A quarter of a cycle later, B will act as a North Pole and B1 as a South Pole; whilst A and A1 will be neutral. However, as explained earlier, the rotor material has a high level of retentivity and therefore will have retained its South Pole at X and its North Pole at Y.

Because of this point X on the rotor will be attracted to the B stator winding and Y is attracted to B1. Since a torque is being exerted on the rotor it will turn to follow the rotating magnetic field and will continue to do so until the error signal falls to zero.

The direction of rotation depends upon the phase of the error-controlled voltage relative to the reference voltage and this, in turn, depends upon the sense of the error. The torque developed depends upon the magnitude of the error. The main virtue of the hysteresis motor is that, with suitable rotor construction, this torque can be kept constant over a wide speed range.

Speed of Single-Phase Induction Motors

The speed of induction motors is dependent on motor design. The synchronous speed, i.e. the speed at which the stator field rotates, is determined by the frequency of the input ac power and the number of poles in the stator. The greater the number of poles, the slower the synchronous speed; the higher the frequency of applied voltage, the higher the synchronous speed.

Remember, however, that neither the operating frequency or pole numbers are variables, i.e. they are both fixed by the manufacturer. The relationship between poles, frequency, and synchronous speed is given in the following formula:

$$\text{Speed (RPM)} - n = \frac{60f}{P}$$

Where: n is the synchronous speed in rpm

f is the frequency of applied voltage in hertz

p is the number of pairs of poles in the stator

As we have seen before, the rotor is never able to reach synchronous speed as if it did, there would be no voltage induced in the rotor; no torque would be developed; the motor would not operate. The difference between rotor speed and synchronous speed, as we have already discussed, is called *slip*. The difference between these two speeds is not usually great, for example, a rotor speed of 3400 to 3500 rpm can be expected from a synchronous speed of 3600 rpm.

Induction Motor Slip

As we have seen, when the rotor of an induction motor is subjected to the revolving magnetic field produced by the stator windings, a voltage is induced in the longitudinal bars. This induced voltage causes a current to flow through the bars, which in turn, produces its own magnetic field. This field combines with the revolving field to produce a force on the rotor to cause it to rotate in the same direction as the rotating magnetic field. Subsequently, the rotor revolves at very nearly the synchronous speed of a stator field, the difference in speed being only just sufficient enough to induce the required amount of current in the rotor to overcome the mechanical and electrical losses in the rotor. On no load conditions, slip can be very small (2-3%).

As the mechanical load on the rotor is increased, the rotor will slow down and the slip increases, thereby inducing larger currents in the rotor to produce the extra torque required. The supply current has to increase to deliver the extra power.

If the rotor were to turn at the same speed as the rotating field, the rotor conductors would not be cut by any magnetic lines of force, no emf would be induced in them, no current could flow, and there would be no torque. The rotor would then slow down. For this reason, there must always be a difference in speed between the rotor and the rotating field. This difference in speed is called slip and is expressed as a percentage of the synchronous speed.

$$\text{Slip} = \frac{N_s - N_r}{N_s} \times 100\%$$

Where: N_s = synchronous speed
N_r = rotor speed

For example, if the rotor turns at 1,950 rpm and the synchronous speed is 2,000 rpm, the difference in speed is 50 rpm. The slip speed is 50 rpm or 50/2000 = 2.5%.

Typically when an induction motor is operating at full load the slip is somewhere between 10% and 20%. If the slip became any greater there is a chance the motor would stall.

How to reverse the Direction of a 3 Phase AC Motor

In order to change the direction of motion of an AC motor, we need to reverse the rotation of the magnetic flux in the stator. This can be achieved by changing the connections to any *two (2)* of the *three (3)* motor terminals. The ease with which it is possible to reverse the direction of rotation constitutes one of the advantages of three (3) phase motors.

Revision

AC Motors

Questions

1. **A synchronous motor requires an additional winding to assist its start as:**

   a) Its starting torque is low

   b) Its starting torque is high

   c) It has no starting torque

2. **A squirrel cage winding has:**

   a) Similar voltage and current values

   b) Low voltage/high current

   c) High voltage/low current

3. **The rotor of an induction motor turns:**

   a) Faster than the stator

   b) Slower than the stator

   c) At the same speed as the stator

4. **For heavier loads the induction motor turns:**

   a) Slower than for lighter loads

   b) Faster than for lighter loads

   c) The same as for lighter loads

5. **In an induction motor, slip is:**

   a) The difference between the speed of the rotating stator field and the rotor speed

   b) The difference between the phase of the current in the stator coils

   c) The phase difference between the current and voltage in the stator

Revision

AC Motors

Answers

1. **C**
2. **B**
3. **B**
4. **A**
5. **A**